Unless Recalled Earlier

DATE DUE

Designs for Life

Molecular biology has come to dominate our perceptions of life, health and disease. In the decades following World War II, the Medical Research Council Laboratory of Molecular Biology at Cambridge became a world-renowned centre of this emerging discipline. It was here that Crick and Watson, Kendrew and Perutz, Sanger and Brenner pursued their celebrated investigations. Soraya de Chadarevian's important new study is the first to examine the creation and expansion of molecular biology through the prism of this remarkable institution. Firmly placing the history of the laboratory in the postwar context, the author shows how molecular biology was built at the bench and through the wide circulation of tools, models and researchers, as well as in governmental committees, international exhibitions and television studios. *Designs for Life* is a major contribution both to the history of molecular biology, and to the history of science and technology in postwar Britain.

SORAYA DE CHADAREVIAN is Senior Research Associate and Affiliated Lecturer in the Department of History and Philosophy of Science at the University of Cambridge. She is co-editor of *Molecularizing Biology and Medicine: New Practices and Alliances 1910s–1970s* (1998) and of a forthcoming volume on 3D models in the history of science. She is advisory editor of *Studies in History and Philosophy of Biological and Biomedical Sciences*.

Designs for Life

Molecular Biology after World War II

Soraya de Chadarevian

CAMBRIDGE
UNIVERSITY PRESS

PUBLISHED BY THE PRESS SYNDICATE OF THE UNIVERSITY OF CAMBRIDGE
The Pitt Building, Trumpington Street, Cambridge, United Kingdom

CAMBRIDGE UNIVERSITY PRESS
The Edinburgh Building, Cambridge CB2 2RU, UK
40 West 20th Street, New York, NY 10011-4211, USA
477 Williamstown Road, Port Melbourne, VIC 3207, Australia
Ruiz de Alarcón 13, 28014 Madrid, Spain
Dock House, The Waterfront, Cape Town 8001, South Africa

http://www.cambridge.org

© Soraya de Chadarevian 2002

First published 2002

Printed in the United Kingdom at the University Press, Cambridge

Typeface Concorde 9.75/14 pt.   *System* LaTeX 2_ε [TB]

A catalogue record for this book is available from the British Library

ISBN 0 521 57078 6 hardback

For Livia and Flavia

Contents

Figures

Tables

Preface

Especially in the summer months, when I approach the neo-gothic archway of the Cavendish building which, from Free School Lane, leads to one of the central laboratory sites of the University of Cambridge as well as to my office, a group of tourists often crowds the entrance. While I slowly push my bike around the group, the local tourist guide unfailingly pronounces the names of James Watson and Francis Crick, two scientists in a long series of Nobel Prize winners who made this place a memorable site. The two researchers, one American, one British, 'unveiled' here the double helical structure of DNA, the 'stuff' from which our genes are made. The next stop of the guided tour is just down the road and across the street to the Eagle, the pub where the two scientists celebrated their breakthrough discovery.

Invariably these encounters intrigue me. This book is my attempt to explain how molecular biology has become a landmark on our cultural map. Like the tourist trail it leads to Cambridge, yet not just to follow in Watson and Crick's footsteps, but to understand, through an in-depth study of one of the world's centres of the field, how molecular biology took shape in the postwar years.

Acknowledgements

Research and writing for this book has been supported by many institutions and individuals. It is a pleasure to thank them here.

Research started on a Unit Project set up by Harmke Kamminga and Andrew Cunningham at the Wellcome Unit for the History of Medicine at the University of Cambridge. Without their initiative and their decision to recruit me for the project I might never have moved from nineteenth- to late twentieth-century history. From 1991 to 1996 (with some brief interruptions) the Wellcome Trust generously funded the project. A brief spell on a 'poste rouge' of the CNRS at La Villette in Paris in spring/summer 1995 brought the chance for some comparative work in the archives. From January till October 1997 funding came from the Hamburg Institute for Social Research. A generous grant from the Laboratory of Molecular Biology and the continuing hospitality of the Department of History and Philosophy of Science at Cambridge allowed me to complete the manuscript. I am grateful to all these institutions and for the help and support of many colleagues and friends. In Cambridge my thanks extend to all staff, although Nick Jardine, Simon Schaffer, Jim Secord and Nick Hopwood have been most involved in the different phases of the work. In Paris John Krige, Dominique Pestre, Jean-Paul Gaudillière, Ilana Löwy and Patrice Pinell offered work space, friendship and many useful discussions. In Hamburg I thank Paula Bradish, Regine Kolleck and Jens Lachmund for endless conversations on my monthly visits. For the later phase of my work at Cambridge special thanks go to Richard Henderson and the History, Archive and Library Committee of the Laboratory of Molecular Biology for unconditional support and complete freedom in completing work on the book, and to Peter Lipton for welcoming me as research associate and affiliated lecturer in the department.

Work on the book would not have been possible without the help of many librarians and archivists who retrieved and granted access to archival holdings (often specially lifting existing restrictions) and without the generous collaboration of numerous participants who agreed to be interviewed and who made their private archives available to me. For their substantial help with archival research I would like to mention in particular Elisabeth Leedham-Green at the University Archives in Cambridge; Mary Nicholas and later Alexandra McAdam Clark and Phillip Thoms at the Medical Research Council; Margaret Brown and later Kirsty Knott at the Laboratory of Molecular Biology in Cambridge; Colin Harris and Steven Tomlinson at the Bodleian Library, Western Manuscripts Room; Denise Ogilvie and Madeleine Brumerie at the Pasteur Institute; Thomas Rosenbaum at the Rockefeller Foundation Archives and Christopher Booth who wrote a decisive letter. More specific acknowledgements are given at particular places in the text.

For granting interviews (in some cases several ones), engaging in long telephone conversations, answering my many questions in writing, providing papers or photographic material from their personal collections (many did all of that) I thank: Uli Arndt, Antony Barrington Brown, David Blow, Sydney Brenner, Robert Bud and his assistants at the Science Museum, Francis Crick, Robert Diamond, Richard Dickerson, Rita Fishpool, Mick Fordham, Michael Fuller, Sandy Geis, John Gray, Freddie Gutfreund, Brian Hartley, Richard Henderson, Jonathan Hodgkin, David Hopwood, Hugh Huxley, Vernon Ingram, Bernard Katz, John Kendrew, Richard Keynes, Annette Lenton, Joan Mason, Hilary Muirhead, Ann Newmark, Leslie Orgel, Peter Pauling, Richard Perham, Max Perutz, Philip Randle, Paul Rayner, Alexander Rich, Martin Richards, Fred Sanger, Eileen Southgate, John Spice, Bror Strandberg, John Sulston, David Wheeler, Maurice Wilkes, Elie Wollmann and Tony Woollard.

Sadly, some of the people just mentioned to whom I owe much gratitude are not here anymore. Among these are, in particular, John Kendrew who supported the project from its very inception, and Margaret Brown who, on her own initiative, had started creating an archive of the institution, the LMB in Cambridge, in which for over twenty years she had served as the director's secretary.

All along, friends, colleagues, students and family, at home and abroad, have helped with comments, reading drafts, offering encouragement and keeping up my spirits. Robert Olby has offered continuous support throughout the project; he also read and commented on substantial parts of the manuscript. Conferences and an exhibition organised with Harmke

Kamminga, Jean-Paul Gaudillière, Nick Hopwood and Bruno Strasser helped in focussing questions and clarifying issues, as did discussions with colleagues at seminars and conferences where I presented my work. Among those not yet mentioned who have followed this work over several years and have helped it along in many ways were especially Pnina Abir-Am, Angela Creager, Lily Kay (now sorely missed) and Hans-Jörg Rheinberger. Nearer to home Jim Secord was always approachable and exceptionally helpful. He also volunteered to read the whole manuscript, as did Nick Hopwood with an earlier draft, and Richard Henderson, Mark Bretscher and John Finch. Their comments and critique as well as encouragement were invaluable. Single chapters or larger parts of the manuscript were read by Jon Agar, Mary Croarken, Jean-Paul Gaudillière, John Krige, Tim Lewens, Ilana Löwy, Sybilla Nikolow, Maria Jesus Santesmases, Simon Schaffer and Mona Singer. They have helped to correct mistakes and sharpen arguments. Any remaining errors are of course my own responsibility. In preparing the photographic material Kirsty Knott of the LMB and Ian Bolton of the Anatomy Department Audio Visual Media Group have been most helpful. Patricia Fara, Annette Faux, Judy Foreshaw, Richard Newbury and Jim Secord helped with the very final work on the manuscript. To all of them my special thanks.

At the Press, my thanks go to Richard Fisher who, as editor of the book, provided the necessary dose of excitement and encouragement for the project and, assisted by Sophie Reed and Teresa Sheppard, successfully guided the manuscript through its different stages; to Frances Brown who gently tweaked some of my not so English sentences into shape and most thoroughly copy-edited the whole text; to Karl Howe who oversaw production; and to Helen Peters who compiled the index. Many thanks also to the three anonymous referees who provided feedback on the book proposal and on a first draft of the book.

Nina de Chadarevian most persistently asked when she would be able to see the book. I thank her for her nagging, Livia and Flavia for distracting me and Michael Cahn for making it all possible.

Permissions

The following institutions granted permissions to quote from their holdings: BBC Written Archives Centre; Bodleian Library, Oxford; Churchill Archives Centre; Medical Research Council; Archives de l'Institut Pasteur; Oregon State University; The Rockefeller Foundation; The Royal Institution of Great Britain; The Royal Society; Sheffield University Library; University of Cambridge; The Wellcome Trust.

Acknowledgements for permission to reproduce photographic material are included in the figure captions. Every reasonable effort has been made to obtain permission to reproduce the figures. In the few cases where the photographers could not be traced, I offer apologies in the hope that they will not be displeased to see their photographs reproduced in this book.

An earlier version of chapter 5 was originally prepared for S. de Chadarevian and N. Hopwood (eds.), *Displaying the Third Dimension: Models in Science, Technology and Medicine* (Stanford: Stanford University Press, in press). Parts of chapter 9 first appeared in M. Hagner, H.-J. Rheinberger and B. Wahrig-Schmidt (eds.), *Objekte, Differenzen und Konjunkturen: Experimentalsysteme im historischen Kontext* (Berlin: Akademie Verlag, 1994), pp. 181–200 and in *Studies in History and Philosophy of Biological and Biomedical Sciences* 29 (1998): 81–105. I thank Stanford University Press, the Akademie Verlag and Elsevier Science Ltd respectively for permission to reprint the material.

Chronology

Events in British and international politics (including science policy) mentioned in this book

1935	Establishment of Tizard Committee for the development of radar
September 1938	'Munich crisis': Britain signs non-agresssion pact with Germany; Central Register of scientists and technical personnel for national work in an emergency set up
1939	Beginning of World War II
1940/1	Battle of Britain
1944	British nuclear physicists join the Manhattan Project
May 1945	Capitulation of Germany
July 1945	Labour Party wins general election; Clement Attlee Prime Minister
August 1945	Atomic bombs dropped on Hiroshima and Nagasaki; end of World War II
1947	Britain's atomic bomb project launched; politics of 'nuclear deterrent'
1950–3	Korean War
1951	Festival of Britain; Conservatives win general election
1956	Suez crisis
1957	Sputnik (first artificial satellite) launched by the Soviet Union
1959	Lord Hailsham nominated as first Minister of Science
1962	Cuba missile crisis
1963	Trend Committee reports on the organisation of civil science
1964	Labour Party wins general election on a 'scientific' platform
1971	Rothschild Report on the organisation and management of government R&D
January 1972	Britain signs treaties of accession to European Union; effective January 1973
1979	Conservatives win general election; Margaret Thatcher becomes Prime Minister
1980	Celltech founded

Abbreviations

Archives

AIP	Archives de l'Institut Pasteur
APSL	American Philosophical Society Library, Philadelphia, PA
BBC-WAC	BBC Written Archives Centre, Reading
Bodl. Lib.	Bodleian Library, Oxford (Western Manuscripts)
CUL	Cambridge University Library
LMB Archives	Laboratory of Molecular Biology Archives, Cambridge
MRC Archives	Medical Research Council Archives, London*
RAC	Rockefeller Archive Center North Tarrytown, NY
RI	The Royal Institution of Great Britain, London
RSL	The Royal Society Library, London

Other abbreviations used

CERN	Centre Européen de Recherche Nucléaire
CNRS	Centre National de la Recherche Scientifique
CSP	Council for Scientific Policy
DGRST	Délégation Générale à la Recherche Scientifique et Technique
EDSAC	Electronic Delay Storage Automatic Calculator
EMBC	European Molecular Biology Conference
EMBL	European Molecular Biology Laboratory
EMBO	European Molecular Biology Organisation
ICI	Imperial Chemical Industries
LMB	MRC Laboratory of Molecular Biology
MRC	Medical Research Council
NIH	National Institutes of Health (USA)
NRDC	National Research Development Corporation

* Most MRC files consulted for this book for which the thirty-year rule has lapsed have meanwhile been transferred from MRC Head Office to the Public Record Office at Kew. These can be identified in the Public Record Office online catalogue using the file references given here. 'Vol.' should be replaced by 'pt', e.g. the file reference E243/109, vol. 1 should be entered as E243/109 pt1 (no comma).

1

...

Introduction

Hardly a common term in the 1950s, molecular biology is now expected to take the dominant role in the twenty-first century that physics played in the twentieth. Our understanding of life, health and disease is as much dependent on knowledge produced by molecular biologists as the fabrication of food and drugs, trials in court, and new ways of waging wars. How, we need to ask, has molecular biology acquired such a dominant position in our society?

To approach this question the book focuses on the Laboratory of Molecular Biology in Cambridge (formerly the Medical Research Council Unit for the Study of Molecular Structure of Biological Systems) which, in the 1950s and 1960s, became an international symbol of the spectacular development of molecular biology. This was the laboratory in which, in 1953, Watson and Crick presented their double helical model of DNA. However, as I will show, this event alone, which in the 1950s attracted far less attention than it does today, cannot explain the explosive growth of the laboratory or the creation of the new science. Rather, the book takes a longer-term view, engaging with events from the immediate postwar years to the late 1970s. The history of the laboratory starts in the mid-1940s, when opportunities created by the postwar reconstruction of the sciences were used to establish new ways of producing knowledge about biological structures and processes in the laboratory. The late 1950s and 1960s saw an extraordinary expansion of activity, the formation of new networks and the use of the science policy arena to promote the new science (only now presented as molecular biology). These events set the stage for later government policies and industrial investments which in turn opened up new opportunities and expectations. Molecular biology, I will argue, was produced as much in the laboratory as in the political and the public arena. Only an in-depth study, as the one presented here, can reconstruct these processes in necessary detail.

A local study

Set up by the Medical Research Council (MRC) in 1947 as a two-man unit dedicated to the crystallographic study of proteins, the laboratory was quickly made an 'obligatory passage point' for the new science of molecular biology. The Queen had only just inaugurated the new four-storey laboratory in 1962 when James Watson, Francis Crick, John Kendrew and Max Perutz were awarded Nobel Prizes for their work in the unit on the structure of DNA and proteins. According to one witness, 'this public ratification of the eminence of the MRC Laboratory was the most important factor in the general recognition of molecular biology as a distinctive scientific discipline' (Fruton 1992, 210–11).[1] The new fame of the MRC Laboratory of Molecular Biology (LMB), together with the opportunity of numerous fellowships to travel to Western Europe, soon attracted a large number of American postdoctoral students to Cambridge. In the years of the expansion of the American universities, good career prospects awaited these researchers on their return. In this way the 'culture' of the LMB was exported to other centres. In Britain itself the LMB so dominated the field that, by the mid-1970s, the 'failure' to 'seed' the subject in universities started to be perceived as a problem.[2]

The pivotal role of Cambridge in the development of molecular biology allows the reconstruction of events and practices that have come to be seen as central to the history of the field and of the mechanisms by which they became disciplinary landmarks.[3] I analyse in particular Perutz and Kendrew's pioneering X-ray analysis of protein structure, including Kendrew's early use of the experimental electronic digital computer at Cambridge; Watson and Crick's work on the structure of DNA and the central role attributed to it as the 'origin' of the new field; early attempts at 'cracking' the genetic code; the crucial role of Fred Sanger's sequencing work for the particular research culture developed at Cambridge which combined structural and genetic approaches; Sydney Brenner's

[1] Joseph Fruton, who himself never accepted that molecular biology was anything else than biochemistry, also suggested that the appearance in 1966 of the *Festschrift* for Max Delbrück, *Phage and the Origins of Molecular Biology*, was influenced by the public esteem gained by the MRC Laboratory (Fruton 1992, 211; Cairns, Stent and Watson 1966). This volume marked the beginning of a whole series of books and articles debating the 'origins' of molecular biology.

[2] 'Cell Board Subcommittee set up to review molecular biology. Unconfirmed minutes of first meeting, 21 July 1975', file A147/14, vol. 1, MRC Archives.

[3] Molecular biology is here taken to mean more than just 'molecular genetics', as indeed was always the case in Cambridge. On the history of the term and its usage in Cambridge see below, especially chapter 7. On the effort to recover research traditions which do not fall under the narrow definition of molecular genetics to a larger 'history of molecular biology project', see Zallen (1992) and Burian (1996).

creation of *Caenorhabditis elegans* as a new model organism for the study of development; and César Milstein and Georges Köhler's invention of monoclonal antibodies. While monoclonals gave rise to a fledgling biotechnology industry, in the late 1980s the plan to sequence the whole DNA of the worm became a pilot for the Human Genome Project and the flagship project of the newly created Sanger Centre, one of the largest sequencing centres in Europe.

However, my choice of a local study is based not so much on the widely recognised excellence of the Cambridge laboratory as on the thesis that widely distributed experimental practices and scientific institutions embody local expertise and negotiations. It is only by studying in detail these local solutions, the resistances they met, and the eventual 'export' of local practices to other laboratories, that one may understand the construction of a new scientific field. Proceeding locally, therefore, need not mean being provincial. The detailed investigation of the Laboratory of Molecular Biology thus offers the possibility of studying the boundaries and connections of local, national and international developments. These structures and mediations get lost in more wide-sweeping accounts. By the same token, the study does not take for granted the excellence of the laboratory or of Cambridge science more generally, but analyses how this one laboratory came to play such a central role in the international establishment of molecular biology – at times despite or even because of local resistances.[4] Cambridge, and especially the Cavendish Laboratory where the MRC unit was first housed, boasted a long and glamorous tradition of research in the natural sciences. However, unlike Oxford, Cambridge voted to contain expansion after World War II. This choice, in addition to the fact that all decisions in the university are made by mixed bodies in which all faculties, as well as the colleges, are represented, made it difficult for new projects to find approval, especially if these were 'no one person's business'. Because of these circumstances, molecular biology at Cambridge developed mainly outside the precincts of the university.[5]

In my analysis of the mechanisms by which the laboratory came to assume such a privileged position in the establishment of the new science, I draw on current approaches in science studies. In particular, I aim to combine a fine-grained analysis of work at the laboratory bench with an analysis of the representational, institutional and political strategies

---

[4] I would like to distinguish this undertaking from the attempt to define the institutional conditions for 'successful science', most often measured in Nobel Prizes. In my understanding, 'success' is socially (and always retrospectively) attributed and historically contingent.

[5] Especially on the Cavendish see Crowther (1974); on the history of Cambridge University more generally see Brooke (1993) and Leedham-Green (1996).

employed to establish the field at the local as well as the national and international level, in competition with other fields and in the face of multiple resistances.[6] Molecular biology, I suggest, was constructed as much at the bench and through the circulation of tools, models and postdoctoral researchers as in institutional negotiations or political committees, in the television studio and in participants' disputes on the 'origins' of the field.

The postwar era

The postwar era in Britain has been depicted in various ways.[7] In political terms it was marked by the loss of empire and the resuming of the 'special relationship' with America, by the onset and hardening of the Cold War and the division of Europe by the Iron Curtain. Economically it was a time of recovery and growth and of low unemployment. Social reforms immediately after the war had introduced a National Health Service, a new system of social security and free secondary education. In addition, economic growth meant material affluence for all classes. However, class divisions remained strong and gender relations remained basically unaltered. The 1960s were marked by rebellion, mainly by the young generation, against these continuing divisions and established political and cultural values. A recent exhibition at the Imperial War Museum has presented a portrait of this era under the motto 'from the bomb to the Beatles', while others have described the two-and-a-half decades following World War II as 'defiant modernism' and as characterised by big technological projects.[8]

The postwar era, as it appears in this book, represents a time of rising science budgets and high public esteem for science.[9] Both were a direct outcome of what was generally perceived as the crucial contribution of scientists to winning the war. In Britain the general opinion was that radar had saved the country from occupation by Hitler's troops. In addition,

---

[6] For the focus on experimental practices see Galison (1987; 1997), Gooding, Pinch and Schaffer (1989) and Pickering (1992; 1995); more specifically for the life sciences, see Latour and Woolgar (1986), Clarke and Fujimura (1992) and Kohler (1994). On the 'place' of knowledge and the export of local practices see Shapin and Schaffer (1985), Latour (1987; 1988) and Ophir and Shapin (1991). On representations see Lynch and Woolgar (1990), Rheinberger (1997) and de Chadarevian and Hopwood (forthcoming). On instituting science see Lenoir (1997). On constructivist approaches in the history of science see Golinski (1998) and the review by Kohler (1999).

[7] On Britain after World War II see Marwick (1982), Morgan (1990), Holland (1991), Hennessy (1992) and Clarke (1996). See also Milward (1984) and Ellwood (1992) on the political and economic reconstruction in Europe.

[8] On the exhibition at the Imperial War Museum see Gardiner (1999). 'Defiant modernism' was the title of a conference held at the Science Museum in London, 25–26 June 1999. See also Bud *et al.* 2000, 158–83.

[9] Science policy became a central political issue only in the late 1950s; see Vig (1968). On science and scientists in Britain after World War II see Edgerton (1996a), Gummett (1980) and Wilkie (1991).

penicillin (a British discovery) and the atomic bomb (developed with the decisive help of British scientists and engineers) had saved thousands of lives, the first by controlling infections in wounded soldiers, the second by ending the war.[10] British scientists had fought for an active role in the war effort. After the war they publicised their contributions and argued for an equally important role of science in postwar reconstruction. 'The problems of reconstruction', Archibald V. Hill, Nobel Prize winning physiologist and high-level scientific administrator and military adviser, noted in his diary, 'will be to an important extent scientific ones.'[11] Politicians were responsive to these views and approved the disbursement of large government funds for research and development. While always small when compared to the military R&D budget, in the early and mid-1960s the annual budget for civil science was growing at an average rate of 13.5 per cent in real terms. Studying the spending for military R&D, Edgerton has suggested that postwar Britain has been as much a 'warfare state' with a solid industrial base as a 'welfare state' (Edgerton 1992, 141).

In Britain, as in France, technological prowess, symbolised above all in an independent atomic bomb project, made up for the loss of empire. The American reconstruction plans for Europe included important measures for the support of science and technology as pillars for security and economic welfare.[12] In the rising tensions of the Cold War, the United States built their own supremacy and that of the West more generally on scientific and technological dominance.

The postwar lustre of science began to fade with the questioning of the role of science and technology in the Vietnam War, loudly voiced on American campuses and throughout Europe in the wake of the student revolts. In Britain, however, civil science budgets continued to rise (if at a lower rate) until the mid-1970s when the oil crisis, general recession and the following devaluation of the pound imposed cuts on government expenditure for science. My study covers this 'long' postwar period.

The making of a new science

The rapid growth of molecular biology after World War II is often assumed to have occurred almost exclusively in the three countries that dominated

---

[10] On the role of penicillin and the myths surrounding it in the reconstruction of Britain and of her self-image see Bud (1998).

[11] A. V. Hill, 'Memoirs and reflections' [unpublished manuscript, p. 568], Hill Papers, AVHL I, 5/4, Churchill Archives Centre, Cambridge.

[12] On Britain see Gowing (1974a; 1974b) and Agar (1998c); on France see Hecht (1998). On the decisive importance of the so-called Berkner Report on Science and Foreign Relations of April 1950 for the formulation of the American policy towards European science see Needell (1996).

the winning coalition: Britain, France and the United States. It has been suggested that an important reason lay in their relative economic strength (Allen 1978, 188). However, economic strength alone does not seem to explain why molecular studies of life processes were privileged over others. To account for the meteoric rise of molecular biology after World War II, the historian Edward Yoxen has suggested that molecular biologists were part of the new scientific establishment which after the war directed the new flow of money towards specific research projects. He has also argued that a biology which conceived of life in terms of a programme fitted the managerial research system which took hold after World War II (Yoxen 1981; 1982). This last thesis, however, fits only a very narrow research agenda, one which to some extent became dominant in the 1970s with the new recombinant DNA technologies and their commercial applications. Focusing on earlier developments, I will argue that molecular biology in Britain (as a distinct scientific enterprise under this name) took form only in the late 1950s – the Cambridge laboratory being the first institution which officially carried that name. In the 1940s and 1950s, much of the research later claimed by molecular biologists (including Watson and Crick's work on the structure of DNA) fell under the heading of biophysics, a larger and more diverse field which attracted considerable support after the war.[13]

By drawing attention to the fortunes and legacies of postwar biophysics I do not intend to create a new 'origin' account or to add a new candidate to the number of disciplines which allegedly contributed to the emergence of molecular biology. My intention is rather to avoid starting with a cognitive (or any other) definition of the field (as most histories of molecular biology do) and to study disciplines as political and cultural institutions.[14]

The opportunity for biophysics after World War II stemmed from a host of new physical approaches developed for the war effort. The hope of

---

[13] For a similar thesis regarding the development of molecular biology in America see Rasmussen (1997a). On the making of molecular biology in France see Gaudillière (1991; in press). For efforts to build up molecular biology in other European countries, including Belgium, Germany, Switzerland, Spain and Italy, see Deichmann (1996, chapter 7), Burian and Thieffry (1997), Santesmases and Muñoz (1997a; 1997b), Strasser (in press) and de Chadarevian and Strasser (forthcoming). On Japan see Uchida (1993).

[14] Robert Kohler in his history of biochemistry also presented disciplines as political institutions (Kohler 1982). While building on this notion, the present study aims to discuss disciplines not just in terms of intellectual programmes and academic politics, but by considering the institution of experimental practices. In his later work Kohler himself moved to consider the material culture and moral economy of experimental practices, but set this approach apart from the study of disciplines and their institutions (Kohler 1994, especially p. 14). On the need to combine the study of disciplines with the study of experimental practices see Lenoir (1997, especially introduction and chapter 3).

turning these technologies, especially those of nuclear physics which had led to the celebrated yet deadly weapon, to peaceful ends gave biophysics its cultural and political appeal. A 'physics of life', with the promise of biomedical applications, fitted neatly into the political discourse of postwar reconstruction. Biologists, physicists and medical researchers alike seized on this opportunity and took advantage of new government funds made available for 'fundamental research'.

Biophysics in postwar Britain comprised at least three different groups: the 'radiation group' which investigated the effects of radiation on the body and ways to protect it as well as biological and medical uses of radioactive isotopes; the 'nerve–muscle group' which exploited new recording devices developed in the context of radar research; and the 'structural group' which used a series of physical techniques and especially X-ray diffraction, decisively aided by the advent of electronic computers, to study complex biological structures. All three groups built on prewar research traditions, but greatly expanded after the war.[15]

The main patron for biophysics in Britain was the Medical Research Council, which had seen its own funds and authority greatly increased as an effect of its role in the wartime mobilisation. The MRC Unit for the Study of the Molecular Structure of Biological Systems fell under this heading, as did, for instance, the Unit for Biophysics at King's College London and the Radiobiological Research Unit at the Atomic Energy Research Establishment at Harwell, all set up in 1947.

In 1957, the MRC unit at Cambridge changed its name to MRC Unit for Molecular Biology. This was a local move which followed a serious institutional crisis for the unit, then still housed in the Physics Department. The only solution was seen in the application for a new and independent laboratory. The plan included new allies and required a new name, then also adopted by the unit. Significantly, around the same time, Kendrew changed the name of the new journal, the editorship of which he had taken on, from *Journal of Molecular Biophysics*, as originally proposed, to *Journal of Molecular Biology*. The journal, edited for many years from Kendrew's college office, is generally credited with having done most to propagate the term. By that time the term 'biophysics', which had served to attract funds after the war, was losing its appeal. Such names, I suggest, are more than mere labels.

In the 1960s, Cambridge molecular biologists skilfully used political channels and connections, many of which dated back to wartime acquaintances, to put the promotion of their new science at the national and the

---

[15] On the centrality and explosive political implications of research on radiation damage and protection in the atomic age see Beatty (1991) and Lindee (1994).

European level on the governmental agenda. In these negotiations, science policy was as much a tool in the hands of scientists as a governmental tool to manage and regulate research. Changes in government policies in the early 1970s, and later regarding the development of biotechnology in Britain, also affected work practices and the position of the Cambridge laboratory.

Britain's special relationship with the United States, a crucial element of Britain's foreign politics which long dominated over European commitments, not only played a key role in Britain's atomic politics, but in many ways also affected the building of molecular biology. In the immediate postwar years, when there were restrictions on foreign currencies for imports, American grants were crucial to buy scientific apparatuses manufactured abroad.[16] On other occasions, however, the MRC, as a government body, was keen to underline that British science could stand on its own feet and expected its leading scientific staff to attract American researchers to Britain rather than to travel to learn from them. On this ground the MRC, for instance, denied Perutz permission to take up a Rockefeller Foundation Travelling Fellowship to visit American laboratories in 1948.[17] Later, scientists used the growing predominance of American biomedical sciences and the fear of a 'brain drain' from Britain to America to argue for more funds. Similar arguments were brought forward in the negotiations for the European Molecular Biology Laboratory (EMBL), which saw Cambridge molecular biologists centrally involved, despite strong opposition from their own peers. As already mentioned, postdoctoral fellowships for American researchers to spend up to five years in Europe as part of their education also played a crucial role in the economy of the LMB and the export of its research culture. The number of available fellowships increased sharply in the wake of Sputnik, the first space satellite launched by the Soviet Union, and America's politics of stepping up the Cold War mobilisation of science and technology throughout the Western alliance. However, despite, or perhaps in response to, America's hegemony, British molecular

---

[16] While the Rockefeller Foundation stopped supporting Perutz (as other European grantees on the natural sciences programme) directly after the war, the Cambridge MRC unit continued receiving grants for additional expenses, including fellowships for its members to travel to the United States, until the mid-1960s. Money also came from other American grant-giving bodies. On the economic situation of Britain after World War II and the convertibility problem of the pound see Milward (1984) and Dore (1996).

[17] Perutz later claimed that, had he gone to America, he might have found out from Pauling that the bond which links amino acids in proteins is planar. This information might have saved the Cambridge laboratory the embarrassment of publishing a structure of the polypeptide chain which was not consistent with stereochemical data. The problem was later brilliantly solved by Pauling through model building (Olby 1994, 267–95).

biologists developed and stressed their own research traditions, or what later became known as the 'British (or Structural) School of Molecular Biology' (Kendrew 1967).[18] In some research fields, like protein X-ray crystallography, British scientists reckoned they held the world lead. When the crystal structure diagrams (including some of proteins) were used to launch a novel design for interior decoration at the Festival of Britain in 1951, the strong national tradition in that branch of science was underlined. The emergence of local research traditions and national 'schools', despite the importance of international exchanges and networks in the making of molecular biology (Abir-Am 1992b), further justifies local and national historical studies. Only they can provide the basis for comparative studies and new 'big picture' accounts, as I will argue below.

Histories of molecular biology

The account presented here addresses some key historiographical issues. I will discuss three main points: the place of 'origin' and discovery accounts; the role of World War II; and the relations of local and 'big picture' accounts.

Participants, historians and science writers have given ample attention to the story of Watson and Crick's elucidation of the structure of DNA (e.g. Kendrew 1967; Sayre 1975; Portugal and Cohen 1977; Watson 1980; Judson 1994; Olby 1994; Edelson 1998).[19] Through popular writings and media presentations it has become one of the most widely known events in the history of science. This book is no exception to the trend. Having become such an integral part of the existing history and iconography of the field and with the story being located at Cambridge, in the institution which lies at the centre of this study, the subject imposed itself. In taking it up, my account continues to depend on the dominant historiography. However, I have tried to approach it in a new way.

[18] The notion has been taken up by historians; see Olby (1994) and for a critique Abir-Am (1985). On 'schools' as historiographical topic see Geison and Holmes (1993).

[19] Molecular biologists have been particularly active in writing the history of their field. In the view of one participant, this can be explained by the 'fantastically rapid' development of the new science, which allowed molecular biologists to look back on their own research and that of their colleagues with an unprecedented 'depth of historical perspective' (Stent in Watson 1980, ix). Most of the accounts are autobiographical, but for exactly that reason claim authenticity. See especially Watson (1968), who inaugurated the trend. For a vocal defence of the figure of the scientist–historian see Fruton (1992; 1999). For a critique of Fruton's position see de Chadarevian (1996b). The standard references in the history of molecular biology are still Olby's thorough study, though only up to 1953, and Judson's highly readable but rather journalistic account. Both books have recently been reprinted (Judson 1994 and Olby 1994).

The retrospective character of discovery and 'origin' accounts and their legitimatory functions have been amply demonstrated. Such analysis has also been applied to the DNA story (Abir-Am 1982b; 1985; see also Forman 1969–70; Olby 1979; Brannigan 1981). While I draw on these studies, my strategy has been to place the double helix back into its local context and to examine the role of Watson and Crick's work in shaping research traditions and institutional developments in the laboratory in which it was performed. I argue that the double helix played only a subordinate role in the negotiations over the future of the Cambridge unit. It was rather in the course of these events and in the following debates concerning the origins and boundaries of the new science that the helix gained its central role (Part II). The uncertainties surrounding the fate of the 'original' model of which only a few pieces survive, I shall suggest, reflect the retrospective construction of the helix's importance. The few surviving plates and model bits were later used to build a model 'the nearest there is' to the original one, for display in the Science Museum in London.[20] Stressing the retrospective construction of the year 1953 as the origin of molecular biology, I do not intend to belittle Watson and Crick's scientific achievement. The making of a science, however, requires more than scientific 'breakthroughs', as the scientists involved seem very well to know. By placing the double helix back into its local historical context, we can retrace these processes and negotiations and reconstruct the work which was necessary to turn the double helix into the icon of a new science.

Scholarship in the history of molecular biology has focused on the funding policies and social agenda of the Rockefeller Foundation in the 1930s and 1940s (Abir-Am 1982a; 1987; Kohler 1991; Kay 1993a), on the development, politics and industrial exploitation of recombinant DNA technologies in the late 1970s and 1980s (e.g. Yoxen 1981; Krimsky 1982; 1991; Bud 1993; Wright 1994; Rabinow 1996; Gottweis 1998a; Thackray 1998) and on the most recent developments regarding the Human Genome Project, which came with a (modest) budget for historical research (Kevles and Hood 1992; Cook-Deegan 1994; Sloan 1999; Fox Keller 2000). In addition, much of this work has been dedicated to developments in the United States. In these studies, World War II, if mentioned at all, is mainly portrayed as an 'interruption' of prewar pursuits or in very general political terms, while our view of postwar science is dominated by studies of the physical sciences (especially nuclear physics) and engineering and the making of 'big science' in the form of the military–industrial–academic complex (e.g. Forman 1988; de Maria, Grilli and

---

[20] The quotation is from the label in the Science Museum.

Sebastiani 1989; Galison and Hevly 1992; Leslie 1993; Galison 1997).[21] Molecular biology in the 1950s and 1960s was not 'big science' in the sense physics was (although to the extent to which molecular biology was part of 'biomedicine' a similar case could be made).[22] Yet, like nuclear physicists, molecular biologists (and before them biophysicists) responded to postwar concerns and effectively used opportunities created after the war and as an effect of the war to further their science. Their efforts, I will argue, set the stage for later developments which gave molecular biology the privileged position it occupies today. Focusing on the changes following World War II does not mean denying continuities with prewar pursuits. However, stressing the opportunities offered by World War II and postwar reconstruction does involve a reassessment of the role of the Rockefeller Foundation in the creation of the new science. The Foundation's 1930s programme, designed by Warren Weaver, aimed at funding chemical and physical approaches in the life sciences, is generally viewed as a decisive factor in the foundation of molecular biology. The institutionalisation of molecular biology in the late 1950s and 1960s, however, cannot be understood as merely subsidiary to intellectual programmes and practices set in place in the interwar years.[23] The focus on Britain not only fills an important gap in the literature, but also moves attention to developments in Europe more generally, despite decisive differences in the way the war affected scientific developments in other European countries.

Finally, how does my local study relate to 'big picture' accounts of the history of molecular biology? Overarching histories of the field continue to be produced.[24] Yet despite an increased focus on details of

---

[21] There is, however, a growing number of booklength studies on the biological and biomedical sciences in the 1940s to 1960s on which my research draws; see for instance Gaudillière (1991), Kay (1993a; 2000), Lindee (1994), Fox Keller (1995a), Rasmussen (1997b) and Rheinberger (1997).

[22] Lily Kay has argued that with the development of large instruments like the Tiselius electrophoresis apparatus, the life sciences, in the 1930s and 1940s, entered the era of 'big science' (Kay 1988). However, as Kay herself has pointed out, postwar commercialisation made 'big apparatuses' more affordable and easy to use. It remains none the less true that investigations at the subcellular level required an increasing quantity of costly apparatus. Molecular biologists partly shared some of the big instruments developed for nuclear physics; this is particularly true of their use of electronic computers in the 1950s and early 1960s and of synchroton radiation, from the 1970s. Both instruments were used for structure determinations of biological molecules.

[23] On molecularisation as a more long-term strategy, involving the state and private funding agencies, the laboratory, industry and the clinic, see de Chadarevian and Kamminga (1998b). Molecularisation, however, gained new momentum with the biomedical mobilisation of World War II. Molecular biology (as a disciplinary formation) was one face of these later endeavours; see especially the introduction and the chapters by Creager, Gaudillière and de Chadarevian in de Chadarevian and Kamminga (1998a).

[24] Recent examples include Morange (1998a) and Corbellini (1999). On the continuing need for 'big pictures' see Secord (1993).

experimentation and instrumentation as well as on 'social factors', they all start with a definition of what molecular biology is and remain 'result' orientated. In these generalised pictures, local contexts, the prime sites of knowledge production, tend to get lost. Is this a necessary price to be paid to gain a wider understanding of historical processes? Questioning the usefulness of traditional big pictures (including the common kind of disciplinary histories), Ludmilla Jordanova has called for a new kind of big picture emerging from in-depth local studies rather than from wide-sweeping generalisations. Only fully contextualised local accounts, she claims, can provide an understanding of the processes involved in know-ledge production at a given time. If such an understanding is achieved, 'the result would be a big-picture history' (Jordanova 1993, 480).[25]

My account of the making of molecular biology at Cambridge is not a history of molecular biology in the grand sense. It is not even a history of one institution. In both respects it is too selective in its choice of topics. However, the topics were selected in such a way as to highlight the pro-cesses involved in the local production of the field. Because of the very fact that it insists on local contingencies, the account given here cannot be generalised. But it points to the multiple arenas and the complex web of negotiations involved in building molecular biology after the war. More lo-cal studies will enrich the picture. But any new picture which emerges will form in the intersections of these local accounts, not by moving beyond them.

I would like to mention one more point which has bothered me through-out. A brief flick through the pictures included in this volume will suffice to confirm that the story here told is dominated by male actors, with most women being relegated to crucial but clearly subordinated tasks back-stage, from washing up, contouring, drawing, and operating machines, to running the cafeteria (molecular biology was hardly different in that from other sciences or public domains). I do at times point to this unequal divi-sion of labour and to unguarded value judgements regarding women and their work, but do not probe any further. Strikingly, protein crystallography was a field which in its beginnings saw several key women practitioners. But after the war, few new women joined. If, as I argue, skills and networks acquired during the scientific wartime mobilisation projects played an im-portant role in the postwar restructuring of the sciences, this certainly put women at a disadvantage. Other authors have focused on the few woman

[25] Jordanova's position resonates with more general discussions regarding the explicative value of qualitative in contrast to quantitative approaches in the social sciences; see e.g. Bude (1988).

actors in the field and the ways their gender has shaped their careers (e.g. Sayre 1975; Fox Keller 1983; 1996; Ferry 1998). More importantly perhaps, we start to understand how key theoretical notions of molecular biology were deeply gendered (Fox Keller 1995a). Though outside the present frame of analysis, these studies seem crucial for an understanding of the cultural inscriptions of molecular biology.

Too early, too late?

Writing on recent science raises a particular set of questions (Söderqvist 1997). By definition, 'recent science' implies that at least some of the actors are still alive. For some people this circumstance, in addition to the missing historic perspective on events which are too close to our own times, imposes serious limitations on historical scholarship. From another point of view, the possibility to elicit information from some of the actors offers the unique chance greatly to enrich our historical records. Oral history projects aim to safeguard this information for later generations of historians. Even among historians of recent sciences, however, the value of 'oral' information is hotly debated (de Chadarevian 1997a). Regarding written sources, the situation is no less complex. While on the one side there is an overflow of records, on the other side many records on which historians are used to relying, such as letters or handwritten memos, have been supplanted by modern technologies of communications, especially the telephone and, more recently, electronic mail (a change with which later historians of the period will also have to come to terms). In addition, many records are still in private hands or unavailable because of the various restrictions regulating access to recent institutional and personal records or simply because they remain uncatalogued. Finally, the overflow of material forces institutions and individuals to reduce their records drastically and also to destroy recent files. The selection will invariably disappoint some historians. Among the MRC records of the LMB, for instance, files relating to the purchase of equipment are difficult to find.

All these conditions have shaped this book. I have used interviews, though mainly as a guide to archival work and to test interpretations. Having become accustomed to being able to rely on these personal exchanges, the death of some participants while research was still in process became a serious loss. Often interviews led to access to personal papers without which this book could not have been written. Papers from private hands as well as from institutions were released in stages, sometimes unexpectedly. Other files remained closed. This often required resetting the agenda and rewriting. Several collections (among these the MRC, Kendrew and Monod Papers) have recently been recatalogued, which required extensive correction of references.

Once a special permission lifted the thirty-year-rule concerning MRC material, I gained access to an abundance of institutional records. Personal records were harder to come by (an exception being Kendrew's and Brenner's rich collections). This imbalance has certainly shaped my perception of the history and may well have induced me to focus more strongly on institutional and policy aspects than I originally planned.

Was it too early to embark on this history, as some colleagues working on earlier periods suggested? More records will become available. The LMB for one is actively engaged in collecting material and setting up its own archive. But by that time oral, written and material sources to which I had access may not be available any more. On several occasions my interest saved some papers or other records from being dumped, while especially the reorganisation of files at the MRC made unavailable some records I had consulted earlier. This book, no more or less than others, is as much a record of its own time as of the period it deals with. The archive will change and other stories, building on new insights, will follow.

Preview

Each of this book's three parts addresses one main question. Part I investigates the impact of the scientific mobilisation during World War II on the place of science and the fortunes of biophysics in postwar Britain. Chapter 2 offers a brief discussion of the involvement of British scientists in World War II. It points to the active mobilisation of scientists and their planning for postwar needs which started well before the end of the war. Chapter 3 investigates the moves which led to the creation of a number of biophysics groups, including the Cambridge unit, after the war. It focuses on the (wartime) careers of those involved in building up biophysics and on the postwar attractions of a 'physics of life' with the promise of medical applications. Chapter 4 focuses on protein X-ray crystallography, the key technology of the Cambridge unit. It discusses especially Kendrew's early use of the experimental digital computer at Cambridge, linking his interest in the new machine to experiences acquired in operational research during the war and reflected in his approach to research and in the organisation of his research work. Chapter 5 investigates the place of models in the experimental practice of protein crystallography and follows their display in public arenas, including early television programmes on science. Models,

I will argue, played a crucial role in promoting a new understanding of life, based on molecular structures and their functions.

Part II considers the role of the double helix in the making of molecular biology. It starts with a review of the debate on the 'origins' of molecular biology, fought out among practitioners in the late 1960s, which retrospectively attributed to Watson and Crick's work a key role in the history of the field (chapter 6). The rest of chapter 6 and chapter 7 place the work on the double helix back into its local context, studying investigative practices and institutional moves following the collaborative effort of the two researchers. I argue that the work on the genetic function of DNA undertaken in the unit, which got going seriously only once Sydney Brenner moved to Cambridge in 1957, did not follow directly from work on the double helix. However, 1953, the year of Watson and Crick's celebrated achievement, marked the beginning of a serious institutional crisis, precipitated by the move of the unit's patron in the Physics Department, the Cavendish Professor of Physics Lawrence Bragg, to London. Fruitless attempts to find another niche in the university finally led to the plans for a new and independent Laboratory of Molecular Biology. Crucial allies in this plan were Sanger and his group working on protein sequencing in the adjacent Biochemistry Department. In the negotiations for the new laboratory the double helix played only a subordinate role. Chapter 8 confronts this local account with the standard story on the place and role of the double helix in the history of the field.

Part III serves a double purpose. On the one hand, it offers a detailed reconstruction of molecular biologists' use of the political arena to promote their science and to turn it into an item on the governmental agenda. The 'Kendrew Report' on molecular biology, issued by the Council of Scientific Policy, the government's central advisory body for science, and the negotiations for EMBL as well as the controversies surrounding these initiatives form the centre of this discussion (chapter 10). On the other hand, Part III points to changes in government policies for science and to increasing competition among molecular biologists, both of which affected work practices and the place and role of the LMB. Two examples serve to illustrate this point: Brenner's worm project, which aimed at a 'complete' description of the worm, eventually including the full sequence determination of its genome, performed in a newly created sequencing centre, and the 'scandal' concerning the 'failure' to patent the technique of producing monoclonal antibodies, developed by researchers at the LMB in 1975, with the subsequent 'push' towards commercialisation under the Thatcher government (chapter 9 and 11). Chapter 11 also discusses the rising political expectations of medical 'applications' from 'fundamental' biological research. Scientists themselves had helped create these expectations when applying for biophysical and biomedical research projects after the war. In the 1970s these expectations led to new policy interventions, new justifications for research and changes in research directions. The interest of government and industries in the field was not the direct and sole effect of new technologies, especially those for cutting, recombining and multiplying DNA, born from fundamental research. Instead, the 'commercial turn' in molecular biology was well prepared by discussions and developments in the preceding decades. The conclusions reflect on the connections to current trends in the history and historiography of the biomedical sciences.

PART I

Postwar Reconstruction and Biophysics

Recalling a series of meetings and discussions in the 1930s which he saw as 'the prehistory of molecular biology', British embryologist Conrad Waddington wondered how much sooner the 'factual information' on which the 'new ideas involved in the Crick–Watson synthesis' depended, might have been discovered, 'if the Second World War had not disrupted the lines of thought which led in direction of them' (Waddington 1969, 321). Similarly, historian of molecular biology Robert Olby sees the protein and nucleic acid 'phase' of molecular biology divided by a 'curtailment and isolation' of research activities owing to the war (Olby 1994, xxii). These are only two of many examples in which scientists and historians of sciences see the war, if it is mentioned at all, as a mere impediment to the development of the sciences, and of molecular biology in particular. Intruding from outside, it hinders the free circulation of people and ideas, diverts minds and funds to war-related projects and, as in the case of Nazi Germany, uproots complete research traditions through forced emigration or extermination.

For some sciences this picture has already been radically revised. An increasing number of studies are dedicated to demonstrating the impact of World War II and the inscription of Cold War logic in the very *form* of the sciences. But in these studies molecular biology, the 'greatest' scientific achievement of the postwar years, is explicitly mentioned as having avoided this dominance (Hacking 1986, 240–1, 258).

In the following chapters I will propose a different picture. A first point arises straight away. In the early postwar years, molecular biology did not yet exist. Even Warren Weaver, the director of the natural science programme of the Rockefeller Foundation, often credited with having coined the term in 1938, in a postwar assessment of the programme conceded: 'Molecular biology is not a generally recognized term.'[1] The Cambridge unit, which is at the centre of this study, carried the long-winded name 'MRC Unit for the Study of the Molecular Structure of Biological Systems' and was housed in the Physics Department at the University of Cambridge. The work of the group, centred on the crystallographic analysis of protein structures, was most often referred to as biophysics, a diverse and loosely defined field which gained strong appeal and attracted considerable funding after the war. It is thus with biophysics I will deal in this first part. Only in the late 1950s did the Cambridge unit change its name to 'MRC Unit for Molecular Biology'. This was a local and political move which I will investigate in more detail in the second part of the book. Molecular biology was only one of the fields built on the legacies of postwar biophysics.

For many scientists and science administrators involved in the construction of biophysics in Britain, the war represented a decisive turning point. They seized just as promptly on opportunities presented by the postwar restructuring of the sciences as their counterparts in physics. The 'physics of life' with its promise of medical applications was sometimes opposed to the 'physics of death' geared to weaponry and destruction, but still drew on the same cultural, economic and scientific resources (Rasmussen 1997a). This point is supported by an increasing number of recent studies, dedicated especially to the development of the biomedical sciences in Cold War America (e.g. Strickland 1972; Yoxen 1982; Kay 1989; 1993, 177–193; 2000; Beatty 1991; Abir-Am 1992b; 1997; Rasmussen 1997b; Creager 1998 and in press; Gaudillière 1998 and in press).

Stressing the opportunities offered by the scientific mobilisation of World War II and postwar reconstruction does not mean denying continuities between pre- and postwar pursuits in the sciences. However, as has been pointed out, it is only by getting to grips with the *dis*continuities marked by World War II that we can understand the specificities of postwar science (Pickering 1993).

In his account of the 'birth' of molecular biology, Max Perutz, first and long-time director of the Cambridge unit, gave one section the heading 'The Science of War'. Briefly introducing the first researchers who joined the laboratory, he remarked: 'What

---

[1] W. Weaver, 'NS Program' [1946]; folder 13, box 2, series 915, record group 3, RAC.

Kendrew, Crick and Huxley had in common was experience of science applied to war, which made them think harder than the average graduate about their future research and realise that the greatest promise of physics and chemistry lay in their application to the understanding of life' (Perutz 1987, 40). Perutz did not comment on this statement any further and went on to attribute the extraordinary success of the laboratory to the excellence of its scientists. What follows is an attempt to substantiate what Perutz might have been alluding to and to provide a contextualised account of the establishment and early history of the laboratory.

After a brief sketch of the involvement of British scientists in World War II (chapter 2), I will move on to consider the war experiences of those involved in building biophysics after the war, their postwar careers, and the moves which, after the war, led to the establishment of several biophysics groups, including the Cambridge unit (chapter 3). In chapters 4 and 5 I will focus on protein X-ray crystallography, the key technology used at Cambridge, emphasising the material and cultural aspects of the work, from the pioneering use of computers to the exhibition of the protein and nucleic acid models at the Brussels World Fair.

2

· · ·

World War II and the mobilisation of British scientists

Shortly after the 'Munich' crisis of September 1938 in which Britain bent to Hitler's claims on the German-speaking parts of Czechoslovakia and signed a non-aggression pact with Germany, the government decided to establish a Central Register of those with 'professional, scientific, technical and higher administrative qualifications' available for national work in an emergency. Following an invitation by the Minister of Labour to nominate a representative to the Advisory Committee of the Central Register, the Royal Society, which had been among the originators of the project, volunteered to supervise the compilation of the scientific section and to pay the costs out of its own budget. The scientists forming the Council of the Royal Society further suggested that the Society could provide not only the machinery necessary to compile the scientific register, but also the expert advice required in deploying the nation's scientific personnel and organisations.[1] A committee chaired by A. V. Hill, Biological Secretary of the Society and already a member of important defence committees, was put in charge of the task. The compilation started in February 1939, and by October, a month after the outbreak of hostilities, the scientific section of the Register counted 6484 names. The list included not only scientists on the staff of scientific institutions or members of scientific organisations as originally suggested by the Ministry, but also young graduates who were at the very beginning of their scientific careers.[2] In October 1940 the Society's copy of the Central Register was moved to the Ministry's offices,

[1] The initiative marked the beginning of a stronger involvement of the Society in governmental affairs; see Rowlinson and Robinson (1992, 1–38).

[2] Meeting of the Council of the Royal Society, 12 January 1939, minute 12 and Meeting of the Council of the Royal Society, 16 February 1939, minute 4, Royal Society Minutes of Council, 1939–1940, vol. 15, RSL and Rowlinson and Robinson (1992, 1). The number of registered scientists kept growing. By 1942 the Register counted 12,000 scientists and engineers of whom 10 per cent were physicists (Bragg 1941/3, 255). For some brief insights into the work of compiling the Register and occasional resistence to 'form-filling' on the part of scientists, see Katz (1978, 113–15).

where scientists continued to assist in its operation.[3] Among the scientists involved in this task during the war was C. P. Snow, scientist, novelist and Fellow of Christ's College, Cambridge.

Some of the wartime projects in which British scientists were involved received much attention after the war. This is particularly true of the radar work, the development of operational research, the work on nuclear fission and the atomic bomb, and, in the medical field, the development of penicillin.[4] These were the fields in which scientists achieved the most spectacular successes (even if in more than one case Britain had to surrender the initiative to the American allies) and which most caught the public imagination. So far, however, we do not have anything like a comprehensive picture of the involvement of British scientists in World War II. Some of the official sources are only now becoming available for study. But this, I suspect, is not the main reason why we still lack historical studies on the involvement of scientists in the war. Sufficient material, from memoirs to official military histories as well as oral histories, has long been in easy reach.[5] Rather, general historians tend to overlook the work of scientists, and historians of science still often see the mobilisation of scientists into wartime projects as a necessary but more or less deplorable aberration from the proper aim of science, which is the disinterested pursuit of knowledge and mainly happens in universities. These attitudes and convictions not only are responsible for a dearth of historical studies on the wartime effort of British scientists, but also have prevented a reflection on how the wartime mobilisation has (permanently) affected the sciences. Among other things, this has also obscured the extent to which Britain continued funding warlike R&D even after the war (Edgerton 1996a).[6] Unlike in the United States, this research happened not on university campuses, but in government establishments to which academic scientists were recruited on a temporary basis, or in industrial laboratories.

---

[3] Meeting of the Council of the Royal Society, 24 October 1940, minute 6, Royal Society Minutes of Council, 1936–1940, vol. 15, RSL.

[4] For instance, in the official publication *Science at War* commissioned by the Scientific Advisory Committee to the Cabinet with the aim of making available to the public an 'authoritative account of some of the more important aspects of the scientific contribution to the war effort', the authors (an established science reporter and a distinguished Professor of Physics) chose to discuss radar, operational research, the atomic bomb and the U-boat war (Crowther and Whiddington 1947). On the place of penicillin and the stories of its discovery in the reconstruction of Britain see Bud (1998).

[5] Clark, writing in the early 1960s on the rise of the 'boffins' during World War II, expressed a similar view (Clark 1962, xvii–xviii).

[6] An exception is again represented by the British Bomb Project, which has been amply documented (Gowing 1974a; 1974b). See also Bud and Gummett (1999).

This book is not concerned with military research during or after the war and in this sense suffers from the same bias. It is, however, concerned with the postwar reconstruction of the sciences in which wartime experiences were cemented. To this end it will be useful to introduce some aspects of the wartime mobilisation of British scientists and its legacies. I will focus on three points: first, the determined actions of British scientists to join the war effort; second, the active role of British scientists in planning post-war needs and in promoting a prominent role for the sciences in postwar reconstruction based on their momentous contributions to the war effort; and third, the increased state support for scientists after the war. British scientists came out of the war with their status and opportunities much enhanced. This, together with newly acquired personal networks and skills and a host of new technological developments awaiting peacetime uses, set the stage for the dramatic postwar development of the sciences. The postwar expansion of biophysics was part of this new scientific culture. Many of the themes introduced in this chapter, therefore, will be filled out with more concrete examples in the following chapters, in which I follow the careers and the work of young scientists and their elder patrons in academia and the scientific administrative apparatus who were involved in building biophysics.

Active mobilisation

With few exceptions British scientists were keen to enrol in the war effort.[7] If it was a survival strategy, it was underpinned by a strong conviction, shared by some politicians, administrators and military leaders, that scientists had an important contribution to make in the defence of Britain. To some extent, this had been a lesson drawn from the Great War. Henry Moseley, one of the most promising young physicists, who was killed at the front at the Dardanelles, had become the symbol of the waste of a complete generation of British scientists. 'Our regret for the untimely end of Moseley', Ernest Rutherford commented on the death of his protégé, 'is all the more poignant [in] that we cannot but recognize that his services would have been far more useful to his country in one of the numerous

---

[7] In his autobiography Zuckerman recalls that to some young scientists like himself it appeared that the 'scientific establishment' was not as enterprising as it could have been in mobilising scientific resources. This conviction led to the publication of the Penguin Special *Science in War*, a pamphlet promoting the full use of scientific resources in the war effort (see below). Qualifying his statement, Zuckerman remarked that 'in retrospect it is remarkable how rapidly the country's academic talent was in fact mobilised' (Zuckerman 1978, 109–10).

fields of scientific inquiry rendered necessary by the war than by exposure to the chances of a Turkish bullet' (Rutherford 1915, 34).[8]

With the rise of fascism in Europe even pacifist left-wing scientists revised their position.[9] Joining the fight against fascism, first in Spain, then against Germany, became a moral obligation. The emergency of the war was now seen as a chance to promote the social uses of science which this group of scientists had long been propagating. In his *The Social Function of Science*, published on the eve of World War II and generally regarded as the manifesto of the group, John Desmond Bernal, crystallographer, first at Cambridge, later at Birkbeck College, London, and prominent member of the scientific left, dedicated a whole chapter to the theme of 'Science and War'. Figures at hand, he strongly denounced the increasing 'misuse' of science to military ends and called on scientists to organise for peace. An important step to this end was to study the causes of war and the part science played in it. However, Bernal also recognised that in the circumstances a boycott of war research would put the democratic countries at a disadvantage to fascist ones, and that even in wartime or in the preparation of the war there was scope for socialist scientists to make an impact, for instance with respect to a fairer protection of the civilian population (Bernal 1939a, 184–6). The Cambridge Scientists' Antiwar Group, of which Bernal was a leading member, had been active in this area, achieving an important revision of Britain's civil defence policies (Werskey 1988, 223–34).[10]

With the beginning of the hostilities, Bernal was among those who most strongly promoted the full use of scientific resources for the war effort. In 1940, the Tots and Quots, a London-based dining club, convened by the Oxford anatomist Solly Zuckerman and assembling liberal as well as left-wing scientists among its members, anonymously published a Penguin Special entitled *Science in War*.[11] The book was produced in less than a month, in response to 'the urgency of the situation' (Anonymous 1940, Publisher's note, p. 1). Most of the planning and editing was done by Zuckerman in close collaboration with Bernal. On the characteristic orange-striped cover

[8] Moseley was able to play this symbolic role despite the fact that he was offered the chance to undertake scientific work, but chose to volunteer for the army as he felt it his duty (Heilbron 1974, 125).

[9] For a personal testimony of such a change see Russell (1968, 191). Bertrand Russell's attitude changed from actively opposing the First World War to supporting the Second.

[10] For a collective biography of the most prominent British socialist scientists and their involvement in the war, including Bernal, see Werskey (1988). On Bernal see also Goldsmith (1980), Hodgkin (1980) and Swann and Aprahamian (1999).

[11] Among the Club members were Bernal, Patrick Blackett, the science writer J. G. Crowther, Cyril D. Darlington, J. B. S. Haldane, Lancelot Hogben, Julian Huxley, Hyman Levy, Joseph Needham and Conrad Waddington. On the history of the dining club and the production of the Penguin Special see Zuckerman (1978, 109–12, 370, 393–404).

the book announced: 'The full use of our scientific resources is essential if we are to win the war. Today they are being half used. This book not only analyses the conditions which have led to this dangerous state of affairs, but also shows where science could be applied with immediate benefit to our national effort' (Anonymous 1940, front cover). In the book the authors argued that if scientists had been more involved in the conduct of social and international affairs, 'wars and other forms of aggression might have been wiped from the face of the earth' (p. 14). The emergency situation of the war called more than ever for the full use of 'scientific intelligence' (p. 13). The book enumerated a series of things science had already done to meet the needs of war and identified the sectors in which more was needed and could be done. This list comprised the conduct of war, the care of the wounded, food supply, industrial production, and techniques of persuasion and efficiency. Weapons research was carefully avoided, but this was of course what many scientists were involved in.

The book was well received and widely read. It was reprinted less than three months after its first appearance and altogether sold more than 20,000 copies.[12] Several of the authors, then still young scientists at the beginning of their careers, subsequently became heavily involved in the business of war. Zuckerman and Bernal as well as naval officer and physicist Patrick Blackett not only became high-level scientific advisers, but, together with Waddington, another Tots and Quots member, also contributed in significant ways to the creation of operational research.

This method of analysis was pioneered in the Royal Air Force Coastal Command at the beginning of the war to render the radar defence system effective, and was later more generally used for the evaluation of equipment and the analysis of war operations. The biggest concentration of operational research scientists was in the Royal Air Force Bomber Command.[13] Scientists were here involved not in the (technical) production of weapons

---

[12] While this represents a large circulation for a programmatic and 'technical' book, it should be remembered that during the war, when reading often represented the only pastime, other Penguins sold up to one million or more copies. The fact that *Science in War* was published anonymously allowed the authors to review their own book enthusiastically and to draw attention to it (e.g. Huxley 1940; Zuckerman 1940).

[13] Note that, with Zuckerman and Waddington, biologists were centrally involved in the development of this new 'branch of the science of war' (Waddington 1973, vii). The explanation often given for this was that biologists were well trained in handling complex phenomena and that they successfully approached operational questions 'from the natural historical point of view' (Crowther and Whiddington 1947, 105). Of the geneticist Cecil Gordon, for instance, it was said that he treated Coastal Command 'as though it were a colony of his pet drosophila', thereby managing to double its flying capacity (Calder 1955, 140). On the history of operational research and its postwar developments see [Air Ministry] (1963), Waddington (1973), Rosenhead (1989) and Fortun and Schweber (1993). On the advisory careers of Zuckerman, Bernal and Blackett see Gummett (1980, 101–4); for more comprehensive accounts of their careers see Zuckerman (1978; 1988), Goldsmith (1980) and Swann and Aprahamian (1999).

but in the field of military strategies and tactics, normally strictly reserved to the military. In fact, scientists working on operational matters never had decision-making but always only advisory functions (they were 'on tap', but not 'on top'). It still became a key example for the 'pure' application of the scientific method, which made war more rational and therefore more humane. As such it gained much publicity after the war (Edgerton 1996a, 22). As Blackett wrote in what was the first methodological exposition of the subject: 'In fact, the scientist can encourage numerical thinking on operational matters, and so can help to avoid running the war by gusts of emotion' (Blackett 1941, reprinted in [Air Ministry] 1963, Appendix I).

During the war, the main argument used to promote the use of scientific methods in strategic matters was that the Germans were pursuing this way all along and that it was a matter of survival to beat them at their own game (Anonymous 1940, 7, 34). After the war, this argument was turned on its head. The deployment of operational research was now used to contrast the 'rational scientific kind of warfare' pursued by the allies to the 'romantic conception of war' held by Hitler. Scientists' introduction of a 'civilian spirit' to the conduct of war was seen as the winning asset (Crowther and Whiddington 1947, 119–20; [Air Ministry] 1963, xx). If there would have been the need, this redefinition of the military as 'civilian' could help scientists justify their involvement in military matters.[14] In reality, however, British scientists felt morally on the right side and, generally, their military work represented a question of pride rather than remorse. The scientists' attitude towards their involvement in the war also prevented any critical reflection on the effects of the wartime mobilisation on the sciences.[15] Bernal represented an exception when, in later years, he wrote:

The only time I could get my ideas translated in any way into action in the real world was in the service of war. And though it was a war which I felt then and still feel had to be won, its destructive character clouded and spoilt for me the real pleasure of being an effective human being. (Bernal, quoted in Hodgkin 1980, 64–5)

Civilians working in close association with the military on technical matters became known as 'boffins'. The term seems to have been first applied

---

[14] On the redefinition of the civilian and the military and the creation of new hybrids in postwar America see Dennis (1994).

[15] For a rare attempt by scientists systematically to assess the effects of the two world wars on the organisation and development of science in Britain see the collection of lectures held at the occasion of a meeting organised at the Royal Society in the mid-1970s. Also on this occasion, however, the British veteran scientists assembled tended to emphasise the positive effects of the wartime mobilisation. Referring especially to biology, Oxford zoologist John Pringle concluded: 'the second world war was, on balance, clearly beneficial' (Pringle 1974, 537).

by members of the Royal Air Force to scientists working on radar.[16] Military officers often used it in a derogative sense, deriding the 'gentleman in grey flannel bags' who before the war had occupied himself with 'something markedly unmilitary such as Biology or Physiology' but on the occasion of war pretended to teach them their own trade (Marshal of the Royal Air Force Sir John C. Slessor, quoted in Waddington 1973, xv). In his novel *The Small Back Room* (1943), Nigel Balchin, himself a Cambridge graduate and during the war a scientific adviser in the Army Council, presented Sammy Rice, one of the boffins or 'backroom boys', as a new and improbable hero. Called out from the safe walls of his laboratory to perform a dangerous but potentially life-saving operation, he indeed wins over his own psychological fears and physical disablement. Above this personal drama, the novel offered intriguing views on the working of the Civil Service and the ways careers were made in the bureaucracy, as well as on the fraught relationships between the scientists in the Civil Service and the military. *The Small Back Room*, turned into a screenplay in 1949, was instrumental in creating a public image of the role of scientists in the war (Balchin 1985).[17]

Despite the fame that operational research acquired after the war, only about 200 scientists (among them several biologists) were engaged in that field in home and overseas commands ([Air Ministry] 1963, 179, fn 1).[18] A far larger number of scientists and engineers worked on radar which, for its greed for scientifically trained people, was sometimes referred to as 'that rapacious beast' (Crowther and Whiddington 1947, 84). The bomb project made a heavy demand on first-rate chemists and physicists at a time when most of them were already engaged in other war projects.[19] Many more scientists, however, were active in less publicised research establishments of the service and supply ministries, of which the biggest were the Admiralty Research Laboratory in Teddington, the Royal Aircraft Establishment at Farnborough and the Armament Research Department of Woolwich Arsenal at Fort Halstead in Kent (after the war taken over by

[16] On various theories regarding the etymology of the term see Clark (1962, vii–viii).

[17] It has been argued that the stereotype of the 'boffin', as featured in several postwar films, still dominates the public image of the scientist in Britain (Jones 1997). A series of novels by Snow, who himself was centrally involved in the organisation of the scientific mobilisation, also dealt with the role of scientists during the war; see for example Snow (1954). Although widely known, they offer less engaging reading than Balchin's novel.

[18] This figure refers to the situation in March 1945. Counting all scientists who had carried out operational research on behalf of the army, Rosenhead quotes a figure of 365 (Rosenhead 1989, 7).

[19] In 1940–1 around thirty-five scientists were engaged in atomic energy research under the Maud Committee in British universities. In 1944–5 around fifty British scientists and engineers, several of them only just naturalised, joined the Manhattan Project. A smaller British team was active in Canada and a few researchers continued work in Britain (Gowing 1964, 53, 190, 243, 268).

the British Bomb Project), as well as in industrial research establishments like ICI.[20] Biologists were drawn in large numbers into the Emergency Public Health Laboratory Service. The Service, consisting of a nation-wide net of local laboratories, was instituted to prepare against the threat of bacteriological warfare, but was more generally directed to the diagnosis, control and prevention of infectious diseases, and also engaged in research projects.[21] Biologists also collaborated in other medical mobilisation programmes like the newly created Blood Transfusion Service. A minimum number of scientists stayed on in the universities to carry on teaching. Remodelling degree courses, they guaranteed a quick output of new scientific recruits for war requirements. Some research which was felt to be of use for the war effort also continued to be carried out in university laboratories. Apart from research in nuclear physics which, up to the exodus of British scientists (including several émigré scientists) to America in 1944, went on at the universities of Birmingham, Bristol, Cambridge, Liverpool and Oxford, this was for instance true for research on poison gases or for research relevant to certain medical problems and nutrition. Under this rubric some work on proteins, viruses and bacteria, including research on anti-bacterial drugs and vaccines, continued during the war. Results were often classified. Among those who continued working in university laboratories were some refugee scientists supported by Rockefeller grants (as for instance Max Perutz, whose work on the structure of haemoglobin was the only research project pursued in the Cavendish in the last years of the war) or a few conscientious objectors (as for instance Fred Sanger, who graduated in 1943 and went on to work on the chemical structure of insulin in the Biochemistry Department at Cambridge).

British scientists, then, were all, in one way or the other, involved in war-related projects and the whole of British science was geared towards the war effort. To a large extent the organisation of the mobilisation was carried by existing institutions, but the two distinctly new bodies created to support the use of scientists in the war effort, namely the Central Register and the Scientific Advisory Committee to the War Cabinet, were set in place to a large extent as a result of pressure exerted by the

---

[20] For a brief overview on the R&D activities of the Service and Supply ministries see Edgerton (1996a, 3–5); more specifically on the Aircraft Establishment and on the Ministry of Supply see Edgerton (1991 and 1992 respectively). On ICI during the war, see Reader (1975, 249–313). The particular chapter carries the title 'The great distraction', despite amply documenting the decisive impact of the wartime mobilisation on the development of the industrial establishment.

[21] Together with the successful development of penicillin and other bacteriostatic drugs, the Service contributed decisively to the cause of a 'scientific', laboratory-based medicine. It worked so successfully that after the war it was decided to continue it on a permanent basis. It was later integrated into the NHS (Mellanby 1943, 352–3; Landsborough Thomson 1975, 258–70).

scientists themselves via the Royal Society (Katz 1978, Appendix I, 136–8). Scientists also initiated many of the more specific projects and were centrally involved in their organisation, though not necessarily in the final decisions regarding the deployment of their products as exemplified by the case of the atomic bomb.[22] Some scientists could use their existing skills. Many, however, were confronted with completely new situations and learned 'on the job'. In many cases this included the experience of new team relations and of collaborations with scientists from other disciplines, as well as contacts with industrialists, military users, civil servants and policy-makers. An important aspect of war-related projects became the skill to persuade other scientists, politicians and the military of its potential or actual importance (Agar and Hughes 1995). After the war, scientists would build on these new skills which for many brought a radical reorientation of their prewar occupations. We will encounter more concrete examples of this in the following chapter.

Postwar planning

Postwar planning for science started well before the end of the war.[23] Scientists played an active role both in publicising their contributions to the war effort and in campaigning for an equally important role of the sciences in the 'battle of peace'. Using newly established links to politicians, administrators and industrialists, they argued that the problems of reconstruction – from economic rebuilding to the organisation of the workforce and the provision of education and healthcare for the people – were, in large part, scientific ones. The newly claimed public role required a new status for scientists and increased funds for scientific education and research.

In September 1941, when German bombers were still threatening London, the Royal Institution hosted a conference in which scientists and politicians of twenty-two nations met to discuss 'the kind of world that must come out of the war' and the place science had to play in it. The meeting was organised by the newly founded Division for Social and International Relations of the British Association for the Advancement of Science and, according to the organisers, it was the first time that 'science and statecraft' met in an open conference (Crowther, Howarth and Riley 1942, 11). The proceedings were broadcast in thirty-nine languages around the world. The programme was accompanied by a

---

[22] For a painstaking reconstruction of the decisionary process leading to the use of the first atomic bombs over Japan see Alperovitz (1995).

[23] The same point is made by Galison regarding the American case (Galison 1988, 50).

transatlantic 'round table' conducted by radio on the themes of the conference. Agreement existed that 'in the winning of the peace as in the winning of the war, men of science are key men', but that to bring their work to fruition and to face the challenges of postwar reconstruction scientists and politicians had to work together (Crowther, Howarth and Riley 1942, 15). How this was going to be achieved was a matter of discussion. British scientists participated with a large contingent, and many of the themes raised in the conference with a strong internationalist ethos continued to be present in national discussions on postwar reconstruction. They concerned the question of scientific literacy (of civil servants, politicians and the public); the relations of science, government and industry and of planning and democracy; and discussions regarding particular areas in need of intervention like food and healthcare for the people.[24]

A few months later, in the same premises, Lawrence Bragg, director of the prestigious Cavendish Laboratory at Cambridge, addressed the audience of the weekly Thursday afternoon lectures on 'Physicists after the War' (Bragg 1941–3).[25] Bragg, an important figure in the mobilisation of physicists and involved in teaching intensive courses for physicists to join scientific wartime projects, stressed the unprecedented extent to which physicists had been drawn into the war. He pointed out that especially for university scientists this meant 'a change in outlook and of occupation of a very revolutionary kind' (Bragg 1941–3, 253). Not only had physicists learnt to apply their knowledge to practical problems and joined engineers and technicians in devising gear which could quickly be put to practical uses. They had also seen more 'of the way the country is run' (Bragg 1941–3, 254). Bragg urged that some of the successful features of the organisation of science in wartime should be kept and cultivated in peacetime. In particular he mentioned the Central Register of scientists, the less bureaucratic way of organising research resources and the closer links between scientists and both state administrators and industrial sponsors. He also recommended that more physicists be trained to avoid the scarcity encountered in the war.[26]

If on the one hand scientists were engaged in promoting the practical uses of science, on the other they expressed concern that fundamental research might suffer in comparison with applied science. The argument

---

[24] The full proceedings of the meeting were published in *The Advancement of Science* 2, no. 5. January 1942.

[25] The lecture took place just two weeks after the Royal Institution mourned the death of its director, William Bragg, Lawrence's father and acting President of the Royal Society.

[26] It should be noted that, notwithstanding the demand for training more physicists, after the war physicists released from the big wartime projects encountered difficulties in finding new occupations (see below).

used was that the striking achievements during the war (the standard examples given after the war were radar, the bomb and penicillin) rested on fundamental research pursued in the preceding decades. Fundamental research then was necessary to secure future useful applications, especially for short-term necessities as presented by wars.[27] Following scientists' concerns, in 1943 the Royal Society set up a committee to advise on the postwar needs of fundamental physics.[28] Committees for other sciences, including chemistry, biology and biochemistry, geology, geophysics, geography and metereology, followed suit. One year later the Council of the Royal Society published a general report based on the work of the various committees for circulation in government circles. As the report revealed, the committees had seized the opportunity for a critical assessment of prewar university research facilities and opportunities as a whole and urged wide-ranging changes. Data and recommendations were based on extensive questioning of a large number of university departments. The basic request concerned an important increase in funds for fundamental research in universities. The report recommended a threefold increase (in real terms) of 'ordinary expenses' which covered the running costs of laboratories and were usually met by the University Grants Committee; an equal increase of parliamentary grants in aid (instituted in 1850) for 'extraordinary expenditure' which were administered by the Royal Society; and the creation of special Treasury grants for exceptional needs involving large sums (£2000 and over). Such special funds were considered vital for the development of new lines of research, especially in borderline subjects, and the Royal Society offered to advise the Treasury in this regard. Interestingly for our concerns, the lack of a 'properly equipped laboratory devoted to research in biophysics' was mentioned in this context, alongside the need for a central institute of general microbiology, for an institute of oceanography and for the provision of cyclotrons. The report also included a series of more specific requests all aimed at facilitating and increasing research activities in universities. They comprised the creation of more research posts and of more research time for academic staff, the provision of student grants for training in research, a substantial increase in the number of laboratory technicians, better workshop facilities, more money

---

[27] The importance attributed to fundamental research lent itself as cover for military research. It became a dominant funding strategy in the Cold War era, especially in the United States where the military invested large sums in 'fundamental' research.

[28] The initiative rested on a letter which Ralph Fowler and Blackett, who both had wide experience in the organisation of military research, had written to the Secretaries of the Royal Society on October 1943, soliciting the formation of such a committee, above all for physics, but also for other subjects; see 'Report on the needs of research in the fundamental sciences after the war', Meeting of 14 December 1944, Appendix A, Royal Society Minutes of Council 1940–1945, vol. 16, p. 334, RSL.

for equipment and materials, and the creation of travel grants. The report was accompanied by an impressive number of tables detailing prewar and estimated postwar expenses for various items.[29]

The report was circulated among fellows and sent to institutions concerned with the organisation of postwar science. It was not sent to *Nature* or any newspaper.[30] It none the less formed an important basis for discussion on the postwar reconstruction of the sciences and found general support.[31] As soon as war ended, it was implemented in various ways, as we shall see in more detail below.

The report called for a substantive amount of central planning. The authors agreed that 'some central guidance on major matters of policy was not only desirable but essential, if the case for increased resources is to be adequately put to the relevant government authorities'.[32] This notwithstanding, the founders of the Society for Freedom of Science, set up in 1941 with the explicit aim of saving British science from any kind of 'socialist' planning, scored the rising pressure by British scientists to secure funds for fundamental and academic research as a success of their campaigns (McGucken 1978).[33] Research in universities, however, represented

---

[29] 'Report on the needs of research in the fundamental sciences after the war', Meeting of 14 December 1944, Appendix A, Royal Society Minutes of Council 1940–1945, vol. 16, p. 334–60, RSL. The individual reports of the postwar needs committees give useful insights into current research concerns. The report of the committee on the postwar needs in biology, for instance, indicates that much emphasis was laid on the development of biochemistry, which was dealt with by a special subcommittee chaired by Charles Chibnall, since 1943 successor to Frederick Gowland Hopkins in the Chair of Biochemistry in Cambridge. However, the report also stressed the importance of studying living animals under natural conditions and the importance of natural history collections and botanic gardens. Subjects to be expanded comprised (plant and animal) genetics, microbiology (including general bacteriology), freshwater biology, ecology and taxonomy. In contrast, the report of the physics committee focused exclusively on structural aspects and hardly touched on 'contents'. This omission notwithstanding, once the war had ended, large funds were invested to build up nuclear physics in universities (see below).

[30] This decision was taken at the Royal Society officers meeting of 22 February 1945.

[31] See for instance [Association of Scientific Workers] (1947, 160).

[32] 'Report on the needs of research in the fundamental sciences after the war', Meeting of 14 December 1944, Appendix A, Royal Society Minutes of Council 1940–1945, vol. 16, p. 334, RSL.

[33] In early 1946, Michael Polanyi, crystallographer at Manchester and, together with the Oxford zoologist John Baxter, founder of the Society, addressed a conciliatory letter to Bernal, whom he had regarded as chief ideologue of socialist planning of British science, suggesting that 'the divergent views' which they had represented 'in the past few years on the subject of freedom in science have now been sufficiently clarified to allow for active cooperation'. Indicating that 'more weighty matters' awaited discussion, Polanyi started by seeking Bernal's comments on a draft proposal for a revision of patent law. Bernal showed himself interested and proposed discussing Polanyi's suggestions with the Association of Scientific Workers, but the correspondence ended here; M. Polanyi to J. D. Bernal, 8 January 1946 and J. D. Bernal to M. Polanyi, 30 January 1946, Bernal Papers, file J.182, CUL.

only a tiny part of the whole R&D budget in postwar Britain. An over-whelming part of the budget was spent on military research, where much more planning and control were exercised. Bernal was among the few who recognised and actively campaigned against the continued mobilisation of scientific research (Edgerton 1996a).

A key military R&D project pursued after the war was the development of nuclear weapons. The decision to continue research in this field in Britain as soon as hostilities had ceased had already been taken in early 1944 (Gowing 1964, chapter 12). The dropping of the first atomic bombs on Japan in August 1945 made this decision even firmer. No member of the newly elected Attlee government was acquainted with the secrets of the atomic bomb project. But on the day Japan surrendered, Attlee proposed that a committee be set up to advise him on general policy for Britain's postwar atomic programme. A few months later the government announced the allocation of important state funds for the creation of a British atomic energy project. The programme was first directed to pursue atomic energy research and development 'in the broadest sense'. Under the aegis of the programme, funds were also made available to equip five British universities (Birmingham, Liverpool, Glasgow, Oxford and Cambridge) with high-energy accelerators, to conduct research in nuclear physics and to train a new generation of physicists and engineers in this field (Krige 1989).[34] Only in January 1947 did the government decide to go ahead with the actual development of atomic bombs (Gowing 1974a, chapter 6). The decision was made public in 1948.

There was very little resistance to this decision.[35] The 1930s had taught British citizens that weakness in defence was morally and politically wrong. Despite the collapse of the empire, most people also still thought of Britain, which had fought so successfully on the side of the allies in the war,

[34] Money for the accelerator construction programme came from the Department of Scientific and Industrial Research (DSIR), but the project could avail itself of the support of the Ministry of Supply. In contrast to the situation in the United States, atomic research in British universities became increasingly detached from the military atomic energy project (Krige 1989).

[35] Blackett, a member of the secret Advisory Committee on Atomic Energy, was among the very few who, at the time, opposed the political and military decision to build Britain's defence on atomic bombs as deterrents. His objections were light-handedly dismissed (Gowing 1974a, 115–16, 171–2, 183–5, 194–206). The Atomic Scientists' Association, formed in 1946 to inform the public and investigate proposals for the control of nuclear energy, also failed to rally against the decision to build the bomb, although Nevill F. Mott, Professor of Physics at the University of Bristol and president of the association, was an outspoken critic (Mott 1986, 80–93). Opposition to Britain's nuclear politics began only in 1957 with the Campaign for Nuclear Disarmament, prompted by the decision to develop the hydrogen bomb (Veldman 1994). As has been argued in another context, it is the scientists' resistance to demobilisation, rather than wartime mobilisation that demands the attention of the historian (see Whittemore 1975).

as a great power and that as such she must have her own bombs. Scientists shared these beliefs with their fellow citizens. It thus 'was not strange' that scientists joined the atomic bomb project (Gowing 1974b, 10). In addition to these political and moral reasons which, if they did not motivate, at least did not impede collaboration, there were professional reasons which made the choice attractive. Many of the scientists joining the project had been part of the teams working on nuclear physics during the war and were keen to continue working in the field. At the research establishment in Harwell not all research was clearly weapon geared and there was a strong motivation to prove the peaceful, in contrast to the destructive, uses of atomic energy. The scarcity of university posts and the reluctance to move to industry, together with a series of incentives offered on the level of work conditions, also made many scientists decide to join the nuclear energy project, in spite of strong misgivings against the Scientific Civil Service (Gowing 1974b, 25–9). As Margaret Gowing, the official historian of the British bomb project, has remarked, 'atomic policy is an important thread in the post-war history of Britain; it is woven into almost every part of that history – international, diplomatic, Commonwealth, military, constitutional, political, administrative, economic, social, scientific, tech-nological and medical' (Gowing 1974a, ix). The role played by scientists as both researchers and advisers in this history affected the roles and oppor-tunities for scientists more generally. Even if the participation of nuclear physicists in national defence and in the civil applications of atomic energy marshalled most attention, other scientists apart from the physicists also profited from the new opportunities offered by postwar reconstruction.

The place of science in postwar Britain

The 1950s have been presented as 'barren years' with respect to science policies (Wilkie 1991, 48). This view contends that despite wide-ranging discussions in the 1930s, and in contrast to the situation that had pre-vailed after World War I, the postwar period saw British scientists deeply divided on issues of scientific policy. Also politicians failed to learn the lessons presented by the role of British science and industry in such key wartime projects as the development and production of penicillin and the atomic bomb. Both were wrongly celebrated as triumphs of (in important parts British) basic science, instead of being recognised as fruits of applied science, engineering and project management, in which Britain lacked initiative and capabilities. As a consequence the support system for civil R&D (which is the only sector considered in this context) remained ba-sically the same as the one set in place during and after World War I, with all its failings and shortcomings especially for Britain's industrial efficiency. Its main pillars were the Department of Scientific and Industrial Research, created in 1916; the Medical and Agricultural Research Councils, set in place in 1918 and 1931 respectively; and the University Grants Committee, created in 1919. The Research Councils (including the Department for Scientific and Industrial Research) were directly responsible to Parliament, while the University Grants Committe, composed mainly of representatives from the universities, administered and independently dispensed a block grant received from the Treasury. Decisive revisions of this system, which lacked central coordination, occurred only in the mid-1960s and early 1970s.[36]

As already noted, taking military R&D (and industrial civil R&D) into account, the overall picture changes dramatically. Postwar Britain easily displayed the largest military R&D complex outside the United States and the Soviet Union. At least in this sector, Britain was thus highly competitive (Edgerton 1996a, 5). But for the civil sector also I would like to revise the view of the 1950s as 'barren years', by emphasising the changes produced by the wartime mobilisation of science rather than the continuities with the interwar years. Only in the wartime mobilisation, or then more dramatically than ever, had scientists proved that their work could serve the national interest. In contrast to the situation after World War I, there was general agreement on the success of the scientific mobilisation and on the key contribution of scientists in winning the war. As we have seen, scientists themselves were active in promoting this view and in contending for themselves a similar role in postwar reconstruction as in the war effort. Herbert Morrison, Lord President of the Council in the new Labour government and in this role responsible for the formulation and execution of government civil science policy, echoed the view of scientists when, in one of the first parliamentary debates on science after the war, he declared: 'Science had contributed enormously to the winning of the war . . . the Government . . . are . . . desirous that science shall play its part in the constructive tasks of peace and of economic development' (H. Morrison, quoted in Rose and Rose 1971, 73). The mood, that is right, was one of complacency, not of critique. This manifested itself in a fall in political engagement of scientists (Hobsbawm 1994, 546). But I would

---

[36] It should be noted that the allocation of large government funds for civil R&D in the fields of nuclear energy and aircraft production lay outside the responsibility of the science agencies just mentioned. On science policy in postwar Britain see also Rose and Rose (1971) and Gummett (1980). Specifically on the Department of Scientific and Industrial Research, until its dismantlement in 1964 the biggest patron of civil research in Britain, see Varcoe (1974). Vig has argued that science policy became a serious topic in Britain only in the late 1950s (Vig 1968). On science policy debates in the 1960s and 1970s and their place in the history of molecular biology see below (chapters 10 and 11).

none the less contend that the high expectations in science, which were seen as unfulfilled in the mid-1960s, were, to a large extent, the product of postwar exultation and of hopes then placed in the sciences.

In postwar Britain the changes regarding the support and opportunities for scientists as well as their place in society were tangible. While government support for civil science had risen very slowly in the interwar years, the postwar years saw a massive expansion. Between 1945–6 and 1962–3 expenditure on civil R&D rose from £6.5 million to £150 million in money terms or by a factor of ten in real terms ([Trend Report] 1963, 10–11). This corresponded to an average growth of 13 per cent per annum in real terms. Throughout the period, expenditure on civil R&D remained small as compared to expenditure on military R&D which in the late 1950s amounted to around 80 per cent of all government spending in science (Rose and Rose 1971, 77). The budget of single institutions, none the less, grew at a remarkable rate. This is particularly true for the Medical Research Council, which was to become the most important funding body for biophysics and later molecular biology (Tables 2.1–2.4). Thus, even if most (but not all) governmental funding agencies had been in place before, they came out of the war with their scope and budgets significantly enhanced.[37] Those responsible for administering these funds, for instance the secretaries of the Research Councils, also gained considerably more influence.

A similar explosion to that in government funds for scientific research could be witnessed in the number of scientific advisers and advisory committees appointed by the services, the ministries and the cabinet. In the ministries alone the number of specialist advisory committees grew from 200 in 1939 to 700 in 1949 (Wilkie 1991, 48). Scientists could serve as advisers on a temporary and part-time basis or could choose a career in what after the war came to be known as the Scientific Civil Service.[38] The most influential scientific advisers of the postwar years in Britain, above all Henry Tizard and Zuckerman, had all made their names during the war (Gummett 1980, 104).[39] In September 1945, Zuckerman, together with Blackett, Bernal, Waddington and the physicist Mark Oliphant, also

---

[37] Among the new government agencies was the National Research Development Corporation (NRDC), aimed at facilitating the commercial exploitation of scientific inventions (see below, chapter 11). For additional tables and information on government expenditure on civil R&D from 1945 to the mid-1970s see Gummett (1980, especially 37–9, 54–9).

[38] For a brief history of the Scientific Civil Service in Britain and the career problems encountered by scientists in the Civil Service see Gummett (1980, especially chapter 3). On advisory 'elites' and problems of scientific advice to government see Gummett (1980, chapter 4).

[39] Note however that, in contrast to Zuckerman, Tizard, his senior by twenty years, had already filled top administrative positions before being called, in 1935, to chair the committee charged with the development of radar. On Tizard's career see Clark (1962; 1965) and Gummett (1980).

Table 2.1. *Government expenditure on civil R&D (1945/6–1966/7). Source:* Council for Scientific Policy, *Report on Science Policy*, Cmnd. 3007 (1966), p. 20.

Department	1945–6 £m	1950–1 £m	1960–1 £m	1961–2 £m	1962–3 £m	1963–4 £m	1964–5 £m (Provisional)	1965–6 £m (Estimated)	1966–7 £m (Estimated)
1. Universities and Learned Societies	1.5	8.3	19.4	21.4	26.6	31.2	37.6	44.0	48.9
2. Department of Scientific and Industrial Research	2.3	4.9	16.0	18.1	21.9	25.4	29.3	–	–
3. Science Research Council	–	–	–	–	–	–	–	28.6	33.9
4. Medical Research Council	0.3	1.7	4.5	5.6	5.9	7.0	8.8	10.3	11.9
5. Agricultural Research Council	0.3	0.8	5.6	6.1	6.4	7.2	8.1	9.3	10.3
6. Natural Environment Research Council	–	–	–	–	–	–	–	3.5	5.4
7. Nature Conservancy	–	0.08	0.5	0.5	0.7	0.7	0.8	–	–
8. Development Fund	0.03	0.3	0.5	0.9	0.8	0.7	0.8	–	–
9. Ministry of Aviation	n.a.	8.0	16.8	19.9	26.8	31.2	39.9	44.9	55.7
10. Ministry of Technology	–	–	–	–	–	–	–	16.1	18.6
11. Agriculture Departments	0.6	2.0	3.7	4.9	4.7	5.1	5.8	6.0	6.3
12. Other Civil Departments	0.9	3.0	6.4	3.4	3.7	4.6	5.5	11.2	6.2
13. Navy Department	0.05	0.2	0.5	0.9	0.8	0.7	0.9	–	–
14. Air Department	0.6	0.4	1.2	1.1	1.2	1.0	1.1	1.2	1.3
15. National Institute for Research in Nuclear Science	–	–	5.2	5.1	6.7	7.8	8.2	–	–
16. Atomic Energy Authority	–	–	–	49.0	48.0	45.0	47.0	48.0	50.0
Total items 1–14	6.58	30.0	75.1	82.8	99.5	114.8	138.6	175.1	208.5
Total items 1–15	–	–	80.3	87.9	106.2	122.6	146.8	175.1	208.5
Grand total (items 1–16)	–	–	–	136.9	154.2	167.6	193.8	223.1	258.5

Table 2.2. *Government expenditure on civil as compared to military R&D (1955/6–1964/5). Source:* Council for Scientific Policy, *Report on Science Policy*, Cmnd. 3007 (1966), p. 18.

Sector	1955–6 Amount £m	1955–6 Per cent	1958–9 Amount £m	1958–9 Per cent	1961–2 Amount £m	1961–2 Per cent	1964–5 Amount £m	1964–5 Per cent
Government								
Defence	65.7	21.9	102.7	21.6	93.2	14.7	91.5	12.1
Civil	10.5	3.5	39.4	8.2	61.9	9.8	72.7	9.6
Research Councils	10.0	3.3	13.1	2.7	23.0	3.6	28.1	3.7
TOTAL	86.2	28.7	155.2	32.5	178.1	28.1	192.3	25.4
Universities and Technical Colleges	14.4	4.8	23.3	4.9	32.4	5.1	55.9	7.4
Private industry, public corporations, research associations and other organisations	199.4	66.5	299.3	62.6	423.5	66.8	508.4	67.2
GRAND TOTAL	300.0	100.0	477.8	100.0	634.0	100.0	756.6	100.0

Table 2.3 (a). *Government expenditure on (civil and military) R&D in Britain as compared to other countries with respect to (a) the population and (b) gross national product (1962). Source:* C. Freeman and A. Young, *The Research and Development Effort in Western Europe, North America and the Soviet Union: An Experimental International Comparison of Research Expenditures and Manpower in 1962* (Paris: OECD, 1965), p. 71.

(a)

	Currency	National currency (millions)	$ US millions official exchange rate	Population (millions)	R&D expenditure per capita $US official exchange rate
United States	$	17,531	17,531	187	93.7
Western Europe[1]	–	–	4,360	176	24.8
Belgium	francs	6,625	133	9	14.8
France	francs	5,430	1,108	47	23.6
Germany	DM	4,419	1,105	55	20.1
The Netherlands	florins	860	239	12	20.3
United Kingdom	£	634	1,775	53	33.5

Note: [1]Belgium, France, Germany, Netherlands, United Kingdom.

(b)

	Currency	Gross expenditure on R&D (millions)	GNP at market price[1] (millions)	Gross expenditure on R&D as % of GNP at market price[1]
United States	$	17,531	557,590	3.1
Belgium	francs	6,625	646,200	1.0
France	francs	5,430	356,300	1.5
Germany	DM	4,419	354,500	1.3
The Netherlands	florins	860	48,090	1.8
United Kingdom	£	634	28,566	2.2

Note: [1]If GNP is taken at factor cost instead of market price the ratios are as follows: United States 3.5%, Belgium 1.2%, France 1.8%, Germany 1.5%, Netherlands 1.7%, United Kingdom 2.5%.

Table 2.4. *Growth of the MRC budget (1913–72).*
Source: A. Landsborough Thomson *Half a Century of Medical Research*, vol. 1 (1973), p. 205.

Year	Recurrent[1]	Buildings, etc.	Total
1913–14			£
1914–15	53,229		53,229
1915–16	49,783		49,783
1916–17	49,203		49,203
1917–18	50,000		50,000
1918–19	54,600		54,600
1919–20	76,000	72,500[2]	148,500
1920–21	125,000		125,000
1921–22	130,000		130,000
1922–23	130,000		130,000
1923–24	130,000		130,000
1924–25	140,000		140,000
1925–26	135,000		135,000
1926–27	135,000		135,000
1927–28	135,000		135,000
1928–29	148,000		148,000
1929–30	148,000		148,000
1930–31	148,000		148,000
1931–32	148,000		148,000
1932–33	139,000		139,000
1933–34	139,000		139,000
1934–35	140,500		140,500
1935–36	165,000		165,000
1936–37	165,000		165,000
1937–38	195,000[3]		195,000
1938–39	195,000		195,000
1939–40	195,000	70,000[4]	265,000
1940–41	195,000		195,000
1941–42	195,000		195,000
1942–43	195,000		195,000
1943–44	215,000		215,000
1944–45	250,000		250,000
1945–46	295,000		295,000
1946–47	415,000	50,000	465,000
1947–48	618,000	130,000	748,000
1948–49	770,000	365,000	1,135,000
1949–50	1,216,000	319,000	1,535,000
1950–51	1,363,671	290,750	1,659,421
1951–52	1,616,500	176,050	1,792,550
1952–53	1,505,917	181,000	1,686,917
1953–54	1,671,146	134,700	1,805,846

Table 2.4. (*cont.*)

Year	Recurrent[1]	Buildings, etc.	Total
1954–55	1,877,441	97,428	1,974,869
1955–56	2,097,000	88,100	2,185,100
1956–57	2,229,500	49,500	2,349,000
1957–58	2,775,500	37,500	2,813,000
1958–59	3,056,600	80,500	3,137,100
1959–60	3,445,050	81,200	3,526,250
1960–61	4,113,060	355,500	4,468,560
1961–62	4,861,950	808,550	5,571,950
1962–63	5,489,000	370,000	5,859,000
1963–64	6,524,000	509,000	7,033,000
1964–65	8,151,000	602,000	8,753,000
1965–66	9,637,536	450,000	10,087,536
1966–67	11,203,488	621,488	11,824,788
1967–62	12,542,896	1,215,108	13,758,004
1968–69	13,358,203	1,872,288	15,230,491
1969–70	16,393,189	1,197,434	17,590,623
1970–71	19,486,837	1,567,731	21,054,568
1971–72	22,260,521	1,200,016	23,464,537

Notes:

[1] The first year's finances were handled by the National Health Insurance Joint Committee. The main item of expenditure was £11,000 as a first instalment of the purchase price (£35,000) of the building at Hampstead for a 'central institute'.

[2] To finance transfer of radium from Ministry of Munitions.

[3] Includes first instalment of special provision of £30,000 per annum for the development of research in chemotherapy.

[4] First provision specifically for capital expenditure on new building (National Institute for Medical Research, Mill Hill).

a member of the Tots and Quots and a major player in the development of airborne radar, had drafted a memorandum addressed to Morrison, the new Lord President of the Council in Attlee's government, which invoked the creation of a 'Science Secretariat' to help direct Britain's scientific effort in peacetime (Zuckerman 1978, 365–6, 425–6). The plan was characteristic of the postwar years in stressing the need for a coordination of scientific resources for the reconstruction of the country and the importance of scientific advice at the highest political levels. Twenty years later, after a long career as high-level scientific adviser, Zuckerman himself came to doubt 'whether scientists, who had rightly been lauded for what they had done to help win the war, have since lived up to what was expected by them, and to the promise which was theirs' (Zuckerman 1978, 372). He was also frank enough to admit that his 'involvement in Committee work at Whitehall' had undoubtedly helped him in the 'effectiveness' of his university work as well as in his efforts to revive the Zoological Society (Zuckerman 1978, 371). A body somewhat similar to the 'Science Secretariat' envisaged by Zuckerman and his friends, but with no executive powers, was created in the form of the new Advisory Council on Scientific Policy. It was composed of representatives of the Research Councils, the University Grants Committee, the Treasury, industry and several academics, all of whom had already played advisory roles during the war. Its first chairman was Tizard, with Zuckerman as his deputy. Tizard also chaired the parallel committee charged to deal with military science, becoming in practice chief scientific adviser to the government. The work and impact of the Advisory Council on Scientific Policy still awaits thorough historical examination (Gummett 1980, 33–6).[40]

An overriding concern of the postwar years was the question of 'scientific manpower'. A small committee, chaired by Alan Barlow, an eminent civil servant, then Second Secretary to the Treasury, and including Blackett, Zuckerman and Snow (the latter as 'Scientific Assessor'), was set up to consider the matter. The committee took the central role of scientists in postwar reconstruction for granted. 'Never before', the committee members wrote in their report, 'has the importance of science been so widely recognised or so many hopes of future progress and welfare founded upon the scientists' ([Barlow Report] 1946, 8). The report recommended that the yearly output of around 2500 scientists should be doubled ([Barlow Report] 1946). The committee assumed that such a task would take five to ten years. In the event it was achieved in only four (Wilkie 1991, 49).[41] To reach this goal, important new government funds were dispensed to the universities by the University Grants Committee, a body in place since 1919, which now saw its power and administrative structure growing considerably. For the first time government funds constituted the greatest part of the income of the universities, marking a fundamental change in government–university relationships (Carswell 1985, 14). At first, Oxford and Cambridge universities, which were anxious to preserve their independence as well as their college system based on selective entrance criteria and an individual tutoring system, were obstructive to the proposed plans. In the longer term, however, they were

---

[40] On the Defence Research Policy Committee see Agar and Balmer (1998).
[41] Student numbers overall, however, only doubled in ten years, with the biggest expansion in the arts, not in the sciences (Timmins 1995, 157).

keen to accept government money and expanded their student numbers accordingly, yet without giving up their elite status (Wilkie 1991, 49). Since the interwar years Oxford, traditionally an arts university, had made a decided effort to build up the sciences (Morrell 1997). After the war it pursued this policy, embarking on an ambitious building programme and rivalling though never really equalling Cambridge, long since a prominent centre for science, but which more strongly restricted expansion.[42]

It should be noted that the new policy did not extend to technical education, which remained the purview of technical colleges. Neither did the aim to expand the number of science graduates automatically translate into corresponding efforts to grant more space to scientific subjects in secondary school curricula. The lack of students prepared to take up scientific subjects threatened to undermine the goals set by the Barlow Committee, as well as later plans to expand further the number of science graduates following the Robbins Report (Timmins 1995, 246). The first serious attempt to renew science teaching in secondary schools was undertaken by an independent body, the Nuffield Foundation, in the late 1950s. The initiative followed the shock imparted to countries of the Western alliance by Russia's launch of Sputnik, the first artificial satellite, which exposed the level of scientific and technical preparation of the West.[43] The Nuffield programme, however, was geared only to the most able students. A more general reform of the school curriculum was taken up only in the mid- and late 1960s (Simon 1991, 314ff). Snow's notion of the 'two cultures', coined in his 1959 Rede Lecture at Cambridge University, reflected and fuelled the educational debates of the time (Snow 1993).[44]

Scientists themselves came out of the war with a new self-image.[45] One indication of this is that the Association of Scientific Workers saw its membership growing from around 1000 members in 1939 to 18,000 in the early postwar years.[46] The Association contributed to the postwar discussions on reconstruction with a book, *Science and the Nation*, published by Penguin in its Pelican Series. 'It seems obvious', the authors argued, 'that if Britain's science could be so effectively planned for war, so likewise it

---

[42] On the sciences at Oxford after World War II see Roche (1994); on the University of Cambridge after World War II see Brooke (1993) and Leedham-Green (1996). For a historical review of Oxbridge in relation to other British universities see Halsey (1994).

[43] On different appropriations of Sputnik in Britain as contrasted to the American reception see Agar (1998c, 118–25).

[44] For a sharp critique of Snow's notion of the two cultures as a distorted picture of Britain's technocratic power in the 1950s see Edgerton (1997).

[45] On the change of the social niche and self-image of the American physicists see Forman (1989).

[46] In the late 1940s and early 1950s the membership dropped again dramatically in line with waning sympathies for the socialist cause (Werskey 1988, 279).

could be for peace' ([Association of Scientific Workers] 1947, v). This was as much a call for scientists and politicians to collaborate in the tasks of reconstruction and international cooperation as an attempt to come to terms with the ambiguous legacy of the atomic bomb. Yet rather than pondering on the responsibility of scientists in producing the deadly weapon, the authors limited themselves to stressing the scientific and organisational effort invested into the project. If the bomb was 'the dramatic and horrifying climax of one line of research for war purposes', a similar scientific and technological effort could be expected 'to yield equally revolutionary advances for constructive social ends' (p. v). Science, the book reiterated, was 'neither good nor bad', but 'a tool or weapon, which society can use for good or evil' (p. 16).

Some scientists had acquired key roles in the war organisation of science and would keep their positions of influence in various advisory functions to the government. Many remained in government service or joined industrial establishments. Others returned to university, to complete their studies or to get back to teaching and research. However, even for those who returned to their prewar occupations, it was not merely a question of picking up where they had left off when joining the war effort, but a time of new departures. In the war scientists had experienced new ways to go about research and had established new contacts with fellow scientists from other disciplines and technical branches, with government officers and industrial producers. Once war had ended, a plethora of new instruments, some of them surplus items from the military which were unloaded in science departments, posed a challenge for new uses. New funds were available and a student generation which had the experience of science at war flooded the science departments in the universities.

Science and public display

The Festival of Britain, which was opened at the South Bank in London in May 1951, epitomises the importance of the sciences in the public and political imagery of postwar Britain. At a time when Britain was losing its imperial power and postwar shortages still dominated daily life, the Festival, originally proposed to commemorate the centenary of the Great Exhibition of 1851, was to present people with a bright utopia of modern Britain. The vision was that of the welfare state, in which all people would enjoy material prosperity. The Festival, seen by some as a 'tonic', by others as a 'narcotic to the people', was a celebration of the faith in this future and of science and technology as a means to achieve these goals (Forty 1976, 26). The main patron of the Festival was Morrison, who steered the proposal through Parliament. In a celebrated essay of the early 1960s, Michael Frayn, then a columnist at the *Observer*, described the Festival as the creation of the 'Herbivores', the 'do-gooders' of the radical middle-classes (Morrison, who had a working-class background, was the exception). In their plans they were opposed by the 'Carnivores', the members of the upper- and middle-classes who 'believe that if God had not wished them to prey on all smaller and weaker creatures without scruple he would not have made them as they are' (Frayn 1986, 308). Frayn also described the Festival as a 'rainbow – a brilliant sign riding the tail of the storm and promising fairer weather' (p. 325). Eight and a half million people visited the South Bank exhibition grounds between May and September that year. While in the exhibits the theme of war was diligently avoided, many of those working on the realisation of the Festival had gained experience in the wartime propaganda and information services. In the press coverage, news from the Festival appeared side by side with reports from the Korean War, which saw Britain's active participation. What the people could not know was that, just one year later, the floating exhibition ship of the Festival, an aircraft carrier of World War II, would be carrying men and material of Britain's top secret atomic establishment to Australia for the test explosion of Britain's first atomic bomb (Hennessy 1992, 421).[47]

The achievements of British scientists as well as the progress of science were celebrated in the Dome of Discovery, a large UFO-shaped building, itself a monument of technological prowess and centrepiece of the main exhibition, *The Land and Its People*, at the South Bank (Fig. 2.1). Yet this was not the only place in which science was displayed. Science figured centrally in the Festival architecture and design, which relied heavily on 'molecular' structures and crystal patterns (Forgan 1998). Science here became a form of art and a fashion setter.

The proposal to use crystal structure diagrams as a source for decoration came from the scientists themselves. The Council of Industrial Design, responsible, together with the Council of Science and Technology, for the development of the South Bank project, enthusiastically took up the idea. The Festival Pattern Group, formed in response, brought together twenty-eight leading British manufacturers. Helen Megaw, a Cambridge crystallographer, played a crucial role mediating between the scientists' requirement of accuracy and the designers' aesthetic requirements (Megaw 1951; Thomas 1951; Forgan 1998). A descriptive panel accompanying the group display in the foyer of the Festival Regatta Restaurant, itself fully furnished with the new system of decorative patterns, stressed the scientific

---

[47] On the Festival see Banham and Hillier (1976); on the display of science at the Festival see Forgan (1997; 1998).

2.1 Cover of the *Illustrated London News* dedicated to the Festival of Britain (1951). Artwork by Terence Cuneo. Note the dome and the vertical structure of the skylon, marking the Festival site on the south bank of the river. From M. Banham and B. Hillier (eds.), *A Tonic to the Nation: The Festival of Britain 1951* (1976), p. 91. Reproduced by permission of the Syndics of Cambridge University Library and the Illustrated London News Picture Library.

2.2 Regatta Restaurant at the Festival of Britain. Note the crystal patterns on wallpaper, curtains, carpet and (less visible) tableware. From *The Souvenir Book of Crystal Designs* (1951), p. 16. Reproduced by permission of the Syndics of Cambridge University Library and of the Trustees of the Design History Research Centre Archives, University of Brighton.

origin of the patterns as well as the strong British tradition in that branch of science (that is, crystallography) (Figs. 2.2 and 2.3).[48] The protein crystallographers were highly gratified to see their patterns used in mass-produced consumer goods, and did not mind the prospect of earning 'some dollars'.[49]

It is true that the Festival which was intended to celebrate the achievements of the Labour government came at a moment when public support for that government was falling and the country was involved in a costly new war. The Conservative government which came to power in September 1951 was opposed to the vision of a socialist welfare state. The 1953 coronation, the next large-scale public event, displayed more traditional values. The New Elizabethan images of science and technology celebrated the heroic discoverer and inventor rather than science as a communal and shared enterprise (Bud 1998). When, later in the decade, the Conservatives appointed Lord Hailsham (later Quintin Hogg) as first Minister of

[48] Crystallography also figured in the Dome of Discovery and at the Exhibition of Science in South Kensington, held in parallel to the South Bank exhibition and dedicated to the more fundamental aspects of science (Banham and Hillier 1976, 144–7).

[49] W. L. Bragg to M. Hartland Thomas (Council of Industrial Design), 11 May 1951, RI, MS WLB 50B/155. On the impact of the Festival Pattern Group on contemporary design see Jackson (1991, 85–94). On the history of a more permanent public spectacle of science than the Festival in postwar Britain, the radio telescope at Jodrell Bank, see Agar (1998c).

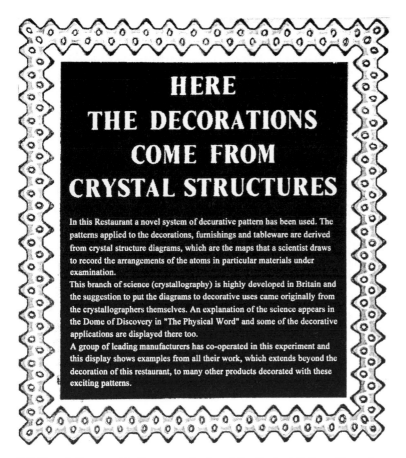

HERE THE DECORATIONS COME FROM CRYSTAL STRUCTURES

In this Restaurant a novel system of decorative pattern has been used. The patterns applied to the decorations, furnishings and tableware are derived from crystal structure diagrams, which are the maps that a scientist draws to record the arrangements of the atoms in particular materials under examination.

This branch of science (crystallography) is highly developed in Britain and the suggestion to put the diagrams to decorative uses came originally from the crystallographers themselves. An explanation of the science appears in the Dome of Discovery in "The Physical Word" and some of the decorative applications are displayed there too.

A group of leading manufacturers has co-operated in this experiment and this display shows examples from all their work, which extends beyond the decoration of this restaurant, to many other products decorated with these exciting patterns.

2.3 Descriptive panel in the group display of the Festival Pattern Group in the foyer of the Regatta Restaurant. Note the border design based on the crystal structure diagram of polythene. From *The Souvenir Book of Crystal Designs* (1951), p. 14. Reproduced by permission of the Syndics of Cambridge University Library and of the Trustees of the Design History Research Centre Archives, University of Brighton.

Science, he stressed that his was a Ministry *for*, not *of* science, because science 'must be done by scientists and not by politicians or Civil Servants' (Lord Hailsham as quoted in Rose and Rose 1971, 83). But despite this *laissez-faire* attitude, state support for science did not diminish during the following thirteen years of Conservative governments.

Scientists had actively collaborated in setting up the science display at the Festival as well as in many other facets of its organisation. This was not the only opportunity scientists seized to present their work to popular audiences. They eagerly wrote for the growing number of popular science magazines which came on the market during or in the years following World War II. These included *Endeavour*, founded by the Imperial

Chemical Industries in 'the dark days' of 1942, 'to record the progress of the sciences in the service of mankind';[50] the *Penguin Science News* which started publication in June 1946 and carried on Penguin's established effort to provide a large public with reliable and affordable scientific information and viewpoints; the *Times Science Review*; and, from 1956, the *New Scientist*.[51] The BBC also slowly increased the time allotted to scientific services. Television, which was more firmly established in 1950s Britian than in other European countries, offered further exciting new possibilities to present science to the public. Already in 1946 the Association of Scientific Workers in their journal were giving considered attention to film as a medium through which the complexities of new scientific knowledge could be explained to the general public and to students. It exhorted scientists to make full use of this exquisitely technical medium of communication to promote their work and to watch the way others used it.[52] In 1957 the BBC launched the series *Eye on Research* devised by Gordon Rattray Taylor, science graduate from Cambridge and wartime reporter for the BBC.[53] He later became editor of *Horizon*, the science programme broadcast on BBC 2, and wrote several books on science-related topics. Both series saw the intensive collaboration of scientists. By that time television had overtaken radio with respect to the transmission of scientific programmes. Some commentators still urged a stronger commitment by the BBC, both radio and television, to science broadcasting, in view of the increasing extent to which science impinged on people's lives ([Anonymous] 1974).[54] Even if British scientists did not directly depend on the public for funding, public interest in science was a legitimation for tax spending.

New departures

In the 1960s the organisation of government research stood at the centre of political debates. When, in 1963, the 'Trend Committee' reported

---

[50] See the subtitle of the journal and the presentation of the new periodical by the president of ICI (McGowan 1942). On the wartime creation of *Endeavour* as a public relations exercise for ICI see Bud (1998, 324–5).

[51] In 1945, Penguin also started the series *New Biology*, which focused on the use of biology in human affairs. On the history of Penguin with brief reference to these new science series see *Sixty Penguin Years* (1995). The *Times Science Review*, started in 1950 as the *Times Survey of the Progress of Science*, ceased publication in 1962. On the purpose and aims of the *New Scientist* see Anonymous (1956) and Dixon (1971).

[52] 'Scientific Film Committee', *The Scientific Worker* (April 1946), 19. See also [Association of Scientific Workers] (1947, 208–10).

[53] On the detailed analysis of one such programme pertaining to the research of the Cavendish protein crystallographers see below (chapter 5).

[54] There is as yet no extensive historical study of science on television in Britain. In Briggs' official history of the BBC, science is hardly mentioned; the same is true for Seymour-Ure's introductory volume on *The British Press and Broadcasting since 1945* (Briggs 1979; Seymour-Ure 1991).

on the government organisation of civil science, it found its structures lacking in many respects ([Trend Report] 1963). In the following years, the Labour government, which had come to power in the 1964 election on a 'scientific' platform, introduced wide-ranging changes in the government apparatus for (civil and military) science. Science policy, however, is not a fixed entity (Jacq 1995). After the war, the organisation of science was discussed, but scientists and administrators gave priority to an increased allocation of government funds for 'fundamental' research and to the expansion of scientific manpower, as well as to the promotion of scientific projects considered important to national security. Succeeding governments held to this promise, even in times of economic stringency. Organisation became a key issue only in the 1960s and once the government apparatus for science had grown to represent a sensible percentage of government expenditure as a whole (see below, chapter 10). The postwar years nevertheless marked a time of new departures and of new opportunities for scientists. We have seen that scientists themselves were active in creating these opportunities, which they used as much for securing a place for themselves in society as for their own disciplinary ends. How the legacies of the scientific mobilisation and the new chances offered to scientists were used for the reconstruction and expansion of biophysical research after the war will occupy us in the next chapter.

3

· · ·

Reconstructing life

In the 1940s and 1950s, biophysics occupied much of the territory later claimed by molecular biologists, most notably Watson and Crick's work on the structure of DNA. The field existed before the war, but it gained new attraction and covered different approaches in the period following World War II. Postwar biophysics encompassed as diverse research traditions as electrophysiology, protein crystallography and radiation studies. The MRC became one of the main patrons of biophysics in Britain. The MRC Unit for the Study of the Molecular Structure of Biological Systems, located in the Physics Department in Cambridge, was among the new institutions created under this heading. Protein crystallography was pursued in the Cavendish before the war. But situating their work under the general banner of biophysics and using resources and opportunities available for biophysical research after the war, the protein crystallographers in the Cavendish managed to put their work on a safer footing. In this chapter I will analyse the reasons for the extraordinary expansion of biophysics after the war and reconstruct the beginnings of the Cambridge unit.[1] I will discuss the question of continuity with prewar pursuits and will revisit the question of the role of physicists in the postwar transformation of biology.

Crick, himself a physicist who turned to biophysical research after his war work and, in 1949, joined the Cambridge unit, has suggested that John Randall, a physicist with an important track record as scientist in the war, played a decisive role in persuading Edward Mellanby, the 'powerful

---

[1] Recently Rasmussen has presented a parallel study regarding the explosion of biophysical research in postwar America (Rasmussen 1997a). While more dramatic – in respect of both the size of the enterprise and the rhetoric employed – the American story shows many common threads to the British one. The differences reflect the stronger weight laid on basic biological research in American health politics and the stronger impact of the bomb on postwar American culture as compared to Britain (Boyer 1985; Weart 1988).

Secretary of the MRC', to make funds available for biophysics after the war. He writes:

During the war scientists had acquired much more influence than they had had before it. It was not difficult for Randall, one of the inventors of the magnetron (the crucial development in military applications of radar), to argue that just as physicists had had a decisive influence on the war effort, so they could now turn their hands to some of the fundamental biological problems that lay at the foundations of medical research. Thus there was money available for 'biophysics'. (Crick 1990, 19)[2]

This reconstruction is interesting if only because, in contrast to other accounts, Crick establishes a connection between wartime exigencies and postwar opportunities. His answer is also instructive in respect of the perception of scientists, in particular physicists, of their newly gained social status and their power in defining postwar needs and policies. Finally, he acknowledges the new emphasis on 'fundamental' medical research and the availability of funds in this field. A closer look at the new fortunes of biophysics after the war will allow us to make these connections tighter.

In this chapter I will take seriously the experience of science at war of those – young scientists and their elder scientific patrons – involved in building biophysics. This will provide a vantage point from which to revisit the widely held thesis that physicists played a leading role in the postwar transformation of biology. Historians have, in their own reckoning, been obsessed with this question (Fox Keller 1990, 389).[3] As with many issues in the field, this one also has been prompted by scientists themselves, who have erected figures like Max Delbrück and Erwin Schrödinger as founding fathers of the new science of molecular biology, building on the prestige and authority of the science, quantum mechanics, to which these scientists had made important contributions before reformulating questions in biology.[4] Instead of playing off physicists' and biologists' contributions in establishing biophysics (or, for that matter, molecular biology), my central argument will be that both physicists and biologists, or, more generally, physical and biomedical scientists, participated in the postwar expansion of biophysics, taking advantage of opportunities created by the legacies of

---

[2] On Randall's role in convincing Mellanby to promote biophysics after the war and the importance of these developments for the history of molecular biology see also Fruton (1992, 213).

[3] See for instance Fleming (1969), Olby (1971), Yoxen (1979), Abir-Am (1982a) (including the 'Responses and replies' 1984) and Kay (1985).

[4] On the appropriation of Schrödinger as founding father of molecular biology see Abir-Am (1985, 104–5). On Schrödinger and Delbrück see also references below.

the scientific mobilisation of World War II. Penicillin and the expectations of medical returns from basic biological research played as important a part in the fortune of postwar biophysics as the attraction of turning physical technologies to more peaceful ends. The very term 'biomedical' was coined in postwar America to justify the diversion of funds from medical research to basic biological, including biophysical, research (Feinstein 1995, 289).[5] In Britain, as already mentioned, the Medical Research Council became the main patron for biophysical research. In particular the postwar restructuring and expansion of medical research, then, was an integral part, as much as a consequence, of new approaches in biology, a theme to which I will return. Stressing the legacies of World War II does not mean denying the impact of prewar developments and especially the role of the Rockefeller Foundation in promoting physical and chemical approaches in the life sciences (Abir-Am 1982a; Kohler 1991; Zallen 1992; Kay 1993a). However, the exigencies of the war and the postwar expansion of the physical as well as of the biomedical sciences, including the call for more fundamental research, functioned as a powerful selector and promoter of (some of) these earlier initiatives.

What biophysics was remained contested. But despite meaning different things to different people, there was an interest in keeping the field together. It was a useful catch-word to attract funds and facilitate certain career moves. It also served to signal a new approach to biology. By the late 1950s, however, different interests prevailed. Molecular biology became one of the fields built on the legacies of postwar biophysics.

The need for biophysics

In 1944, joining others in making plans for the development of the sciences after the war, Hill, prominent physiologist and science adviser, Biological Secretary of the Royal Society (1935–45) and Member of Parliament for the University of Cambridge (1940–5),[6] drafted a document

---

[5] On the debates leading to the allocation of huge funds for medical research in lieu of the creation of a national healthcare system in postwar America, a politics which continued through Nixon's cancer campaign in the 1970s to more recent times, see Strickland (1972). The policy led to the growth of a biomedical complex, consisting of research establishments with their political lobbies and tight connections to the pharmaceutical industry which has become difficult to displace, as events in the first Clinton administration have shown. The biomedical complex developed parallel to the military–industrial–academic complex. Biomedical funding by military agencies points to connections between the two interest areas.

[6] Hill, who had a Cambridge degree, but from 1923 held a chair at University College London, was nominated as candidate by the Cambridge University Conservative Association, despite his objection that he 'could not properly be described as a

entitled 'The need for an Institute of Biophysics' which he sent to the President of the Royal Society and some other high officials.[7] The aim of the Institute proposed by Hill was to provide training and promote research in the application of physical methods to biological questions. A memorandum on the same subject dating back to 1937 as well as a document relating to the plan of creating a 'Cavendish Laboratory for biologists and doctors' as a memorial to Rutherford, attached as appendices, were to prove that the plan had long been in his and others' minds.[8] The war, Hill explained, had called for other actions, but much had happened in the last years to strengthen the case for biophysics. A variegated, but most impressive list of technological achievements in the physical and medical field followed. It comprised the development of radioactive and other isotopes; radio, and in particular radar; the electron microscope; supersonic and vacuum techniques; electronics; chemotherapy; insecticides and repellents; viruses; neurology; and the physiology of flying. All these technological advances, Hill suggested, represented 'a challenge to biophysical application, research or methods'. The institute he proposed would allow physicists who had become interested in applying physical methods to biology to train and to open up career opportunities in this new field of 'fundamental importance'. More and more physicists, said Hill, 'two of them recently within a week', inquired about such possibilities.

Technological innovations and a new interest in the field were not the only features which distinguished Hill's new proposal from his prewar plan. In both cases he stressed that London (where he ran his own Laboratory of Biophysics) would offer the best location for the proposed institute. But while in 1937 he put most weight on the interactions with other university departments, after the experience of the war, interactions with government establishments and industry were mentioned as well. The plan had also gained in ambition. Not any longer the University of London but only the Royal Society, which during the war had taken a more active role in the organisation of scientific research, was deemed capable of seeing the need and taking the necessary steps to the realisation of the

Conservative' (Katz 1978, appendix II, p. 141). The privilege of the University of Cambridge to nominate a representative to Parliament was abolished in 1945.

[7] A. V. Hill, 'The need for an Institute of Biophysics' [accompanying letter to E. Mellanby dated 7 December 1944], file FD1/7099, MRC Archives. On Hill's wartime activities, especially regarding his participation in the Tizard Committee for the development of radar and in setting up the Central Scientific Register, the Scientific Advisory Committee to the War Cabinet and scientific liaison offices for the Commonwealth and with the United States see Katz (1978), esp. 113–22, 136–40. On Hill's scientific career see below.

[8] See A. V. Hill, 'An institute of biophysics' (1937) and letter from A. S. Eve to A. V. Hill, 11 December 1937; attached as appendices I and II to A. V. Hill, 'The need for an institute of biophysics' [1944], file FD1/7099, MRC Archives.

proposed institute. This included directly approaching the Treasury. The cost of building, equipping and staffing an Institute of Biophysics would be considerable. Equipment was expensive and laboratory assistants, mechanics and electricians would be needed on a generous scale. 'To try to carry out the plan in a minor way', Hill intimated, 'would be no good'.

Hill made it clear that his interest was only to get the institute off the ground. Before the war he had hoped to expand his own research group at University College London, dedicated to physico-chemical investigations of muscles and their functions, into an Institute of Biophysics. But after a six-year interruption from active research owing to his public activities during the war, he needed all his energy to find his own way back to research. He therefore did not feel in a position to take over the directorship of a large research enterprise.[9] Using all the weight of his authority, however, he reiterated the timeliness and the importance of creating such a centre, both for the advancement of fundamental science and of medical applications and to keep up with American developments in the field.

The circulation of Hill's statement on biophysics coincided with the final drafting of the Royal Society 'Report on the needs of research in the fundamental sciences after the war', based on a series of reports prepared by special subject committees (see above, chapter 2). In the document biophysics, along with biochemistry and oceanography, was mentioned as a 'borderline subject' which called for '"extraordinary" expenditure' to be met by special Treasury grants.[10] Interestingly, neither the physics committee nor the biology committee, whose work had preceded the drafting of the final document, had mentioned biophysics in their reports. Very probably Hill in his function of Biological Secretary had been able to introduce the subject in the final rounds of deliberations leading to the document.

The initiative did not end here. Very soon Hill identified in Randall a possible candidate to lead an institute of the sort he had envisaged and strongly promoted his application to the Royal Society for special Treasury funds.[11] Despite Hill's backing, Randall's application took a complicated institutional journey and was the subject of complex negotiations. In the fluid postwar situation policies and competencies in allocating increased budgets had still to be established. The interdisciplinarity of Randall's

---

[9] A. V. Hill, 'The need for an institute of biophysics' [1944], file FD1/7099, MRC Archives.

[10] 'Report on the needs of research in the fundamental sciences after the war', Meeting of 14 December 1944, Appendix A, Royal Society Minutes of Council, 1940–5, vol. 16, pp. 334–60, RSL.

[11] In his memorandum Hill had suggested D. W. Bronk, Professor of Medical Physics at the University of Pennsylvania, as first director for the proposed Institute of Biophysics; see 'Need', p. 3. This suggestion was not further pursued.

project, as well as the unprecedented size of the grant requested, posed further problems. The bureaucratic passage of the application transformed Randall's project as much as it created the administrative channels and the institutional space for biophysics. In these and other ways Randall's application prepared the way for other applications in the field, and especially for Bragg's project to set up a protein crystallography unit at Cambridge.

The 'Randall incident'

When in November 1945 the Physical Secretary of the Royal Society circulated a letter to all physics departments in Britain inviting applications for 'extraordinary expenditures' to be met by special Treasury grants, Randall, Professor of Natural Philosophy at St Andrews, seized the opportunity. Encouraged by Hill, he asked for a grant to cover the costs of a biologist and a technician, as well as some apparatus, to develop his biophysical work.[12]

A physicist by training, Randall had shown an interest in applying physical methods to biological material since before the war. In this connection he had been in touch with Bernal, the crystallographer William Astbury and the British cytogeneticist Cyril Darlington. However, only after having acquired fame for his crucial contribution, together with Henry A. H. Booth, to the development of microwave radar in the early years of the war did he see the chance to pursue his interest more seriously. In 1943, Randall had left Oliphant's physical laboratory at Birmingham to fill a teaching position in the Cavendish at Cambridge, and one year later had moved to St Andrews where he had started biophysical research on a small Admiralty grant.

Together with Randall, a few other physicists responded to the invitation of the Royal Society. Patrick Blackett from Manchester applied for a cosmic ray spectrograph; Eric M. Lindsay, the director of the Armagh Observatory, for a telescope for studies of galactic structures; and Arthur M. Tyndall from Bristol for research in nuclear physics.[13] The Royal Society board dealing with the applications approved all the requests except the one by Randall, for which it deferred a decision. It welcomed Randall's plan of 'an advance of physics . . . into the biological field', but felt 'unable to express an opinion whether the biological side of the investigation would be covered by the appointment of a young biologist' as

---

[12] Royal Society Council Agenda and Papers, Meeting of 17 January 1946, minute 14 and document U, RSL. On Randall's application see also Wilkins (1987, 510–12) and Olby (1994, 326–31).

[13] Royal Society Council Agenda and Papers, Meeting of 17 January 1946, minute 14 and document S, RSL.

suggested by Randall.[14] On Hill's incitement, Randall rebutted submitting a new and much more substantial application. If in his first application he had asked for £1000 annually for a period of five years, he now applied for a grant three times that size.[15] The new proposal included the salary costs for a physicist, Maurice Wilkins, a member of the British team in the Manhattan Project, who had started work under Randall. The most substantial increase, however, regarded the projected costs for 'special apparatus'. The research project itself had become both more ambitious and more specific. It comprised a much broader range of physical, especially optical, approaches to the study of cells. But Randall now also proposed to focus the investigations on 'the physical factors affecting mitosis and cell division', a project clearly informed by Darlington's work. The matter was handed over from the Physical to the Biological Secretary, a position now occupied by the botanist and serving director of the Royal Botanic Gardens at Kew, Sir Edward Salisbury, and a committee was formed. Although himself representing quite a different biological tradition, Salisbury dispassionately pursued the matter which had been so near to the heart of his predecessor. But opposition to Randall's scheme there was.

Among those asked to serve on the committee was David Keilin, a highly regarded parasitologist and biochemist, Quick Professor of Biology and head of the privately funded Molteno Institute in Cambridge. Keilin, who had himself turned to using physical instruments to study molecular cell functions, was supportive of other biophysics projects, like Perutz's crystallographic studies of haemoglobin (see below). However, in his view a reliance on physical methods alone was obviously not enough to draw up a biological research programme. In a letter to Salisbury he made it clear that he was little impressed by Randall's knowledge of the subject he proposed to study and that he could not discover any original idea or suggestion of a new approach in his proposal. Instead of directing a unit, Keilin suggested, Randall should come to Cambridge to learn some biology and test his ideas on mitosis and chromosomes with researchers who were active in the field. If, however, he only wished 'to dabble in biology, perhaps under the influence of the admirable small book by Schrödinger', money spent on him not only would be wasted, but would also harm the subject.

[14] Royal Society Council Agenda and Papers, Meeting of 17 January 1946, minute 14 and documents S and U; Royal Society Supplementary Council Agenda, Meeting of 14 February 1946, minute 2, RSL.

[15] Royal Society Council Agenda and Papers, Meeting of 11 April 1946, minute 3.c, RSL. The full document could not be located. But see the final document, 'Programme of biophysics research to be carried out by Professor J. T. Randall, F.R.S., in King's College, University of London' [July 1946], file FD1/7096, MRC Archives. At this point the total expenditure had risen to £22,000.

Even if the Society 'may now be in a strong position to risk a large sum of money on research', Keilin objected, it was still not right to divert the few good workers in the field from useful work by offering them attractive salaries and asking them to join a '"Five Year Plan" of a wild goose chase'. There were better ways to promote the subject.[16]

Keilin's harsh statement led to a recomposition of the committee to include some of the people Keilin had mentioned in his letter as experts in the field. Those co-elected were the Cambridge botanist David G. Catcheside, Darlington, Honor Fell, cell biologist and director of the Strangeways Laboratory in Cambridge, and Joseph S. Mitchell, Professor of Radiotherapeutics in Cambridge. The enlarged committee, however, endorsed Randall's application, expressing doubts only about St Andrews as a location for the project – money alone could not 'create' a good place – and inviting Randall further to consider possible collaborations with biologists.[17]

Meanwhile, Randall had been offered the prestigious Wheatstone Chair of Physics at King's College London. The most serious obstacle to his proposal for research in biophysics seemed removed. He submitted a final proposal organised around physical methods and their application to subcellular structures, especially chromosomes.[18] Anticipating that the position in London would make it easier for Randall to get the necessary feedback from biologists, even Keilin adopted a more conciliatory tone,[19] and the Royal Society approached the Treasury for funds. However, new difficulties lay ahead.

The responsible secretary of the Treasury, Alan Barlow, harboured doubts about whether the Royal Society was the appropriate body to oversee a project like Randall's. In a confidential letter to Edward Mellanby, the secretary of the MRC, he enquired if his Council was willing to consider the application. Mellanby, who had steered the Council through the war years, had seen its standing grow considerably owing to the role played by the medical research agency in the wartime mobilisation (Fig. 3.1). As Landsborough Thomson, long-serving deputy secretary of the MRC and its official historian, laconically noted: 'Council had emerged from each

---

[16] D. Keilin to E. Salisbury, 6 April 1946, file FD1/7098, MRC Archives. Keilin's remark on Schrödinger's *What is Life?* is interesting in relation to the importance the book acquired in the founding myths of molecular biology (see below).

[17] 'Memorandum regarding the application of Prof Randall', file FD1/7096, MRC Archives.

[18] 'Programme of biophysics research to be carried out by Professor J. T. Randall, F.R.S., in King's College, University of London' [submitted 22 July 1946], file FD1/7096, MRC Archives.

[19] D. Keilin to E. Salisbury, 23 August 1946, and D. Keilin to E. Mellanby, 17 February 1947, file FD1/7098, MRC Archives.

3.1 Edward Mellanby, Secretary of the MRC (1933–49).
Reprinted with the permission of the Medical Research
Council 2000.

of the wars with its range and resources greatly enlarged and its reputation greatly enhanced. The applauded achievements were attributed to the Secretaries' (Landsborough Thomson 1973, 215).[20] On receipt of the note from the Treasury, Mellanby at once perceived the opportunity to settle

[20] For a disturbingly triumphant view on the wartime achievements of the MRC see Mellanby (1943). The most important areas of intervention were seen in the organisation of the Emergency Public Health Laboratory Service and of the Blood Transfusion Service, the development of bacteriostatic drugs, and the advisory functions to the government, especially with regard to food and nutrition. Another area on which the MRC prided itself was physiological and psychological research into the safety, efficiency and comfort of fighting personnel, and especially of the crews engaged in actions on aircraft, tanks and submarines. On the early history of the MRC and its research policy see Austoker (1989). In the postwar period the MRC became notorious for not issuing policy statements; see editorial 'Growth like Topsy's' in *Nature* 215 (1967), 567. Internally, policy issues were certainly discussed. Confidential documents as well as more public statements reveal an institution determined to capitalise on its contribution to the war effort – unmistakably represented as a great success and 'tremendous stimulus' – and to raise its profile; see files PF100/19/3 and FD5/110,

a question of principle. The Treasury, he agreed with Barlow, should not make decisions on single applications made by the Royal Society, which was not equipped to administer large grants. As far as Randall's application was concerned, it certainly lay in the field of competence of the MRC. In an attempt to keep control over the project, the Secretary of the Royal Society explained to the Treasury that Randall's main objective was an understanding of fundamental cellular processes, which could have medical as well as other applications. To deal with it simply as a medical research project meant imposing undesirable biases.[21] At stake in this discussion was clearly not just Randall's application, but the authority of various bodies and their competence in defining research aims and controlling research budgets which needed to be renegotiated after the changes produced by the war and postwar restructuring. What was classified as fundamental and applied research and the place of medical research in this context were also at stake.

The Secretaries of the three Research Councils were called to the Treasury to discuss the problems arising from the 'Randall incident'. It was agreed that it was undesirable that there existed parallel channels for Exchequer's assistance to the same project and that the best way to solve the problem was to restrict the responsibility of the Royal Society to those few subjects for which no Research Council felt directly responsible, namely astronomy, some forms of mathematics, and certain problems in biology, including taxonomy. In the particular case of Randall's application it was left to the Secretaries of the Agricultural and Medical Research Councils to decide who should be responsible for it.[22]

121–3, 127, MRC Archives. The two areas most consistently mentioned as in need of particular attention after the war were industrial psychology and the application of advances in nuclear physics to medicine. Policy statements regarding research in Britain more generally also mentioned the need to raise the status and budget of the MRC; see e.g. 'Research and development expenditure in Great Britain', Third Report of the Select Committee on Estimates (1947), file FD5/121, MRC Archives. Which research areas the Council supported depended to a large extent on the applications made by individuals and on the interests and initiatives of its Secretaries, who all had a scientific background, if not, as in the case of Mellanby, an active scientific career. Until the mid-1960s the Secretaries were directly responsible to the Lord President of the Council and strongly dominated the organisation over which they presided. The general 'philosophy' was to grant generous support to 'excellent' people, allowing them to pursue their own research agendas. After the passage of the National Insurance Act (1946) and following discussions with the Ministry of Health, the MRC also became responsible for clinical research, a field close to the heart of Harold Himsworth, who succeeded Mellanby as Secretary of the MRC in 1949.

21 A. Barlow (Treasury) to A. Landsborough Thomson (MRC), 29 August 1946; A. Landsborough Thomson to A. Barlow, 2 September 1946; A. Barlow to E. Salisbury, 5 September 1946; E. Salisbury to A. Barlow, 5 September 1946; E. Mellanby to A. Barlow, 1 October 1946, file FD1/7096, MRC Archives.

22 Note by E. Mellanby, 11 October 1946, file FD1/7096, MRC Archives.

When the Secretary of the Agricultural Research Council showed himself impartial, Mellanby was quick to seize the opportunity and assume responsibility for initiating the project, even while agreeing that its later implications might well be rather agricultural.[23] By November, Randall's project was approved and a five-year grant agreed, subject to the conditions that a senior biologist join the team and that major equipment would be shared with other MRC researchers. A special advisory committee was formed to oversee the development of the biophysics programme.[24]

Randall's unit opened in March 1947. The unit was attached to the Physics Department of which Randall held the chair. On his encouragement, many physicists of the university department also embarked on biophysical research. With around fifty scientists and technicians working on biophysical projects, Randall's group at King's College London was among the biggest centres for biophysical research in Britain in the 1950s (it was topped only by the Radiobiological Research Unit at Harwell, which grew into a huge enterprise).

In a letter to Keilin, Mellanby admitted that he was 'under no illusions' about Randall's scheme, but that 'the method whereby it came into our hands' made it necessary to give the proposal full support.[25] In front of the Royal Society, Mellanby justified the decision of his Council by adopting a very large definition of what medical research meant. 'Remember', he wrote to Salisbury, 'medical research means research of value to human beings and it is not simply a matter which concerns ill health.' He also accused Salisbury of chasing a will-o'-the wisp in his ideas about the difference between fundamental and applied research. The point was not whether the results were of value 'to men or flowers or animals'. A fundamental discovery, according to Mellanby, was 'something which is really new, opening up a fresh vista'. Fundamental research was work in pursuit of such knowledge. 'In medical research and I think probably in biological research generally', he proclaimed, 'it is no more unlikely that a fundamental discovery will come out from research on a practical problem than a practical discovery will come out from research on a fundamental problem.'[26]

Mellanby, himself an active researcher in the field of vitamins trained in Gowland Hopkins' biochemical laboratory at Cambridge, here not only reiterated the general terms of references of the MRC which were negotiated

---

[23] J. Fryer to E. Mellanby, 16 October 1946, and E. Mellanby to J. Fryer, 21 October 1946, file FD1/7096, MRC Archives.

[24] MRC Council Minutes, Meeting of 22 November 1946.

[25] E. Mellanby to D. Keilin, 21 February 1947, file FD1/7098, MRC Archives.

[26] E. Salisbury to E. Mellanby, 15 November 1946, and E. Mellanby to E. Salisbury, 20 November 1946, file FD1/7096, MRC Archives.

with the newly created Ministry of Health in the interwar years and accorded the Council the field of 'basic' medical research. Rather, he interpreted current debates on the need for fundamental research in a way that secured his Council a key role in the postwar resettlement and consolidated the influence the Council and its Secretary had gained through their contribution to the war effort. The 'Randall incident' had offered Mellanby an opportunity to affirm his ambitions. But even if political motivations dominated, the support for Randall's project marked an important engagement of the MRC in the field of biophysics. A few months after the approval of Randall's project, the MRC received and approved an application by Bragg to set up a protein X-ray crystallography unit at the Cavendish Laboratory in Cambridge. The unit was grouped under the newly created heading of 'biophysics'.

Molecular structure

In April 1946, Bragg also applied to the Royal Society for funds made available for 'border-line subjects' to provide assistance for Perutz in his investigations of crystalline proteins by means of X-ray analysis. Bragg, who with his father William Bragg had instituted the subject of crystallography in Britain, argued that Perutz's work was of 'national importance' because very few centres were engaged in the structural analysis of proteins, a field in which Britain was the undisputed leader. The progress of the work depended crucially on the help Perutz could get with the routine measurements and computations involved in the investigations. The grant Bragg was asking for was rather modest, a sum not exceeding £800 a year for a period of two years. The application, however, was not successful. Instead Bragg was advised to make his application to the MRC or the Department of Scientific and Industrial Research (DSIR) in the first instance.[27] Bragg had been in touch with the DSIR before, and in the event successfully applied for funds there. But more serious problems loomed: Bragg's ambition to secure a university position for Perutz and thus to establish the work, which he had been supporting since before the war, on a more permanent basis seemed doomed to fail.

Perutz, a graduate in chemistry from Vienna, had come to Cambridge in 1936 to learn crystallography from Bernal who, a few years before, together with Dorothy Crowfoot (later Hodgkin), had obtained the first X-ray picture of a crystalline protein (Fig. 3.2). The following year Perutz settled on haemoglobin as research object for his Ph.D. When, later that year, Bernal

---

[27] Meeting of 11 April 1946, Council Agenda and Papers, document L and Royal Society Minutes of Council 1945–1948, vol. 17, minute 16.

3.2 John D. Bernal (back row, centre) and his crystallography group in Cambridge around 1935. From G. Werskey, *The Visible College: A Collective Biography of British Scientists and Socialists of the 1930s* (1988), fig. 4. Reprinted with permission of Free Association Books.

moved to London to take up the chair of Physics at Birkbeck College, London, Perutz decided to remain in Cambridge and continue research under Bragg, who in the same year succeeded Rutherford to the Cavendish Chair in Cambridge. Bragg became keenly interested in extending the method of X-ray analysis he had pioneered with his father to the challenging problem of protein structure. With the annexation of Austria by Hitler in March 1938, Perutz became a refugee, but Bragg came to the rescue. Drawing up a project for research on the structure of haemoglobin, which saw his own and Keilin's collaboration, he applied for a Rockefeller grant to employ Perutz as his assistant.[28] This allowed Perutz to continue

---

[28] L. Bragg to W. E. Tisdale, 28 November 1938, folder 561, box 42, series 401, record group 1.1, RAC. For the Rockefeller Foundation the main attraction of Bragg's scheme lay in the planned collaboration with Keilin, a long-time grantee of the Foundation, and the hope of preparing the ground for further cooperative projects between physicists and biologists at Cambridge, which represented the centrepiece of the Foundation's programme in the natural sciences. Following his first meeting with Bragg in London, Wilbur E. Tisdale, the Rockefeller Foundation representative for Europe, noted in his diary: 'I do not believe [Bragg] has any real interest in biology. I do know that he has a deep interest in the study of crystals which are of biological interest, and it is very

his research during the war, albeit with some interruptions (see below). However, Bragg strongly felt that he 'must not trespass on [the] generosity' of the Foundation for too long and, with the war drawing to an end, he was hopeful of receiving university funds for Perutz's work, 'particularly in view of the greatly increased funds which we all hope will be put at the disposal of academic research'.[29] Bragg managed to obtain an ICI fellowship for Perutz, but his aim to secure a university post for his protégé looked increasingly bleak. To Bragg this must have seemed the only way to guarantee Perutz's future and to establish the subject of protein crystallography at Cambridge. However, despite his position as Cavendish Professor, Bragg did not wield much power in university politics and the university was not keen to institute a lectureship in a 'no man's subject'. Kendrew, a graduate in chemistry with a distinguished career as scientist in the war, who had joined Perutz in the autumn of 1945, was also on short-term funding.[30] The story goes that at this critical juncture Keilin suggested to Bragg to ask the MRC for support (Perutz 1980, 327; 1987, 40). Yet the timing is crucial here. Keilin was well acquainted with Perutz's work. Comparative aspects of the structure and function of haemoglobin had long fascinated him (Mann 1964) and very early on he had offered Perutz bench space for his crystal preparation in the Molteno Institute (Fig. 3.3). Keilin also had a long-standing connection with the Medical Research Council, having

evident that he completely reverses past traditions of keeping [the Physics Department] isolated from the others... I feel that if we can make a smallish grant to start a cooperation between X-rays and biology at Cambridge, the atmosphere will be cleared also for neutron work relations, and that eventually this scientific center of the British Empire will follow in at least some respects those cooperations which in other centers are proving of value'; see WET (W. E. Tysdale) diary entry, 23 November 1938, folder 561, box 42, series 401, record group 1.1, RAC. The entry also reveals Bragg's interest in attracting Astbury to Cambridge. Perutz acknowledged Hill for suggesting to Bragg that the Rockefeller Foundation might give Bragg a grant to employ Perutz as his assistant; see M. Perutz to A. V. Hill, 4 November 1962, file AVHL 3/75, Churchill Archives Centre, Cambridge.

[29] L. Bragg to F. B. Hanson (Rockefeller Foundation), 20 April 1945 (the sentence 'must not trespass' was underlined in red on the receiving side) and L. Bragg to H. M. Miller (Rockefeller Foundation), 28 September 1945, folder 565, box 44, series 401, record group 1.1, RAC. See also L. Bragg to H. M. Miller, 3 August 1944, folder 564, box 44, series 401, record group 1.1, RAC. Equipment and travel grants from the Foundation for protein crystallography at Cambridge, however, continued well into the 1960s. The dollar grants, most of which were directly spent on equipment purchased in America, were particularly important in the early postwar years when the convertibility crisis of the pound and restrictions on foreign imports made it virtually impossible to make such expenditures from Britain. Among the instruments imported from America in the 1940s were X-ray tubes and several Buerger precession cameras.

[30] Kendrew had resumed a college fellowship at Trinity from before the war which gave him two years of funding. On Perutz's early career see his own account (Perutz 1980; 1987). The piece represents a personal account of the history of the Cambridge unit. For a discussion see below, pp. 170–1.

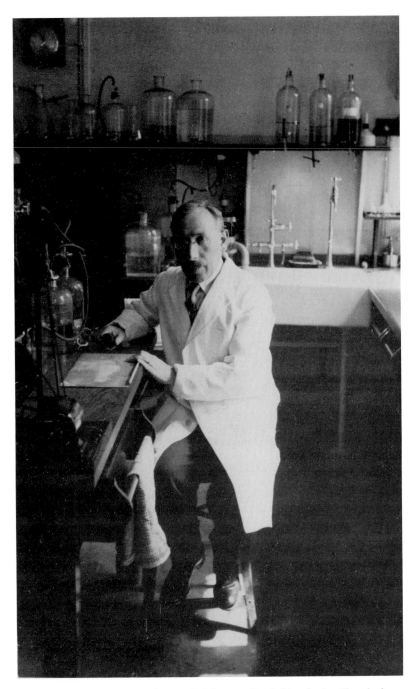

3.3 David Keilin, Quick Professor of Biology at Cambridge, in the Chemical Room in the Molteno Institute, also used by the protein crystallographers for their chemical work (late 1930s). From the Keilin Papers, Add. 7953, file E.12, A 23, Cambridge University Library. Reproduced by permission of the Syndics of Cambridge University Library.

only just concluded his term as member of the Council. The suggestion to turn to the MRC for funding, however, came only after his involvement with Randall's application and the approval of the project by the MRC. Bragg himself, not used to dealing with the MRC, had probably not appreciated the possibility. He personally certainly perceived the haemoglobin project as a crystallographic challenge rather than as an investigation of biomedical interest.

Plans were drawn up for a research unit on the molecular structure of biological systems by means of X-ray diffraction. The current work on haemoglobin was to be extended to other proteins and new X-ray techniques developed to study the fine structure of nerve and muscle cells and other cell structures. Support for a period of five years was sought for two full researchers (Perutz and Kendrew), two assistants and some special apparatus. Bragg, who would further house the group in the Cavendish, drafted the proposal and personally discussed it with Mellanby. Keilin wrote in support, stressing that the project had developed as a collaboration between the Cavendish and the Molteno Institute and was of 'fundamental biological importance'.[31]

The MRC Unit for the Study of Molecular Structure of Biological Systems with Perutz as director was officially set in place in October 1947. If Mellanby had acquired Randall's project for reasons of prestige, this was the confirmation of a serious engagement of the MRC in the field of biophysics (Figs. 3.4 and 3.5).

Certainly, the MRC had independent reasons to support the crystallographic work at Cambridge. During the war years, Dorothy Hodgkin and her collaborators, using X-ray crystallography, had provided conclusive proof of the structure of penicillin, where traditional chemical methods of structure analysis had failed. It was an essential step in the frenetic efforts to produce penicillin synthetically (Swann 1983). Even if in the end this avenue was not further pursued at the time, Hodgkin's accomplishment had provided an important recognition of X-ray crystallography as analytical tool for the structure determination of complex biological molecules.[32] Protein crystallography represented a new challenge, but the prospects of biomedical returns were equally big: haemoglobin, the

[31] L. Bragg to E. Mellanby, 26 May 1947; D. Keilin to E. Mellanby, 31 May 1947, file FD1/426, and 'Proposed unit at Cambridge for research on molecular structure of biological systems. Memorandum from Sir W. Lawrence Bragg, F.R.S., Cavendish Laboratory, University of Cambridge', Council Paper no. MRC 47/449, presented at meeting of 17 October 1947, MRC Archives.

[32] For a strong recognition of this achievement see e.g. Chain (1964, 142). Among many other publications on the penicillin story see Liebenau (1987). On Hodgkin see also Ferry (1998).

3.4 Snapshot of Max Perutz (1946). Copyright Cavendish
Laboratory Cambridge.

molecule at the centre of Perutz and Kendrew's efforts, was of eminent
medical importance.

National pride and elements of competition with the United States,
with which the scientific wartime collaboration on unequal terms had
left Britain in an ambiguous position, also played a role. Crystallography
was fashioned as a field created by British scientists. If American phi-
lanthropists had saved the haemoglobin project over the stringencies of
wartime research budgets, time had come to take responsibility for it and
not to surrender the fruits of research to the American initiative.[33]

[33] Arguments of this kind continued to prevail. When in 1948 Perutz, now Director of the
newly established MRC unit, asked for leave to take advantage of a Rockefeller
Travelling Fellowship, Mellanby took the view not only that he should remain at his
post, but also that he 'should be of sufficient standing to attract Americans to his
laboratory, rather than that he should go and seek their help'; E. Mellanby to M. Perutz,
1948; cited after Olby (1994, 291). Similarly, when in the late 1950s the NIH offered the
unit a significant sum of money to support the crystallographic work, the Secretary of
the MRC retorted that as long as the Council did not refuse them any money they asked
for, they did not need to take American money; interview with J. Kendrew, 14 July 1992.

3.5 Main entrance to the Cavendish Laboratory on Free School Lane (1946). *Source*: Computer Laboratory, University of Cambridge.

But in addition to these reasons, the Cambridge proposal fitted the new engagement of the MRC in biophysics and its declared aim to win physicists and chemists to apply their technologies to work on biological and medical problems. The new Cambridge unit was consistently listed under the new heading of 'biophysics' in the scientific programme of the MRC, and even if the term did not appear in its official denomination it was commonly used to refer to the group and its work by all parties involved.[34] Perutz became a member of the Biophysics Committee, whose purpose was extended to advise the Council in promoting research work 'over the whole field of biophysics in relation to medicine'. In practice the work of the committee consisted of a series of visits to the member institutions to increase scientific exchange and support. A brief glance at the list of institutions indicates the range of pursuits included under the general heading of 'biophysics': apart from Randall's unit at King's College London, Perutz's unit in Cambridge, Hill's Biophysics Research Unit at University College London and the newly created Biophysics and Optics Division at the National Institute for Medical Research, it comprised various London- or Cambridge-based groups working in the field of radiobiology and radiotherapeutics.[35] This was a field the MRC had seized upon soon after the war, but which was now subsumed under biophysics. In the same year in which Randall's and Perutz's groups got started, the MRC also opened the Radiobiological Research Unit at the Atomic Energy Research Establishment at Harwell and financed the installation of two 30 megavolt electron synchrotons, one at the Department of Radiotherapeutics in Cambridge, the other at the Institute of Cancer Research in London (Landsborough Thomson 1975, 62).

The MRC was an important, but not the only patron for biophysical research in postwar Britain. Neither were all applications for biophysical support successful. Astbury, crystallographer at the University of Leeds, whose research focused on fibrous proteins, for instance, failed to receive funds from the MRC.[36] Through the mediation of Tizard, Bernal was successful in securing funds from the Nuffield Foundation, a private charity created in 1943 by Sir William Nuffield, founder of Morris Motors, for building up his Biomolecular Research Laboratory at Birkbeck

[34] This holds true up to the end of the 1950s, when the unit had already changed its name to MRC Unit for Molecular Biology. To such an extent was this use ingrained that a university committee set up to consider the future of the unit in the late 1950s had to enquire specifically from the MRC what the official name of the unit was; see 'Report of the General Board of their Committee on the future of the M.R.C. Unit of Biophysics, and associated matters', third draft, handwritten corrections and note, file GB120 (1), University Archives, CUL. On the change of name of the unit see below (Part II). 'Biophysics' was introduced as new scientific heading in the MRC programme of 1948; see 'Outline of research programme', *Report of the Medical Research Council for the Years 1948–1950* (London: Her Majesty's Stationery Office, 1951), p. 48. See also 'Developments in biophysics', *Report of the Medical Research Council for the Year 1954–1955* (London: Her Majesty's Stationery Office, 1956), pp. 28–32.

[35] The committee was discharged in 1954. It is through a report of December 1952 by Randall to this committee that Rosalind Franklin's X-ray pictures became available to Watson and Crick via Perutz.

[36] On the possible reasons see Olby (1994, 326–7).

College (Hodgkin 1980, 59–62; Clark 1972, 61). On a more modest scale, the Nuffield Foundation also supported Randall's department and Hodgkin's crystallographic group at Oxford. Biophysics and nuclear physics accounted for the bulk of the Foundation's grant in natural sciences in the first quinquennium. Further support for biophysical research in university departments came from the University Grants Committee which, following the Barlow Report of 1946, commanded a much larger budget than before the war (above, chapter 2).

War, mother of all things?

Was 'biophysics' really something new, a new product of the postwar era? Or was it only a more fashionable term for endeavours pursued also before the war? If so, why was it fashionable? Or what was new?

The career of Hill, who so actively promoted biophysics after the war, forms a useful point of departure to consider questions of continuity or discontinuity with prewar endeavours. Hill, born in 1886, first trained in mathematics at Cambridge. Disillusioned with his studies and strongly encouraged by his tutor at Trinity, the physiologist Walter Morley Fletcher, he decided to enter the field of his teacher. Traditionally a medical degree was a necessary qualification to enter physiology, but not so at Cambridge where there existed a strong experimental tradition and physiology was taught as part of the Natural Science Tripos.[37] Hill embarked on his second course of study, choosing chemistry and physics to complement his study of physiology and fulfil the requirement of three subjects. In 1910 he started research in the Physiological Laboratory at Cambridge, taking up the study of muscle contraction and nerve action, two classical subjects of physiology, which were already extensively investigated in the laboratory. Hill brought to the investigations new mathematical tools, physico-chemical concepts and refined physical instrumentation. He made a career in physiology, moving first to the chair of Physiology in Manchester (1920), and from there to University College London (1923). In 1922 he and Otto Meyerhof were awarded the Nobel Prize for Physiology or Medicine for their contribution to the analysis of physical and chemical changes

---

[37] On the history of physiology at Cambridge see Geison (1978). As first Secretary of the Medical Research Committee, the predecessor of the MRC, Fletcher later became instrumental in setting up the Biochemistry Department at Cambridge with Hopkins, his earlier collaborator in the Physiological Laboratory, as first professor (Kohler 1978). The making of biochemistry into an independent discipline remained an important example for Hill in his later efforts to turn biophysics into a fully fledged discipline. On the Tripos system in Cambridge see below.

underlying muscular contraction. In further recognition of his work he was appointed Royal Society Research Professor in 1926.

Hill's particular approach to physiology was already termed 'biophysics' in the interwar years. Indeed, his research unit at University College London, supported by a grant from the Rockefeller Foundation, was known as the 'Laboratory of Biophysics'. A series of lectures he gave in Philadelphia in 1930 were published as *Adventures in Biophysics* (Hill 1931). However, it was only in the period of postwar reconstruction that he saw the possiblity of establishing his own brand of biophysics as well as biophysics more generally on a larger and firmer basis. He not only supported Randall's ambitious research project in biophysics, but became a 'general point of reference' for scientists moving into the field.[38] For his own laboratory he successfully negotiated first a series of appointments and scholarships and finally, in 1951, the establishment of an independent Department of Biophysics at University College London. Hill also became an influential figure in the formation of the American biophysical community. Francis O. Schmitt, biophysicist at the Massachusetts Institute of Technology, engineered that Hill's article 'Why biophysics?' was republished in *Science* just in time to appear before the constituting meeting of the American Biophysical Society, where it set the tone of the discussion (Hill 1956; Rasmussen 1997b, 192).

Hill's physical and mathematical approach to physiology was not the only brand of biophysics in the interwar years. William B. Hardy's physico-chemical studies of colloids at Cambridge, for instance, also ran under the name of 'biophysics'. In Hardy's 'ideal Biological College' biophysics occupied the central floor, with molecular physics and cell mechanics being situated in the basement and the attic floor respectively (Hardy 1928, 16). The creation of a separate Department of Biophysics was an integral part of grand plans for restructuring Cambridge science with a substantive grant provided by the International Education Board, a Rockefeller philanthropic organisation. But the Professor of Physics, Ernest Rutherford, pressed to co-sponsor the plan, declined to have biologists within the precincts of the Physics Department and this part of the scheme had to be abandoned (Kohler 1991, 182–8).

The International Education Board was closed down in the course of a general overhaul of the Rockefeller funding programme in the natural and medical sciences, but support for biophysics was again inscribed in the new Rockefeller Foundation programme in natural sciences devised by Warren Weaver in the mid-1930s (Kohler 1991, 233–46, 282–3). Weaver,

---

[38] Among several others Crick especially has credited Hill for his encouragement to enter biophysics; Crick (1990, 19); F. Crick to A.V. Hill, March 1959 and 3 November 1962, file AVHII, 4/18, Churchill Archives Centre, Cambridge. See also Szilard (1968, 98).

however, had problems delineating the field. It seemed to include everything in which physical methods were applied to biological material, and
in this generality did not strike him as very useful, despite the fact that
bringing physical and chemical methods to bear on the biological sciences formed the core of his programme (Kohler 1991, 299). In Britain,
Astbury and the mathematician Dorothy Wrinch, working respectively
on the X-ray analysis of fibrous molecules and on chemico-mathematical
models of protein structure, did receive generous funding from the
Rockefeller Foundation natural sciences programme. But the more ambitious plan by members of the 'Biotheoretical Gathering' to set up an
interdisciplinary institute in Cambridge for the study of morphological
structure in conjunction with mathematical, biochemical and biophysical
approaches, including X-ray crystallography, failed to receive support and
had to be abandoned. The scheme was submitted by Cambridge biochemist
Joseph Needham and Wrinch, then tutor at Oxford, together with a memorandum by Bernal on the prospects of X-ray analysis in biology. It was
the fruit of intellectual exchanges the three researchers shared with other
members of an interdisciplinary discussion group, also referred to as the
'scientific Bloomsbury', convened by the theoretical embryologist Joseph
D. Woodger and aimed at redefining the relationships between the sciences along more participatory lines (Abir-Am 1987). Especially through
Bernal, there were links between this institutional project and postwar
programmes in 'biophysics'.[39] Protein crystallography, however, was only
in its infancy in the 1930s. Technological resources available after the war
radically changed the research practices of protein crystallographers, as
we will see in the next chapter.

Between 1936 and 1938 the Rockefeller Foundation also funded three
meetings initiated by Niels Bohr and attended by physicists and biologists to discuss problems of gene and chromosome structure. Among
the participants at the meetings held in Copenhagen, Klampenborg (near
Copenhagen) and Spa (Belgium) were Darlington, Astbury, Bernal and
Waddington.[40] The ideas discussed at these meetings spread beyond the
circle of participants. They were, for instance, taken up by Randall, whose
early biophysical programme was very much in line with Darlington's experimental cytology and the Klampenborg programme (Wilkins 1987, 511
and this chapter, above). But as we know, Randall moved to biophysics
only after his substantial contribution as physicist to the war effort. The

---

[39] For a closer discussion of the suggested link between Needham, Wrinch and Bernal's
    institutional project of the 1930s and the proposal for a laboratory of molecular biology
    at Cambridge, submitted by Perutz, Crick and Sanger in the late 1950s, see below,
    pp. 234–5.
[40] See Olby (1994, 113–15) and for a participant's account Waddington (1969).

German physicist Max Delbrück, impressed by Bohr's ideas on the complementarity principle and the new nature of physical laws in biology, had pioneered the move from physics to biology several years earlier, when fleeing Germany and moving to Caltech (the California Institute of Technology in Pasadena) as a Rockefeller Fellow in 1937. He soon settled on the problem of gene replication and on bacteriophage as research tool. Yet the phage course, which attracted a large number of researchers, including many physicists, to the field staked out by Delbrück, started only in 1945.[41] Schrödinger's *What Is Life?*, which leaned heavily on the quantum mechanical description of the gene offered by Delbrück in the famous 'Three-Man-Work' of 1935, also appeared only ten years later (Timoféeff-Ressovsky, Zimmer and Delbrück 1935; Schrödinger 1992). Its main impact was with disillusioned physicists towards the end of the war. Among British biophysicists, Wilkins and Crick in particular claimed to have been influenced by Schrödinger's highly readable booklet.[42]

Physics of life versus physics of death

A number of factors concurred in the fortunes of biophysics after the war. In his 1944 pamphlet for biophysics Hill anticipated some of them. The document therefore offers once more a useful starting point. In the first place Hill wrote his memorandum to argue for and in the expectation of a general overhaul of the research system and the disbursement of new government funds for 'fundamental' research. But many different research areas could potentially profit from these measures (as indeed they did). These general changes, therefore, cannot account for the particular attractiveness of biophysics after the war, even if they were essential for the postwar expansion of the field.

More specifically, Hill pointed to the host of new acoustical, optical, electrical and informational technologies, developed for the exigencies of war, in search of new uses. Peacetime applications of these instruments or their by-products, which would give them a new legitimation, as well as men skilled in their use, were highly valued. In some cases the instruments employed for biophysical research were actually surplus items from the services or 'liberated' from Germany (Katz 1978, 124). John Gray, researcher in neurophysiology at the National Institute of Medical Research and,

---

[41] On Delbrück see Kay (1985), Fischer and Lipson (1988) and Olby (1994, 227–40).
[42] On Schrödinger's place in the history of molecular biology see Fleming (1969), Olby (1971; 1994, 227, 240–7), Yoxen (1979) and Fox Keller (1990). The interest of physicists in biology and the interactions between the two communities in the interwar years deserves a more systematic study.

after 1952, Reader in Physiology at University College London (Hill's 'old' department), remembers: 'The parts from the Services were going for nothing. They filled the cupboards in our lab. We built up our electronics from them.'[43] The power of physical instrumentation rather than specific biological questions often provided the legitimation for early research projects in biophysics. For some this presented a point of concern. Crick, himself searching for an entry into biology, critically remarked on his first visit to Randall's Biophysics Unit in 1947: 'Exactly what biophysics was, or could usefully become, was less clear. At King's they seemed to feel that an important step would be to apply modern physical techniques to biological problems...Exactly what they hoped to discover with these new instruments was less clear' (Crick 1990, 19–20).

The war mobilisation produced not only new instruments, but also people keen to apply their instrumental skills to biophysical problems. Much has been made of the role of the atomic bomb to explain the dissatisfaction of physicists with their subject and their attraction to biology. I will say more about this point below. But Hill was writing before the atomic explosions over Japan. In his view the war experience generally was decisive to make both physicists *and* biologists more appreciative of each other's work and acquire the boldness and the skills necessary to work in biophysics. In his words:

many physicists, some of standing and experience, others enterprising youngsters who have seen many strange applications of physics during the war, have become interested in the exciting new possibilities and the probable practical importance of applying the latest resources of physics, theoretical and experimental, to biology. On the other side, many first-rate young biologists, during the last $5\frac{1}{2}$ years, have been devoting their talents very effectively to physical war-problems, and have acquired techniques and knowledge and shown initiative in experiment which could find undoubted scope in biophysics.[44]

Coming back to the same point in later years, Hill stressed that 'the resource and initiative with which many biologists mixed up with physicists,

[43] Interview with John Gray, 1 December 1995, Cambridge. See also Hodgkin (1992, 261). The same point regarding the transfer of instruments from the military to academic research laboratories after World War II is made by Galison for the American context, especially with respect to physics (Galison 1988). On the war-related development of technologies used in biophysics and later molecular biology, including the electron microscope, isotopes, mutagens, cryptoanalysis and computers, see also below, chapter 4.

[44] A. V. Hill, 'The needs of biophysics' (memorandum of 21 February 1945), quoted in Katz (1978, 122–3). This document seems to represent a later version of the one from which I quoted before. Unfortunately I was unable to locate it, either at the Royal Society or among Hill's papers in the Churchill Archives Centre, Cambridge. Nor does Bernard Katz keep a copy.

in telecommunication and in operational research, and the like' convinced the latter that 'biology, after all, can be quite a respectable subject' (Hill 1956, 1234). (Despite these local conversions, though, the picture of the biologist with his specimen box or as beetle collector survived well into the 1960s.)

The war not only brought scientists from different disciplines to work on common projects, but also served to establish new contacts in industrial and political circles. Hill expressed the hope that 'all these connections (established during the war with governmental and industrial research laboratories) with science on a wider stage would be useful in the plans one has in mind'.[45] The following extract from Hill's 'Memories and reflections', a privately circulated typescript, will give a taste of how Hill himself made use of his wartime connections for getting things done. He referred to it as the ' "club tie" trick'. The event occurred in 1945:

> I was anxious to get a few first-rate men who had worked at radio and radar for the services, and turn them into biophysicists. I asked two friends, who were high up in the radar business, if they could find suitable chaps for me. Three men were proposed who in the end were appointed. They were absolutely first-class, among my best investments. But one, who was a Captain in Signals, was refused permission because (as was said) there was a great shortage of officers at the time. I knew that he was not, in fact, working as an officer at all, but in a laboratory, so again I went to the highest authority... Two days later my friend arrived.[46]

The bomb definitely changed the face of physics and of science more generally. It was received as a most extraordinary, but also threatening, achievement. In the words of project leader Robert Oppenheimer, it had made scientists 'become Death' (Kevles 1987, 333).

The threat that emanated from the new weapon was exploited in the decision of Britain's postwar Labour government to base the defence of the country on atomic bombs as deterrents, a move hardly opposed by British scientists. They shared the virtually unchallenged view that Britain as (still) a first-class power needed first-class weapons (Gowing 1974b,

---

[45] A. V. Hill, 'The needs of biophysics' (memorandum of 21 February 1945), quoted in Katz (1978, 121).

[46] A. V. Hill, 'Memories and reflections', article 91 (typescript), file AVHL I, 5/4, Churchill Archives Centre, Cambridge. In 1945, Hill also offered Katz, who had worked with him before the war and then became involved with radar, a senior research appointment in his laboratory. This confirms that radar work struck Hill as useful preparation for a career in biophysics. To Katz, who was still busy in the Radio Physics Laboratory in Sydney, he wrote: 'You will all find cathode ray oscillography like second nature when you return to your physiology'; A. V. Hill to B. Katz, 12 February 1945, file MDA A 6.4, RSL.

497; cf. above pp. 32–3). But this view went hand in hand with a strong urge to turn the same scientific and technological developments as well as equivalent efforts in other scientific fields to the more direct benefit of the people. If the atomic bomb represented 'the dramatic and horrifying climax of one line of research for war purposes', an equally intense scientific and technological effort could be expected 'to yield equally revolutionary advances for constructive social ends' ([Association of Scientific Workers] 1947, v). The reaction to the atomic bomb here fitted with more general themes of postwar reconstruction and the making of the welfare state, while biophysics, the physics of life, became part of the efforts to redeem the physics which had led to the deadly weapon.

Nuclear physics itself opened up a huge field of medical and biological research comprising the study of the effects of radiation on the body and ways to protect it, the therapeutic use of radiation, and the use of heavy and radioactive isotopes as tracer elements in biological research and medical diagnosis. Echoing similar remarks which abounded in the literature of the time, Mellanby described it as a 'pleasant irony' that the 'medical by-products' of the atomic bombs, and especially the isotopes which where produced by the same accelerators built to yield fissile material, would themselves 'directly or more likely indirectly, save many thousands more lives than all the weapons of war destroyed'.[47] Studies on the effects of radiation as well as first experiments with isotopes dated back to the inter-war years (Kohler 1977), but this branch of biophysics virtually exploded after the war, offering medics, biologists and physicists new professional outlets.[48] Indeed, the military connection could never be completely ignored. Studies on the effects of radiation on workers handling radioactive material were conducted as part of the Manhattan Project (Hacker 1987). The survivors of Hiroshima and Nagasaki formed a unique test

---

[47] E. Mellanby, 'Some aspects of medical research in Great Britain during the last seven years', paper presented to the Norwegian Medical Society [undated; probably 1948], file FD5/110, MRC Archives.

[48] Joseph Mitchell, appointed Professor of Radiotherapeutics at Cambridge in 1945, represents an example of a medic who made a career in the new field. After completing his clinical studies and having acquired some background in physics, Mitchell started research in radiology before the war. During the war he was summoned to Canada to study the radiobiological hazards for workers handling radioactive material in connection with the atomic bomb project. On his return to Cambridge he continued this and other lines of research aimed at harnessing radiation sources for the treatment of cancer. Joseph Rotblat came to the field as a physicist. He was one of the British scientists who joined the Manhattan Project, but quit the Los Alamos weapon laboratory when it became clear that the Germans were not developing an atomic bomb. After the war he became Professor of Physics at St Bartholomew's Hospital Medical College in London, where he worked on radiobiology. He actively opposed the proliferation of nuclear weapons and became a leading figure in the Pugwash movement. He was honoured with the Nobel Prize for Peace in 1995.

population to pursue the investigations on a much larger scale. In the context of the Cold War and nuclear rearmament these studies gained strategic importance (Beatty 1991; Lindee 1994). In Britain the largest centre for radiobiological research was the MRC unit at the Atomic Energy Research Establishment in Harwell.

However, postwar biophysics profited as much from the medical mobilisation epitomised in the development of penicillin as from the disenchantment and new technological tools deriving from the bomb. Penicillin occupied an equally pivotal place in the plans and hopes of postwar reconstruction as the bomb (Bud 1998) and provided a main rationale for the allocation of important new government funds for basic medical research (including biophysics). Other, less publicised medical mobilisation projects also centred on molecules. These included the investigation and production of other antibacterial substances, antimalaria drugs, blood fractionation products, vitamins, proteins, viruses and substances used in chemical warfare (including research on their toxic, mutagenic and therapeutic effects) (de Chadarevian and Kamminga 1998b).[49] The search for strategic molecules boosted the development of technologies for the visualisation, isolation and large-scale production of molecules and macromolecular assemblies and forged new alliances between researchers, industrialists and clinicians. Postwar molecular studies could draw on these resources, as well as benefiting from the new funds made available for 'fundamental' research.

Postwar biophysics in Britain, then, comprised at least three quite separate groups: the 'nerve–muscle group' represented by Hill and a new generation of researchers able to apply the electronic skills gained in radar research during the war to devise apparatus for the amplification and measurement of minute electrophysiological signals; the 'radiation group', whose work gained particular significance in the atomic era; and the 'molecular structure group', including X-ray crystallographers, who took advantage of new instrumentation made more widely available through increased government funds.[50] All three groups built to some extent on prewar research traditions, but the opportunities of postwar restructuring, including new sources of funding for 'fundamental research', new technological opportunities, the recruitment of a new generation of physicists and biologists prepared to take up problems at the interface of biology and

---

[49] So far none of these medical mobilisation projects in Britain except the penicillin project has attracted significant historical research. For an overview of medical research during the war see Mellanby (1943) and Committee of the Privy Council for Medical Research (1947). For recent studies of American medical mobilisation programmes see Rasmussen (1997b), Creager (1998) and Gaudillière (1998).

[50] J. Gray, letter to the author, 17 October 1995. On the increased reliance of protein crystallography on electronic computers see below (chapter 4).

physics, as well as the cultural currency of a 'physics of life' in contrast to a 'physics of death', concurred in a fantastic expansion of some but not all threads of biophysical research laid out in the interwar years.[51]

Practitioners held different views on what counted as biophysics. We have heard of Crick's reservations about Randall's research programme. When Crick in 1949 decided to join the protein crystallographers in the Cavendish unit, Hill, his mentor in his move to biophysics, lamented the crystallographers' lack of interest in functioning living systems.[52] Several years later, Hill insisted: 'The crystallography of material of biological origin is not in itself biophysics... Using... radioisotopes, or working on muscles and nerves, does not confer any biophysical status. It all depends on the motive, the idea, on the method and manner of approach' (Hill 1956, 1233). While internal battle lines were drawn, giving a positive definition of the field as a whole proved still more difficult. The editorial of the newly founded review journal *Progress in Biophysics and Biophysical Chemistry* opened with the declaration: 'The editors have had some difficulty in deciding what is the proper field to be covered by reviews of recent progress in biophysics. Excluding biochemistry on the one hand and physiology on the other, there lies between a vast and rather amorphous field of study of which the frontiers and lines of demarcation are anything but well defined' (Butler and Randall 1950, vii).[53] Even Hill had to concede: 'The term *biophysics* is coming today into common use, but as yet no clear definition of it has emerged' (Hill 1956, 1233).

If on the one hand there were problems of delineating the field, on the other there was an interest in keeping the different communities together. As chairman of the Board of Studies for Biophysics of the University of London, Randall saw his job as 'reconciling the interests of the "nerve/muscle group", the "molecular structure group" and the "radiation group"'.[54] The review journal prided itself on the 'diversity' of its 'biophysical menu', hoping to offer something for everyone and to provide a 'meeting ground' for all those working in biophysics broadly understood

---

[51] Certain sub-fields which were vigorous in the 1920s and 1930s, for instance colloid biophysics, quickly lost currency as a result of technological advances brought about by the same colloidists; see Creager (1998). At Cambridge the Department of Colloid Science closed down with the retirement of its professor in 1966.

[52] A. V. Hill to F. Crick, 11 March 1949, file AVHL II, 4/18, Churchill Archives Centre, Cambridge.

[53] After 1963 the same review journal continued publication under the title *Progress in Biophysics and Molecular Biology* (with the last three words appearing in smaller print). As the editors noted, this change of title did not imply a change of policy, but rather an acknowledgement of the fact that the journal had always published many articles within the field now widely known as molecular biology. As if justifying this move, they added that it had always been difficult to define the field and that 'the editors have never wished to be unnecessarily restrictive' (Butler and Huxley 1963).

[54] J. Gray, letter to the author, 17 October 1995.

(Butler and Katz 1957, vii). And when in 1960 the British Biophysical Society was founded, all three groups were included, although the radiation group was less well represented and over time became less active.

However, despite the wide currency of the term and the efforts to establish it as an academic subject, biophysics remained always more of a merger than a clearly defined field. This contributed to the fact that, with time, the attraction of the term waned. By the time the Society was founded, Kendrew, the Society's first Honorary Secretary, had agreed with his American colleague, Paul Doty, that 'molecular biology' rather than 'biophysics' or 'molecular biophysics' more appropriately described the subject of the journal he was going to edit (see chapter 7). Molecular biology was only one of the fields built on the legacies of postwar biophysics.[55]

War recruits

Not the move of physicists to biology *per se*, we have seen, but the war experience of both physical scientists and biologists and the attraction of a physics of life that promised medical applications were crucial for the postwar fortunes of biophysics. A closer look at the careers, practices and institutional connections of those involved in building biophysics at Cambridge will allow us to pursue this point and to investigate in more detail how the scientific mobilisation of World War II set the course for developments in the physical and biological sciences and hence on biophysics.

At first view, the Cambridge biophysics group appears as a physicists' enterprise. The group relied heavily on Bragg's patronage and on the skills and workshops available in the Physics Department. All of Perutz's first collaborators, namely Kendrew, Hugh Huxley, the first research student in the unit, and Crick, were physicists or chemists with the experience of science at war (as was Wilkins, Randall's first collaborator). However, as Keilin's role in supporting Perutz's early crystallographic work and the crucial financial backing of the MRC already indicated, biophysicists at Cambridge depended as much on biomedical resources and patronage. This dependence extended to the exchange of research material and the use of functional knowledge for molecular analysis and the interpretation of results.[56]

---

[55] On the bursting of the 'biophysics bubble' in America and the dispersion of biophysical investigations into such diverse fields as biochemistry, cell biology, the neurosciences and nuclear medicine see Rasmussen (1997a).

[56] Dependence did not mean subservience. As has been pointed out, molecular biology could establish itself despite the resistance of the traditional biological discipline, by using the new institutional spaces created by the postwar expansion of the sciences (Abir-Am 1985, 108, 110).

Bragg exposed himself to the stinging critique of fellow physicists for backing 'weird' subjects like biophysics and radioastronomy instead of building up nuclear physics, which represented the grand tradition of the Cavendish (Dyson 1970).[57] He supported the construction of a new linear accelerator on the outskirts of Cambridge for research into particle physics, but overall opted for a more flexible development of the department, allowing various promising lines of research to develop. In reorganising the Cavendish after the war, and in view of an increasing number of research students, he split the laboratory into several subject groups (nuclear, low temperature and mathematical physics, metals, crystallography and radioastronomy) headed by a senior researcher. Each group operated fairly independently, having its own workshop, assistant and secretariat. This assured a more effective organisation of the lab and at least within each group a more productive communication among researchers. In such a set-up the biophysics unit, which was separately funded, could also find its place (Crowther 1974, 269–90) (Fig. 3.6).[58]

The dependence on the Cavendish workshops presented one of the main reasons why the group remained in the Physics Department when it had long grown into a 'cuckoo's egg' and the new head of the department, Nevill Mott, wanted to get rid of it (see Part II). When the biophysicists, now turned molecular biologists, finally moved to a purpose-built laboratory in the early 1960s, they saw to it that it was equipped with electrical and mechanical workshops that matched the Cavendish ones. They were operating before any other division moved to the new site (Perutz 1962a, 209; 1996, 661; and below, chapter 9) (Fig. 3.7).

The early careers of the first members of the unit illustrate the experience of science at war shared by a generation of scientists.[59] Despite the very

---

[57] Similar accusations were moved against Randall and his 'circus' (see below, p. 249n). On the making of radioastronomy in postwar Britain see Edge and Mulkay (1976) and Agar (1998c).

[58] On Bragg's directorship, the reorganisation of the Cavendish after the war and the place of the MRC unit in the laboratory see Crowther (1974, 269–90).

[59] On 'generation' as an analytical tool in history see the contributions to the special issue on 'Generations' of *Daedalus* in the twentieth-anniversary year of the journal (Fall, 1978). Annan (in the same issue) has described the generation who came of age or graduated from university between the end of World War I and the end of World War II, or more precisely 1952, when the last veterans of World War II had taken their degrees, as 'our generation' (Annan 1978). What this generation shared, it seems, was the active experience of World War II. Annan was well acquainted with the molecular biologists at Cambridge and, on at least one occasion, lobbied the university administration on their behalf in an attempt to get the plans for a laboratory of molecular biology to be built at Cambridge moving; see N. G. Annan to R. E. M. Macpherson (Financial Board, University of Cambridge), 2 May 1959, file Himsworth, LMB Archives. For a recent study in the history of science which takes a generational approach in a more systematic way than I propose here, see Nyhart (1995).

3.6 W. Lawrence Bragg at his desk at the Cavendish
(*c.* 1950). Copyright Cavendish Laboratory Cambridge.

personal trajectory each of them followed, their stories reveal common threads.

Perutz's early career differs from that of his colleagues and of most British scientists of his age in that as refugee and grantee of the Rockefeller Foundation he could complete his Ph.D. and continue his research during the war – if with some interruptions. Yet in the production of this difference, his career, too, was deeply structured by the events of war. With the end of the war the Rockefeller grant also ended, and Perutz, like the scientists released from war service, had to make new choices.[60]

Perutz submitted his doctoral dissertation on the structure of haemoglobin in spring 1940. A few months later, with mounting fears of a German invasion, he was rounded up with other 'enemy aliens' and interned in Canada. As a result of pressure from colleagues and friends, he was

[60] For recent studies on German émigré scientists, including reference to Perutz, see Ash and Söllner (1996).

3.7 View of the Cavendish workshop in the Austin Wing also used by the protein crystallographers. Copyright Cavendish Laboratory Cambridge.

released after several months and returned to Britain to resume his research. When in 1943 his participation was requested on the secret war project HABAKKUK, he received British citizenship. The plan, conceived by Geoffrey Pyke, a man with a chequered career who had won the trust of Mountbatten, Chief of Combined Operations, was to build an island of ice as an aircraft base in the Atlantic. The ice was to be made stronger and to freeze faster by adding wood pulp. Perutz, who had studied glaciers before, was called to the project as expert on the crystalline structure of ice. Eventually, the project was abandoned, and in January 1944 Perutz returned once more to his research in the Cavendish.[61] By that time his project on

[61] For a personal account of his wartime experiences see Perutz (1989, 101–36). While keen at the time to help the British war effort, he does not spare critique of some of the 'impossible' wartime projects on which, in his view, huge amounts of resources and manpower were wasted. Conditions at the Cavendish and progress of Perutz's haemoglobin research during and just after the war can be gleaned from Bragg's correspondence with the Rockefeller Foundation and from Foundation officers' diary entries on their visits to the laboratory; see folder 561–4, box 43 and folder 565–70, box 44, series 401, record group 1.1, RAC.

the structure of haemoglobin was the 'only piece of pure research' still pursued in the Cavendish.[62] Blood and its products, however, were intensively studied in other laboratories because of its relevance to wartime medicine, and Perutz was part of an active network of Cambridge researchers working on different aspects of haemoglobin. The informal group was held together by the physiologist Joseph Barcroft, a pioneer in the field and inspired teacher (de Chadarevian 1998a).[63]

Kendrew, who joined Perutz in 1945, had graduated in chemistry in 1939. He started doctoral research on the rotation kinetics of molecules in the Department of Physical Chemistry. When war began, the Cambridge Recruiting Board told him to continue his research, but with most of his colleagues leaving for war-related projects, he became increasingly dissatisfied. On his own initiative he wrote to Wilfrid B. Lewis, a lecturer in the Cavendish, of whom he knew that he was involved in 'some secret war work', enquiring if he could join.[64] As a result, in February 1940 Kendrew was recruited to work on the development of airborne radar at the Air Ministry Research Establishment in Dundee. In the same year he was moved to operational research duties, a field then in its infancy in which he remained engaged for the duration of the war, taking on increasing responsibilities. From Royal Air Force Coastal Command he moved to Cairo to build up the Operational Research Section at Headquarters RAF Middle East, and from there to Air Command South East Asia, where he served in India and Ceylon as Officer in Charge of Operational Research and as Scientific Adviser to Mountbatten, then Allied Air Commander-in-Chief. He ended the war with the honorary rank of Wing Commander.[65]

[62] L. Bragg to H. M. Miller (Rockefeller Foundation), 31 January 1944, folder 565, box 44, projects series 401, record group 1.1, RAC.

[63] A volume, published in commemoration of Barcroft, documents the extent of haemoglobin research pursued, much of it under his direct inspiration (Roughton and Kendrew 1949). Barcroft had been active in haemoglobin research since before World War I. During the Great War he was appointed Chief Physiologist at Porton Down, the Chemical and Biological Defence Establishment, where he worked as the only civilian scientist on medical aspects of gas poisoning. Barcroft found that certain poison gases as well as shock led to oxygen deficit in the circulating arterial blood. After the war Barcroft chaired the Medical Research Council Haemoglobin Committee, which was founded at his instigation in 1919 to encourage and sponsor research on haemoglobin. It remained active for around a decade. During World War II, research on blood components including haemoglobin as well as on blood groups and other compatibility factors received new impetus in connection with a powerful transfusion service, set up in Britain as in the United States. On the transfusion service in Britain see Mellanby (1943); on the scientific, medical, political and commercial importance of research into blood and its components during World War II see Creager (1998). On haemoglobin research after the war see also de Chadarevian (1998a).

[64] Interview with J. Kendrew, Cambridge, 14 July 1992. Before the war, Kendrew had done some radio work with Lewis in the Signals Unit of the Officers Training Corps. The work was contracted by the War Office, but was not secret.

[65] On Kendrew's wartime contribution to operational research see [Air Ministry] (1963), esp. 74, 109, 111, 125, 169, 177, 179.

According to Kendrew's own account, it was Bernal who most influenced his decision to take up protein crystallography after the war. Kendrew had met Bernal as an undergraduate in Cambridge, but the decisive conversation took place 'in a jungle in Ceylon', while both were serving as advisers to the Air Commander-in-Chief (Hodgkin 1980, 59).[66] In his spare time while running bomb trials, Bernal convinced Kendrew that it should be possible to use X-ray diffraction to solve the structure of proteins and that this was a most worthwhile problem to embark on. Stopping over in California to discuss military matters on his eastern route back to Europe, Kendrew was presented to Caltech structural chemist Linus Pauling, himself involved in a costly wartime project on the structure and function of immunoglobulins, the proteins involved in immunological reactions (Kay 1993a). Conversations with him confirmed Kendrew's resolution.

While there is no reason to doubt that the two encounters indeed took place, it is likely that they assumed their decisiveness only retrospectively.[67] There is clear evidence that for several months after his return to England Kendrew seriously contemplated and extensively discussed the possibility of remaining with the Scientific Civil Service. With others, he was convinced that science had an important role to play in postwar reconstruction and felt that scientists who like him had gained 'experience of administration and of working alongside Government' had the 'strongest moral obligation' to remain in government service. At the same time, however, he was frustrated by the lack of any central organisation for science in government and by the inadequate purpose and advice offered to scientists in his position. He finally decided that it would take time before definite openings in government employment would arise and that it was preferable, at least temporarily, to return to university for 'intellectual refurbishment'.[68] This decision was once more put in question when, following his resignation, several offers of permanent appointments in government service were made to him, and it was intimated to him that it was in both his personal and the national interest to remain in the service rather than return to academia.[69] Important offers to return to government service,

---

[66] See also interview with J. Kendrew, Cambridge, 6 May 1993, and video interview K. Holmes with J. Kendrew, 18 June 1997, LMB Archives.

[67] In his first letter to Pauling after taking up his position in the Cavendish, Kendrew vividly referred to their meeting in 1945, but did not present it as decisive for his decision to return to research. See J. Kendrew to L. Pauling, 10 December 1946, correspondence files, Pauling Papers, Oregon State University Library.

[68] J. Kendrew to C. Gordon, 15 August 1945, Kendrew Papers, MS Eng. c.2394, B.51, and J. Kendrew to C. P. Snow, 8 October 1945 [draft version], Kendrew Papers, MS Eng. c.2606, R.7, Bodl. Lib. For more information on Kendrew's activity during the time spent at the Ministry of Aircraft Production in London see below, p. 120.

[69] J. Kendrew to Appointment Department, Technical and Scientific Register, 18 November 1945, Kendrew Papers, MS Eng. c.2606, R.7, Bodl. Lib.

including the invitation to participate in building up operational research in the Board of Trade, continued to reach Kendrew even after he had taken up his fellowship at Cambridge.[70]

If Kendrew's decision to return to academic research was anything but straightforward, his dissatisfaction with chemistry seems to have dated from before the war. In an exchange of letters following Kendrew's Nobel Prize in 1962, he wrote to Bernal: 'I'm always grateful that it was you, during the war, who more than anyone awoke the biological interest in a chemist tired of chemistry.' However, the shift was not seen as such a radical one in disciplinary terms after all. Having won the chemistry prize, Kendrew was comforted by the thought that 'after being a renegade chemist all these years I now seem to be a respectable one once more'.[71]

Bernal anticipated that Kendrew would collaborate in the team he planned to rebuild at Birkbeck College (and apparently later was disappointed that he did not do so). But when Kendrew found out that he could take up his fellowship at Cambridge, which he had abandoned to join the war effort, Bernal himself advised him to enquire if Bragg could set him up in the Cavendish.[72] Bragg arranged for him to work with Perutz, being officially supervised by William H. Taylor, head of the crystallography division in the Cavendish. Taylor was working on the structure of silicates. Like most professional crystallographers, he regarded protein crystallography as a hopeless undertaking, but still accepted the formal agreement regarding Kendrew.[73]

Huxley joined Perutz and Kendrew as first research student, when they were just set up as an MRC unit.[74] Huxley had started his last year of undergraduate studies in physics at Cambridge in 1943, when he began feeling 'very restive at playing no direct part in, nor even being very near to, the great wartime events that were taking place' (Huxley 1996, 7). He

---

[70] See C. Gordon to J. Kendrew, 17 January 1946 and C. Gordon to J. Kendrew, 22 January 1947, Kendrew Papers, MS Eng. c.2606, R.7 and R.11, Bodl. Lib.

[71] J. Kendrew to J. D. Bernal, 16 November 1962, Bernal Papers, file J 109, CUL. Besides Bernal and Pauling, Kendrew credited Waddington with interesting him in biology. They met while they were both involved in operational research during the war; interview with J. Kendrew, Linton, 18 March 1993.

[72] J. Kendrew to L. Bragg, 15 September 1945 [draft], Kendrew Papers, MS Eng. c.2606, R. 7, Bodl. Lib.

[73] Interview with J. Kendrew, Cambridge, 14 July 1992. On the whole there was very little interaction between the crystallography division in the Cavendish and the MRC unit.

[74] Also closely associated with the unit from 1947, though officially a research fellow in the Department of Colloid Science, was Herbert ('Freddie') Gutfreund, an Austrian émigré, with a Ph.D. in biophysics. He became an important link to Sanger and later was instrumental in alerting Vernon Ingram, then at the Rockefeller Institute to learn protein chemistry, that the MRC unit was looking for a biochemist. Ingram was hired in 1951.

decided to interrupt his studies and join the Royal Air Force as a radar officer. On his wartime experience Huxley wrote:

Basically, I suppose I wanted to have some adventure first. I was not successful in getting myself into Europe at the opening of the Second Front, but I did have an extremely interesting time doing flight trials on experimental radar systems, at Malvern and with Bomber Command . . . as well as spending some time on an operational bomber station (p. 7).

A *Life* report on the dropping of the atomic bombs on Japan, featuring a gallery of photographs of the physicists behind the bomb, Huxley's 'great heroes', followed by a line of photographs of some of the survivors from Hiroshima, had a sobering effect on the young Huxley and made him drop his wish to pursue a career as research scientist in nuclear physics (p. 7).[75] Once demobilised in 1947, he none the less returned to his physics studies, hoping eventually to do scientific research involving physics, 'but far away from its wartime uses', possibly in some form of medical research (p. 7). Through one of his supervisors he heard about Perutz and Kendrew's biophysical investigations in the Cavendish. This facilitated his decison. The Physics Department was a more congenial workplace to Huxley than a clinical environment (another option he had considered). Even if he knew nothing about proteins and had not been 'particularly enamoured by what I had learned about crystallography', Huxley chose to do a Ph.D. in that topic, later switching from protein crystallography to X-ray diffraction studies of muscle fibres. For Huxley this marked the beginning of a lifelong interest in the molecular mechanism of muscle contraction. In the mid-1950s, he became the first worker in the group to take up electron microscopy to complement his X-ray diffraction studies.[76]

The last war recruit to join the unit was Francis Crick. After graduating in physics at University College London in 1937, Crick had started research towards a Ph.D., building an apparatus to study the viscosity of water under pressure. He later described it as 'the dullest problem imaginable'

---

[75] *Life* magazine featured a number of investigative articles and distressing photographic reportages on the bomb and its effects in the aftermath of the explosions in Japan; see especially *Life*, 20 August 1945, pp. 17–20; 17 September 1945, pp. 37–40; 24 September 1945, pp. 27–31; 8 October 1945, pp. 27–35. None of these articles on its own fits Huxley's description completely, but it is likely that the pictures from various articles coalesced in his mind, forming the grim nexus that triggered his 'conversion'.

[76] For an autobiographical account of Huxley's research career see Huxley (1987; 1996). Also interview with H. Huxley, Cambridge, 4 December 1997. Huxley learnt electron microscopy as a postdoctoral student in Francis Schmitt's laboratory at MIT. On his return to Cambridge, plans were drawn up for Huxley to share in the use of an electron microscope to be installed in the Zoology Department. But before the instrument was up and running, Huxley left for University College London. A few years later the unit first acquired its own electron microscope.

(Crick 1990, 13). During the war a land-mine destroyed the apparatus. For Crick this was rather a relief. A few months after the beginning of the war, he was given a civilian job in the Admiralty. He worked first in the Admiralty Research Laboratory in Teddington and later at the Mine Design Department near Havant, mainly on the development of magnetic and acoustic mines. When war ended he was given a job in the Naval Intelligence Division in London. He did 'the obvious thing' and applied to become a permanent scientific civil servant. But by 1947 he had decided to leave the Civil Service and return to research (Crick 1990, 14–15).[77] After a seven-year break and with no special qualification to boot, this meant a completely new beginning. Crick resolved to pursue the question he was most interested in. This he determined to be the borderline between the living and the non-living. Anti-religious reasons apparently played a role in this choice (Crick 1990, 17). On the advice of Harrie Massey, the distinguished theoretical physicist under whom he had worked during the war, he contacted Hill, expressing 'a strong, though uninformed inclination to some form of bio-physics'.[78] Hill arranged an introduction to the Secretary of the MRC. After further explorations which he described in detail in his scientific autobiography, Crick settled to work in the Strangeways Laboratory in Cambridge, receiving an MRC studentship (Crick 1990, 15–23). The laboratory, originally founded as a research hospital, but now completely dedicated to scientific investigations, was loosely attached to the university. Research focused on tissue cultures. Crick was given the task of investigating the viscosity of cytoplasm, a problem not too different from his first research project in physics. It was a long way from the 'big problems' Crick aspired to tackle (Fig. 3.8). But the position gave him the chance to read extensively in his new area of interest and to talk to other Cambridge scientists. After an apprenticeship of two years he decided to move to the protein crystallography group at the Cavendish. To some extent this represented an escape from more biologically orientated work back to physics. To Hill, who had supported his move to biophysics but who had advised him against taking up protein crystallography, he justified his decision: 'The problem really does fascinate me, and I feel that after having given up my Civil Service job in order to do work in which I was really interested, it would be better to do this, even if the work is difficult.'[79]

---

[77] Crick's services with the Admiralty are listed in detail in the curriculum vitae attached to his first 'Application for M.R.C. studentship', both enclosed in letter by A. V. Hill to E. Mellanby, 9 July 1947, file P3/125, vol. 1, MRC Archives.

[78] F. Crick to A. V. Hill, 1 May 1947, file AVH II, 4/18, Churchill Archives Centre, Cambridge.

[79] F. Crick to A. V. Hill, 7 May 1949, file AVH II, 4/18, Churchill Archives Centre, Cambridge.

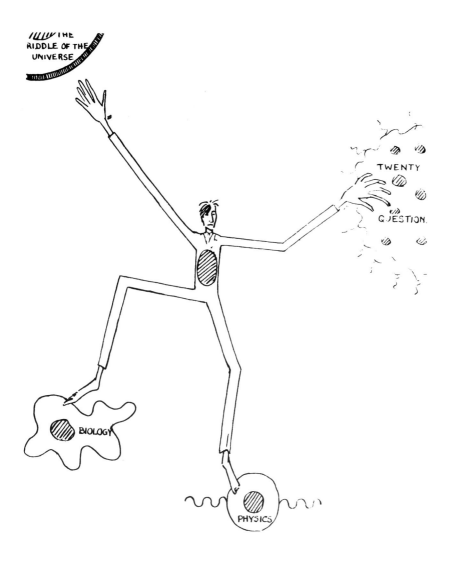

THE DENCRITIC CELL

3.8 Crick as seen by his colleagues at the Strangeways Laboratory. From the Medical Photographic Library, CMAC:PP/FGS/C.29. Frederick Gordon Spear, personal papers. Courtesy of The Wellcome Library, London.

For Kendrew and Crick, as for many other scientists of that generation, the war offered the chance of a new beginning and of a more conscious choice of research topic. An unexciting project grown out of a supervisor's research interest was exchanged for a much more ambitious one – often outside the original area of specialisation. After the war, disciplinary divisions in the universities still ran deep and the qualifications gained before the war were seldom completely thrown overboard, but equally important was the freshly won experience that one could apply oneself to new problems and learn new techniques, as most researchers had done during their war work. In an interview of Cambridge Nobel laureates, the radioastronomer Anthony Hewish whose physics studies were interrupted by three years of radio work in the war – 'a marvellous interjection' – confirmed: 'I did far better when I got back to university, than if I would have gone straight through, I think.'[80] And Watson, the first American to join the Cambridge unit in 1951, remembered being struck by the 'self-confidence' of British scientists which he interpreted as stemming from their participation in the war effort.[81] Another observer remarked:

Scientists who came back to university from war work had a particular way to approach research; in war research goals were set, deadlines were tight, but resources were no problem; the only thing that mattered were the research goals. This attitude stayed. They were very research orientated, not interested in teaching, colleges etc. They knew how to argue for their work and some of them became very arrogant.[82]

Doing a Ph.D., a qualification only introduced in British universities in the interwar years and not yet a prerequisite for an academic career, became a common route to requalify for research.[83] The relation between teacher and 'mature' candidate, however, often resembled more that between colleagues, especially when, as in the case of Kendrew, the candidate was only a few years junior to his teacher (Perutz) and had had a distinguished career as scientist in the war. Crick, too, considered that his achievements in the Admiralty had somewhat 'redeemed' his 'not-very-good degree' in physics (Crick 1990, 15–16). When entering biophysics, he first settled on a MRC studentship, but later was advised by Bragg and Perutz to follow a Ph.D. to improve his future career prospects. When in 1952 he still had not acquired his Ph.D., this became a serious cause of concern for Bragg.[84]

---

[80] *Cambridge Science Nobel Laureates*. Cambridge Audio Tape presented and recorded by Tamsin Palmer, American Friends of Cambridge University (Cambridge, 1996).

[81] J. Watson, address given at a memorial meeting in honour of Sir John Kendrew, Cambridge, 5 November 1997.

[82] Interview with M. Richards, Cambridge, 5 December 1995; see also Richards (1972).

[83] On the history of the Ph.D. in Britain see Simpson (1983). Simpson's account focuses on the introduction of the Ph.D. around 1920, but figures appended in a 'postscript' confirm the steep increase in the number of Ph.D.s awarded at British universities after World War II (1983, 165–6). The trend was facilitated by increased funds and improved opportunities for research in universities after the war.

[84] L. Bragg to A. V. Hill, 18 January 1952, Hill Papers, file AVH II, 4/18, Churchill Archives Centre, Cambridge.

Biophysicists and the transformation of biology

Kendrew, Huxley and Crick all taught themselves some biology. For Huxley this took the form of 'reading through Perutz's reprint collection during long night vigils over water-cooled X-ray generators' (Huxley 1996, 2). Crick read extensively, being most impressed by Baldwin's *Dynamic Aspects of Biochemistry*. All attended some lectures – the course in biochemistry, co-taught by several members of the Biochemistry Department, being best remembered.[85] But none of them registered for a full-time degree in biology or joined a biological department. The 'move to biology' was in all cases a very guarded one. As has been pointed out, physicists moving to biology took pride not only in having been physicists or chemists, but especially in *not* being biologists (Fleming 1969, 163). If what they were doing was biology, it certainly had little to do with what they perceived as mainstream biology. In their view most of biology was old-fashioned or antiquarian and not worth engaging with. As one participant put it to me: 'Biology, I am sorry to say, was a subject for girls. Bright people were not attracted by it.'[86]

This remark may be indicative of contemporary rhetoric. It may also explain why there were so few women in the Cambridge unit. But it hardly describes the actual position of the biological sciences, especially not at Cambridge, where women were admitted to full membership only in 1948 (it took another quarter of a century before male colleges slowly opened their gates for women students) and the biological sciences (as everywhere else) were dominated by men.[87] In contrast, special features of science

---

[85] Kendrew had studied biochemistry as half-subject as an undergraduate. At that time the course was still taught by Hopkins in collaboration with prominent visitors from abroad. Kendrew remembered it as a 'very remarkable' and 'extremely stimulating' course, which first made him decide that he did not want to go on being a chemist – though at the time he did go on to specialise in chemistry; interview with J. Kendrew, Linton, 18 March 1993.

[86] Interview with S. Brenner, Cambridge, 30 June 1993. See also Huxley (1996, 6). The implication clearly was that molecular biologists had changed this state of affairs.

[87] In 1968–9 women still represented a meagre 11 per cent of the undergraduate population and just 14 per cent of the postgraduates (Brooke 1993, 526–33). A cursory view through student lists of the first decade after the war indicates that in their final year as many women chose chemistry or physics as botany or zoology and a higher number consistently opted for mathematics, though numbers are too small for any significant statistics. See students lists in *The Historical Register of the University of Cambridge* (Supplement volumes 1941–) (Cambridge: Cambridge University Press, 1952–). This picture contrasts strongly with the situation today when (molecularised) biology has become the preferred choice of women science students (Mason 1993). Before the war Gowland Hopkins' Biochemistry Department in Cambridge was known to offer places to an unusual (if always small) number of women researchers (Weatherall and Kamminga 1992, 46–8) – but so was crystallography as a subject. After the war the situation changed rather to the disadvantage of women in these subjects (see above,

teaching at Cambridge worked against a complete division of the sciences. In addition the war had left its mark also in the biological disciplines.

In Cambridge all biological subjects (including botany, zoology, genetics, anatomy, physiology, pharmacology, biochemistry, pathology, psychology and veterinary medicine) were taught in the Science Tripos. Medics also took the Natural Science Tripos.[88] This allowed but also *required* students (unlike anywhere else, including Oxford) to study a combination of physical, chemical, biological, medical and mathematical sciences before specialising in any one subject in their final year and being socialised into this one department. Combining biological and physico-chemical subjects was not unusual. As Kendrew took up biochemistry as half-subject next to physics, chemistry and mathematics, others, for instance the physiologists Alan Hodgkin, Andrew Huxley and Richard Keynes, all read physical chemistry and mathematics as undergraduates. The organisation of the Science Tripos thus worked against a sharp separation of the incoming students and required teaching staff from various departments to elaborate common curricula. In a 1948 review of the Natural Science Tripos this particular feature of the science curriculum 'which has given a special outlook to the Cambridge scientist' was widely praised. It was seen to allow students to defer specialisation until they could be expected to make more informed choices and to provide 'an excellent preliminary training for men [*sic*] who undertake work in those borderline subjects from which new advances in the natural sciences often arise'.[89]

Certainly in botany and zoology the taxonomic and morphological tradition was still strong, but experimental traditions were as securely established and a sizeable number of researchers across the biological disciplines were interested in applying new physical technologies to biological problems. Indeed, a number of these actively interacted with the biophysicists in the Physics Department. An important arena for these exchanges was a science club, known as the Hardy Club. Clubs were a common feature of Cambridge academic life. Many scientists were members of more than one of them.[90] But the history and composition of the Hardy Club is of particular interest here.

Introduction). On women at Cambridge see also Hunt and Barker (1998), and more specifically on women in Cambridge science, though dealing with earlier periods, Gould (1997) and Richmond (1997).

[88] Medics were fully integrated in the Tripos in 1934 and remained so until 1968 when medical studies were organised in an independent Tripos. On the discussions taking place in the 1930s reform see Weatherall and Kamminga (1992, 54–7).

[89] Reporter no. 3611, October 1948, p. 340. Quite in contrast to this spirit, in the negotiations for a new laboratory of molecular biology in the late 1950s, interests of entrenched disciplines seemed to prevail over the support for a new (interdisciplinary) venture.

[90] Crick for instance was part of the Hardy Club as well as a member of two physics clubs, the Kapitza Club and the $\nabla^2 V$ Club; interview with F. Crick, Cambridge, 25 May 1993.

The club was launched by a group of young zoologists and physiologists in 1949. It took its name from the Cambridge colloid scientist William B. Hardy, mentioned earlier, who was trained as a zoologist and worked in the Physiological Laboratory. The aim of the club was to read papers and discuss problems 'in the general field of biophysics', but the name was avoided in the title so as not to offend people who were not invited, probably above all members of the Department of Colloid Chemistry who traditionally laid claim to the field. Membership was by invitation only. Those elected were all fairly junior in their academic careers, but the majority had 'spent the war being physicists'.[91] Some knew each other from before the war; some first met during the war and then again afterwards. There was no woman in the group, though some of the members were related to each other via women relatives.[92] A remarkable number of the original members later became Fellows of the Royal Society, to 'say nothing about the Nobel prizes, the Order of Merit and things like that'.[93]

Among the first members were Horace Barlow, Hodgkin, Huxley and Keynes from Physiology, Murdoch Mitchison, Mark Pryor, Victor Rothschild and Michael Swann from Zoology, Kendrew from the Cavendish Laboratory, Crick from the Strangeways, Peter Mitchell from the Biochemistry Laboratory and J. R. Robinson from the Department of Experimental Medicine. A few other members were soon coopted, including the protein biochemist Kenneth Bailey, Philip George from Colloid Science, Perutz, and later Hugh Huxley and the zoologist John Pringle. Watson also participated at the meetings when he came to Cambridge (Watson 1987, 63), as did Sydney Brenner, once he joined the MRC unit. The club met in Fellows' college rooms (that was one reason to keep it small). There were one or two meetings per term, with one of the members, sometimes a visitor, presenting a talk, followed by discussion. Topics were wide ranging, though the structure and function of nerves, muscles, membranes and proteins were the most recurrent themes.

---

[91] The Hardy Club Minute Book, p. 1, Churchill Archives Centre, Cambridge, and interview with R. Keynes, Secretary of the Club, Cambridge, 1 April 1993. A description given by Alan Turing's biographer of the Ratio Club, a cybernetics discussion group started in the same year and meeting in London, could fit as well the Hardy Club: 'In some ways it was an attempt to revive the democratic association of young scientists which had characterised the war years. They excluded those of professorial rank... Many of the Ratio Club had worked at TRE [the Telecommunication Research Establishment]... It was just a faint ghost of the "creative anarchy"' (Hodges 1983, 412).

[92] For example, Barlow and Keynes were cousins and Pryor was married to another of Keynes' cousins. Close familial ties were not unusual among Cambridge scientists, where some families represented true dynasties reaching back many generations. On growing up in one of these Cambridge families see Raverat (1952).

[93] Interview with F. Crick, Cambridge, 25 May 1993.

3.9 (a) The Eagle (1937). From C. Jakes, *Cambridge: Britain in Old Photographs* (1996), p. 38. Copyright Cambridge Antiquarian Society, Cambridgeshire Collection.

Crick and Kendrew in particular singled out the Hardy Club as a very useful forum. That was the place where they kept in touch with biologists and biology and presented their own work.[94] More informal but no less institutionalised occasions of exchange were the celebrated lunches in the Eagle, presided over by Crick, and coffees at the Bun Shop. Both pubs were just around the corner from the central laboratory site. Among the regular participants at the lunches were Herbert Gutfreund from Colloid Science, Daniel Brown and Leslie Orgel from Chemistry and Bailey from Biochemistry. The Bun Shop was regularly visited by all scientists. For most of them the laboratory and the Bun Shop were more important than college life (Figs. 3.9 a and b).

The Hardy Club ran successfully up to the late 1950s or early 1960s, when meetings became more sporadic and membership was less clearly defined. Some of the original members, including Richard Keynes, long-time secretary of the club, took up positions outside the university or moved elsewhere. More importantly, with time there seemed to be less

---

[94] Interviews with J. Kendrew, Linton, 18 March 1993, and with F. Crick, Cambridge, 25 May 1993.

3.9 (b) The Bun Shop. From C. Clarke, *British Medical Journal* 301 (1990), 1447. Photograph by W. Bodmer. Reprinted with permission from the BM Publishing Company.

common ground to share. Members split up into different groups, and especially physiologists and biophysicists now turned molecular biologists seemed to speak different 'languages'.[95] In this respect the dying of the Hardy Club was but a symptom of wider rearrangements of the biological sciences (see Parts II and III). In the first decade after the war, however, common interests between biophysicists from various quarters in Cambridge prevailed, and, importantly, several of them were as successful in their endeavours as the protein crystallographers.[96] Hodgkin, for instance, was among the few biologists who could directly use his wartime skills acquired in work on radar to build powerful amplifiers for his electrophysiological experiments on squid axons. Together with Huxley and Keynes, he ran a very successful research group which attracted generous funds from various bodies, including the Rockefeller Foundation, the

[95] Interview with R. Keynes, Cambridge, 1 April 1993.
[96] Why in the long run, at least apparently, 'molecular' approaches prevailed and even highly successful fields of biology, like electrophysiology, felt the pressure to 'go molecular' remains an open question at this point. I consider this in Part III of this book.

Nuffield Foundation and the Royal Society. He became a Fellow of the Royal Society in 1948 (the only member of the Hardy Club who had this distinction when the club started) and shared the Nobel Prize with Huxley in the year following the award to Crick, Watson, Wilkins, Kendrew and Perutz for their work on the structures of DNA and protein. In a brief passage in his autobiography, which was written around an earlier account dealing exclusively with his wartime contributions to the development of radar, Hodgkin reflects on how this work experience as physicist had changed his attitude as biologist:

I found it much harder to give tutorials... than before the war. This was partly because I had forgotten a good deal and partly because I had ceased to believe in some of the principles that had once seemed to hold physiology together... I suppose that after five years working as a physicist I had little use for biological generalizations and always wanted to concentrate on the physicochemical approach to physiology. (Hodgkin 1992, 262)[97]

Biochemists were not well represented in the Hardy Club. But their field also underwent great expansion after the war. In Cambridge itself the department had lost some of its aura since Hopkins' pioneering years, but many of his collaborators remained, and Charles Chibnall, who had succeeded Hopkins in 1943, assembled a strong group of protein chemists, including Bailey and Sanger. While Hopkins' strategy had been to build up a Biochemistry Department which was free from any service role to medicine, Chibnall, himself a plant biochemist with strong links to the Agricultural Research Council, saw the future of biochemistry in its application to the medical sciences, which he expected to explode after the war. In order for the department to respond to this situation, he resigned his chair in 1949 to make place for Frank G. Young, an animal biochemist well connected to the medical research community. Young soon started negotiations for an extension of the laboratory, though these turned out to be long and thorny.[98]

Genetics at Cambridge represented a special case. The chair was held by Ronald A. Fisher, internationally renowned for his genetic and mathematical investigations of the theory of natural selection. However, his mathematical approach to the subject did not attract many students, and the department, accommodated in an extension to his private house at some

---

[97] In the view of some of my interlocutors, physiologists like Hodgkin and Huxley became as 'arrogant' as the Cavendish biophysicists with respect to other biologists.

[98] On biochemistry in Cambridge from Hopkins' days to Frank Young see Chibnall (1966), Kohler (1978: and 1982), Randle (1990), Weatherall and Kamminga (1992) and Kamminga and Weatherall (1996).

distance from the central laboratory sites, was in a somewhat desolate state.[99]

In the late 1950s, the biologists at Cambridge initiated a review of their teaching in the Tripos, introducing a new one-year (i.e. half-subject) course in cell biology. The course assembled 'the basic facts and principles of cytology, cell physiology, biochemistry and genetics common to all types of organisms' and was specially designed to attract physics and chemistry students to biology.[100] In particular, the genetics section of the proposed course included new work on the molecular structure and function of genes, but the molecular biologists, as non-teaching officers, were not involved in the discussions.[101] By that time, plans for a MRC Laboratory of Molecular Biology on the outskirts of Cambridge, only loosely connected to the university, were already well advanced.

To a large extent, then, biologists themselves initiated the postwar renewal of biology, even if this process involved the integration of techniques and results developed by people outside the traditional biological disciplines. But biologists and medical researchers also played a much more direct role in shaping the research programme of the biophysicists at Cambridge.

As Perutz himself has always acknowledged, he settled on haemoglobin as research subject for his thesis on the advice of the Austrian biochemist Felix Haurowitz. The first crystals were given to him by the Cambridge physiologist cum biophysicist Gilbert S. Adair. Later Keilin, Professor of Biology at Cambridge, offered Perutz bench space for his biochemical preparations until the group got its own biochemical research facilities in the Cavendish. When Kendrew joined Perutz to do a Ph.D. in protein crystallography, it was the physiologist Joseph Barcroft who suggested a comparative analysis of foetal and adult haemoglobin. Perutz himself had been at a loss as to what to suggest. Barcroft also supplied the crystals of foetal sheep haemoglobin for Kendrew's research (Perutz 1980). Sickle cell

[99] At Cambridge, botanists before zoologists and geneticists included molecular genetics in their courses. As early as autumn 1953 students of botany were introduced to Watson and Crick's double helical model of DNA; D. Hopwood to author (November 1995). On Crick's application for the chair of genetics after Fisher's retirement in 1957 see below (chapter 6).

[100] 'Half subject cell biology', memorandum prepared by E. F. Gale, R. D. Keynes, J. W. S. Pringle, J. A. Ramsay and J. M. Thoday for discussion by the Faculty Boards; cf. Minutes of the Faculty Board of Physics & Chemistry, January 1957–October 1963, Meeting of 27 May 1960, minute 62. The course was first offered in October 1961. The proposal was part of a larger review of the Science Tripos.

[101] Perutz held a lectureship in biophysics from 1953, but when his promotion to reader was refused in 1957, he resigned from his position and returned to the MRC payroll.

haemoglobin as well as other haemoglobin variants from the collection of Hermann Lehmann, clinical biochemist at Cambridge, became important research tools for the structural as well as the genetic research work in the unit. Once Perutz had a structural model of haemoglobin, Lehmann's clinical and chemical knowledge of haemoglobin variants became crucial for relating structure to function (de Chadarevian 1998a, and below, pp. 348–9).

Watson, the first American postdoc and biologist who joined the unit had, in Perutz's own words, 'a tremendous impact' on the group, focusing their attention on the 'most basic problems in biology'. Before Watson's arrival they were 'chemists and physicists who had received inspiration from biochemists and physiologists interested mainly in the function of proteins, but never asking themselves where proteins came from' (Perutz 1980, 328). Brenner, who joined in 1957, strengthened the biological direction of the work of the unit. Like Watson he belonged to a younger generation who had started their studies after the war. He held a medical degree from South Africa and had worked on the resistance of bacteria to phage infection in Cyril Hinshelwood's chemical laboratory in Oxford (Brenner 1997). Finally, when the new Physics Professor, Nevill Mott, was not prepared to house the unit any longer on his premises, the group joined forces with Sanger, who had been working on protein sequencing in the Biochemistry Department in Cambridge, to ask the MRC to build a new independent laboratory. Winning Sanger's support was a crucial step for setting up the proposal (Perutz 1980, 329). Having him on board determined the way molecular biology was defined and practised in Cambridge (Part II; also de Chadarevian 1996a). When no suitable place for the new laboratory building could be found, Mitchell, in his new role as Regius Professor of Physic at Cambridge, came to the rescue. He suggested the building of a new Laboratory of Molecular Biology adjacent to and sharing common facilities with the new Department of Radiotherapeutics on the site of the planned Postgraduate Medical School and the new hospital (see chapter 7).

As we will see later, Mitchell had opportunistic reasons to support the case of the MRC unit. But this only strengthens the point that biophysicists depended on the active interest of biologists or medical researchers to further their case. At the same time biologists and medics used biophysicists and the funds they attracted to pursue their own disciplinary and institutional agendas. These disciplinary projects could well be in conflict with those of other biologists. Indeed, a selected number of physiologists, zoologists, biochemists and pathologists, several of them members of the Hardy Club, protested against the removal of the MRC unit from the (geographical and institutional) centre of the university engineered by Mitchell. But their protest came too late (below, chapter 7).

A brief look outside Cambridge confirms the same pattern. From the beginning, the funding of Randall's enterprise was made dependent on the condition that he could secure adequate biological supervision, a role fulfilled by Honor Fell, zoologist by training and director of the Strangeways Laboratory in Cambridge.[102] Several biologists joined his staff, among them Jean Hanson, a skilled zoological microscopist. Later, Randall recognised the importance of a more formal link with the life sciences. He renounced his chair in Physics and engineered the move of the (by now independent) Department of Biophysics to the Biological Faculty (Wilkins 1987). Biologists at Leeds rejected Astbury's crystallographic work as not biological enough, and on this ground, in the late 1940s, vetoed the name of Molecular Biology for his research unit. Similar reasons may explain Astbury's failure to secure funding from the MRC after the war (Olby 1994, 326–7). The lack of institutional and financial support hampered the development of his work.

In this chapter I have delineated the extraordinary expansion of biophysics after the war and have analysed the reasons for this phenomenon. Postwar biophysics fitted the agenda of postwar reconstruction. I have pointed to the attraction of a 'physics of life' versus a 'physics of death' and to new institutional opportunities offered for basic medical research. These opportunities were eagerly taken up (if not created) by physical scientists and biomedical researchers, keen either to build on their wartime experiences or rather to leave them behind and start afresh. The next chapter looks more closely at the research practices of protein crystallographers and at the transformations introduced through war-related technological developments. My main concern will be with Kendrew's early use of the electronic digital computer at Cambridge, the reasons for his prompt interest in the experimental machine when it became operational in 1949, and the place of computer development in protein crystallography. If the separation of institutional negotiations and personal careers in this chapter and technological developments in the following is largely artificial, it will help to clarify the different ways the transition of World War II worked in the case of biophysics and of protein crystallography in particular.

---

[102] On Fell see Mason (1996) and Vaughan (1987).

4
...
Proteins, crystals and computers

In the decades following World War II the appropriation of a host of new instruments transformed the life sciences. The list includes ultra-centrifuges, spectroscopes, electrophoresis apparatuses, electron micro-scopes, and heavy and radioactive isotopes. The production of simpler commercial models and the allocation of new government funds for scientific research, including large sums for equipment, made these instruments, many of which existed only in a few prototypes before the war, available to a large number of researchers (e.g. Kohler 1977; Elzen 1986; Kay 1988; Creager 1993; Rasmussen 1996).[1] This is particularly true for those countries which, after the war, could afford and decided to invest in scientific research, like the United States and, to a minor degree, Britain or France, while for scientists in other countries it was more difficult to participate in these new trends. With the use of expensive machines, life scientists took their share in the Cold War funding of research.

Historians have stressed how wartime needs and military funding were used to promote the development and use of some of these instruments. Shifting the focus from the university laboratory to the industrial setting, they have shown that researchers in the development departments of commercial firms played an active role in 'making machines instrumental' or turning them into laboratory tools (Rasmussen 1996; also Lenoir and Lécuyer 1995).

This chapter focuses on the appropriation of the electronic digital computer by protein crystallographers. The postwar development of electronic

---

[1] On the role of the Rockefeller Foundation natural science programme in supporting the development of laboratory technologies as a means of fostering collaborations between physicists, physical chemists and biologists on 'vital processes' in the 1930s see Kay (1988), Kohler (1991, 358–91) and Zallen (1992). For a more critical view on the 'technology transfer' promoted by the Rockefeller Foundation see Abir-Am (1982a). Kay has argued that the technological programme targeted at investigation of the subcellular level of living organisms from the beginning involved 'intervention' of a kind that has become prominent with genetic engineering (Kay 1993a).

computers built directly on military projects and technological innovations of World War II. Even when directed to civilian uses, developments always remained tied to military interests (Cohen 1988; Edwards 1996). At the Mathematical Laboratory at Cambridge Maurice Wilkes and his team pioneered the construction of a stored program electronic digital computer, and the protein crystallographers were among its first and most ambitious users.[2] Kendrew and Perutz have always acknowledged that their Nobel Prize winning achievements of the full structure determination of myoglobin (the oxygen carrier in muscle and the first protein ever for which the atomic structure was known) and haemoglobin (the most complicated molecule to be 'solved' at the time) crucially depended on the development of electronic computers. Indeed, Perutz reckoned that they had been 'very fortunate' that the development of computers had 'always just kept in step' with the expanding needs of their X-ray analyses (Perutz 1964, 662). However, Kendrew started using the first electronic digital computer at Cambridge when mechanical methods of calculation were still feasible. In the early 1950s, when Kendrew already routinely used the computer, Perutz did not 'trust' the new device and preferred to rely on the more familiar but still fairly new Hollerith punched card machines.[3] The move from mechanical to electronic calculation devices, then, was not as straightforward as might be assumed. A closer investigation of Kendrew's early use of the computer will shed light on how this transition was achieved. It will also show the way in which the work of early users like Kendrew was of interest to the computer engineers, keen to improve the performance of their machines, and how, in turn, computers increasingly shaped the direction and pace of work in protein crystallography. Kendrew's interest in electronic computers and computer programming, like his approach to protein crystallography more generally, I suggest, in various and subtle ways refers back to his occupation with radar and operational research during the war. This connection underlines the impact of approaches and skills gained on wartime scientific projects on the very practice of scientists, here biophysical scientists, after the war. Through their use of powerful electronic computers for their calculations, protein crystallographers participated

---

[2] In contrast for instance to a parallel project pursued at Manchester University, the Cambridge project was not funded by the military, but the military keenly followed its development (see below). On the recent interest in studying *uses* in the history of computing see Agar (1998a).

[3] Hollerith punched card machines were first introduced into crystallographic practice for the structure analysis of penicillin during the war. They became more widely available to researchers in the field in the late 1940s (see below).

in the money, patronage and technological developments of the Cold War era.[4]

To chart postwar changes in the practice of protein crystallographers, it will be useful to start with a brief account of protein research and protein crystallography in the interwar years.

Mad pursuit

In the 1920s and 1930s proteins were expected to hold the key to all life processes, including inheritance. The chemical constitution of proteins was an area of intense study and debate, fuelled by strong commercial interests. When the Swedish physical chemist Theodor Svedberg, to his own surprise, found in 1926 that haemoglobin formed a sharp band in his new ultracentrifuge, it gave a big boost to the molecular over the colloid theory of proteins. The ultracentrifuge, originally designed for the analysis of colloid particles, turned into a key instrument for molecular weight determination. On the basis of his sedimentation studies, Svedberg also claimed that all proteins were composed of sub-units of the same molecular weight (Svedberg and Fåhraeus 1926; Svedberg and Pederson 1940).

Even before Svedberg's sedimentation studies, researchers following the lead of the German chemist Emil Fischer were investigating the amino acid composition and chemical structure of proteins as well as the overall size and shape of the molecules. Efforts also focused on the development of new physical and chemical techniques to analyse, synthesise and characterise proteins and their components in view of their biological, medical and economic importance.[5]

When in 1934 Bernal, working in the Mineralogy Department in Cambridge, found that pepsin crystals in their liquid of crystallisation produced sharp X-ray diffraction images with hundreds of spots extending to reflections corresponding to interatomic distances, he was 'full of excitement . . . thinking of how much it might be possible to know about the structure of proteins if the photographs . . . could be interpreted in every detail'

---

[4] Computers were not the only war-born technology which protein crystallographers adopted for their research and which changed their work practices. Electronic gear, much of it military surplus equipment which filled the cupboards of the Cavendish as well as other physics laboratories after the war, was used to stabilise X-ray tubes and develop new intensity measuring devices. For a vivid description of the radical change of crystallographic practice following World War II see U. Arndt, 'The development of X-ray crystallographic techniques in molecular biology' [unpublished manuscript, c. 1987].

[5] On the 'protein paradigm' in biology see Kay (1993a). In contrast to the situation sketched by Kay, in the Cambridge biophysics unit the study of the structure and function of proteins remained a central concern after 1953 and became integrated into the approaches to molecular genetics (de Chadarevian 1999).

(Hodgkin and Riley 1968, 16).[6] However, the X-ray diffraction pictures of the digestive enzyme produced by Bernal were much more complex than any structure analysis ever attempted, and although the mathematics to interpret the pictures was well established, their application was anything but straightforward.

The principles of X-ray crystallography were established by William and Lawrence Bragg in 1913, building on earlier observations made by German physicists that X-rays were diffracted by crystals. The Braggs postulated that the diffraction patterns recorded on photographic plates resulted from the reflection from sets of parallel atomic planes in the crystals (or more exactly from the electrons in these planes of atoms). Later Lawrence Bragg showed that each spot in the diffraction pattern could be represented as a term in a Fourier synthesis, thus providing the mathematical tools by which electron densities and hence the atomic structure of crystals could, in principal, be deduced from the diffraction pattern.[7]

Following the Braggs' groundbreaking work, X-ray crystallography was applied to inorganic and organic molecules of increasing size and complexity. But in contrast to big strides in the X-ray analysis of inorganic molecules, progress in the analysis of organic molecules had been slow.[8] In 1934, the largest organic molecule of which the atomic structure had been solved was a benzene derivative with just twenty-four atoms (Olby 1985a, 175). To interpret the diffraction pictures (i.e. to apply the mathematical tools), both the intensities and the phases of the reflections needed to be established. But while the intensities could be directly measured from the diffraction pictures, the phases had to be established by trial and error methods. With the number of atoms the problem of 'guessing' the phases grew dramatically and the calculations involved were gigantic.

Bernal based his hopes of interpreting the complex protein diffraction pictures on the development of new crystallographic techniques as well as on certain expectations regarding the structure of proteins. Among the

---

[6] On the retrospective and ritualistic transformation of this 'non-event' into the origin not only of a new speciality but of molecular biology more generally see Abir-Am (1992c). On 'myths of origin' in crystallography and the call for an anthropological analysis of these myths see also Forman (1969–70) and Ewald's response in defence of the scientist's approach to history (Ewald 1969–70).

[7] On the history of X-ray crystallography see Ewald (1962), Olby (1985a; 1994, 1–70, 249–66) and Perutz (1990). Glusker and Trueblood (1985) offer a useful introduction to crystallography.

[8] An exception was represented by the X-ray analysis of fibrous proteins like silk, cotton, wool and keratin which was intensively pursued from the 1920s in the context of (textile) fibre research, especially by researchers in the Kaiser Wilhelm Institut für Faserstoffchemie [chemistry of fibres] in Berlin and by William Astbury's group in Leeds (Olby 1985b).

new tools, the Patterson function and the method of isomorphous replacement deserve special mention. The Patterson function, introduced in 1934, the same year as Bernal's breakthrough, by Lindo Patterson, then at the Kaiser Wilhelm Institute for the Chemistry of Fibres in Berlin, allowed calculation of the distances between atoms (rather than their exact position) on the basis of measured intensities alone (that is disregarding the phases of the reflections). This represented an important simplification of the Fourier analysis.[9] In contrast, the isomorphous replacement method was based on introducing heavy atoms into the molecule with the aim of using the produced differences in the diffraction patterns to calculate the phases. (To determine the positions of the heavy atoms in the molecules this method also relied on the Patterson function.) The method promised to determine atomic structures 'directly', that is without referring to previous chemical knowledge. In 1936, J. Monteath Robertson for the first time successfully applied this method in his analysis of the structure of the phthalocyanines, a series of industrially produced organic pigments consisting of fifty-eight atoms arranged in a porphyrin ring-like structure (Olby 1985a). But apparently the possibility of using the isomorphous replacement method had been discussed in Bernal's lab well before this date (Hodgkin and Riley 1968, 25–6). A new mathematical tool, known as Beevers–Lipson strips after its two inventors, designed to help with the massive calculations involved in the Fourier syntheses, represented another important innovation. The cardboard strips, arranged in two wooden boxes, each contained a row of cosine and sine figures. The strips could be lined up and the figures in the columns directly added, thus saving the time that would be necessary to look up the right figures in the tables and write them down, a step which also easily led to errors (Fig. 4.1).

Besides these technological innovations, it was the expectation that proteins possessed a regular structure that made Bernal and his few followers believe that the highly complex diffraction pictures were indeed interpretable and that the problem of protein structure was 'solvable'. Bernal saw this widely held (though not uncontested) view confirmed in the clear spacings of his first diffraction picture. In their communication on the subject, he and Dorothy Crowfoot (later Hodgkin) dispelled alternative suggestions and postulated a symmetrical arrangement of the 'constituent parts' of the proteins (Bernal and Crowfoot 1934, 795).[10] They also supported the view that there was a general plan of protein structure

---

[9] On Patterson and pattersons see Glusker, Patterson and Rossi (1987).

[10] Astbury, a pioneer in the X-ray analysis of fibrous proteins, using dried crystals, had not succeeded in obtaining clear X-ray pictures of the globular proteins Bernal was investigating. On the basis of this (and other) observations, he postulated a less ordered structure for this class of proteins. In contrast to Bernal, he also suggested that globular proteins, like fibrous ones, consisted of a single coiled chain (Olby 1985b).

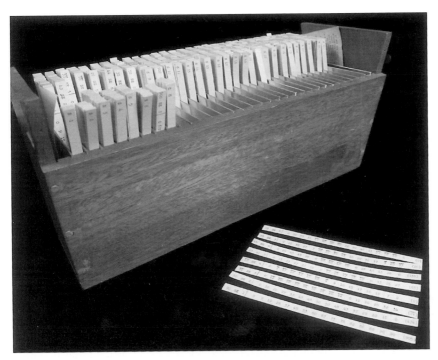

4.1 Set of Beevers–Lipson strips owned by the LMB. The cabinet was locally built (mid- to late 1940s). Photograph by Lesley McKeane (2000). Courtesy of Medical Research Council Laboratory of Molecular Biology, Cambridge.

and that the solution of one protein structure would give clues to the structure of proteins more generally. These expectations dominated X-ray crystallography up to the late 1940s and the famous 'pill box' or 'hat box' model of haemoglobin which pictured the polypeptide chains running in parallel bundles (Perutz 1948–9). Without these expectations, the few crystallographers who did take up the challenge set by Bernal might not have done so. Most 'professional' crystallographers considered protein crystallography in any case a hopeless undertaking and a 'waste of time', a view that prevailed well into the 1950s.[11] Bragg explicitly made this point when he wrote:

For a long time the idea that the [haemoglobin] molecule contained some kind of regular structure of protein chains...was a guiding star which encouraged the investigations. As events turned out, it was a false star...If this had been realized at the time the problem would have seemed so hopeless that the quest might well have been discouraged, but fortunately this was not the case. (Bragg 1965, 4–6)

[11] Interview with J. Kendrew, Cambridge, 14 July 1992.

Protein crystallographers were not alone in advocating a symmetric or regular structure of protein molecules. On the basis of extensive studies of the amino acid composition of proteins, Max Bergmann and Carl Niemann from the Rockefeller Institute in New York suggested that amino acids were arranged in a regular manner along polypeptide chains, forming repeats that corresponded to the size of Svedberg's sub-units. This hypothesis was reinforced by the cycol theory of Oxford mathematician Dorothy Wrinch. Combining ultracentrifugal, chemical and crystallographic data, Wrinch proposed that proteins consisted of a regular arrangement of ring-like sub-units counting 288 amino acids, Bergman and Niemann's repeat unit (Wrinch 1937).[12]

Bernal initially supported Wrinch's theory, but later became increasingly critical of it, sharing growing concerns that the work, based on 'doubtful chemical assumptions', would only bring X-ray crystallography into disrepute (Bernal 1937–9, 552–3; 1939b; Olby 1985b; 1994, 13–14). The most decisive attack against both Wrinch's and Bergmann and Niemann's hypotheses came from the British protein chemist Albert Neuberger, who based his critique on detailed chemical data (Neuberger 1939). However, only Sanger's sequencing studies of insulin (the blood sugar regulating protein of the pancreas) in the early 1950s conclusively confounded any speculation of a regular arrangement of amino acids in proteins.

At the same Royal Society meeting at which Neuberger launched his critique against symmetrical or periodical models of protein structures, Bernal called for more coordination between protein X-ray crystallographers and other researchers in the field. He envisaged the creation of a 'central bureau for protein research' which would 'facilitate exchange of information and material' as well as 'assist in an ordered attack on the whole problem' (Bernal 1939b, 38).[13] By that point he may well have become less optimistic that X-ray crystallography alone could provide the decisive clues on the structure of proteins.

There certainly was something of a gap between Bernal's hopes and expectations and what was really achieved. Soon after their first publication on the subject, Crowfoot left Cambridge for Oxford, leaving nobody behind who was experienced in crystal growing. At Oxford, she continued her investigations of protein crystals, focusing on insulin and lactoglobulin. But it very quickly became clear that the Patterson maps of the

---

[12] On Wrinch's life and scientific career see Abir-Am (1989).

[13] See also Bernal (1937–9, 556) and, in a more general context, Bernal (1939a, 257, 338). Bernal, however, did not follow up his suggestion with any concrete step for calling such an organisation into being. He was by that time increasingly involved in left-wing politics and as scientific adviser on civil defence (see below).

complex protein molecules were extremely difficult to interpret and the attachment of heavy atoms into the protein molecules proved anything but straightforward (it took twenty years before Perutz was able to demonstrate that the method actually worked in proteins) (Hodgkin and Riley 1968, 20–6).[14]

When Perutz in his first year at Cambridge was on the lookout for a 'biological problem', Bernal had no crystals of biological interest at hand (Perutz 1987, 38). When Perutz finally settled on haemoglobin as the topic for his dissertation, the first crystals were given to him by Gilbert Adair from the Physiological Laboratory at Cambridge. Soon afterwards Bernal left Cambridge, where he was not offered any career prospects, to take up the Physics chair at Birkbeck College, London, an institution dedicated to continuing adult education and therefore well suited to Bernal's social engagement. From his new post he continued to promote the fledgling field of protein crystallography, but spent most of his own formidable energy on left-wing political activities and government advisory functions in preparation for the war. In 1939 his *The Social Function of Science* appeared, which became the manifesto of socialist thought applied to science (Bernal 1939a and above, chapter 2). Fortunately enough for Perutz, Bernal's departure coincided with Lawrence Bragg's appointment as Rutherford's successor to the Cavendish Chair in Cambridge. Bragg became excited at the prospect of extending the method he himself had first applied to propose a structure for common salt some twenty-five years before, to the fantastically complex structure of proteins.

What were Perutz's expectations when embarking on the X-ray analysis of haemoglobin as a dissertation project? Perutz was to dedicate most of his scientific career to the study of the structure and function of the haemoglobin molecule. But he had certainly not anticipated this at the time. A brief look at his dissertation, submitted in March 1940, shows that his aims were limited, even if the hope of achieving a significant result ran high. To prepare and mount the haemoglobin crystals, determine the dimensions and hydration of the crystals and study the effects of denaturation was enough to acquire a Ph.D. Perutz's dissertation also included the discussion of a two-dimensional vector analysis of a haemoglobin derivative calculated by Dennis Riley of Hodgkin's group in Oxford. Comparison with Hodgkin's Patterson analysis of insulin led Perutz to confirm 'that protein molecules are composed of small subunits, arranged in a regular manner' (Perutz 1940, 60).

---

[14] On the continuing problems in applying the method to insulin see Ferry (1998).

It should be noted at this point that the British crystallographers were not the only ones who attempted to apply X-ray analysis to the study of protein structure. In particular, Linus Pauling at the California Institute of Technology was active in the field. In contrast to his British counterparts, he did not attempt a straightforward crystallographic analysis of proteins, but early on settled on a model building approach based on detailed knowledge of the structure of amino acids and small peptides combined with stereochemical reasoning (Bernal 1968, 372).[15] After the war a keen competition developed between Caltech and Cambridge. From his visit to Oxford in 1947/8, Pauling wrote to Robert Corey, his closest collaborator at Caltech: 'I am beginning to feel a bit uncomfortable about the English competition. They have a gift for driving straight at the heart of a problem, and getting its solution by hook or crook.' From Pasadena Corey confirmed Pauling's feelings: 'I am terribly impatient about getting into the protein work with a force that will really give the British some competition. Right now they are miles ahead of us on personnel and experience.'[16] From then on, the 'complete solution of the crystal structure of a protein' became a much more clearly defined goal for the group at Caltech, but the emphasis remained on a stepwise approach.[17] Pauling's detailed chemical and structural knowledge of the protein building blocks allowed him to deduce the two basic conformations of peptide chains, known as α-helix and ß-sheet. The announcement of the α-helix conformation in 1951 caused consternation among Cambridge crystallographers (Olby 1994, 289–93). But after the 'remarkable Pasadena Conference' in 1953, at which 'Perutz showed how protein structures could be solved by X-ray diffraction, and Crick and Watson described the DNA double helix' (Huxley 1996, 10), Perutz reckoned that 'the focus of structural research in biology had shifted to Cambridge' (Perutz 1987, 40).[18] Watson and Crick,

[15] On model building among British crystallographers in the early 1940s see Hodgkin and Riley (1968, 27).

[16] L. Pauling to R. Corey, 18 February 1948, and R. Corey to L. Pauling, 25 February 1948, Pauling Papers, individual correspondence, file C, 13.10, Oregon State University Library. I thank Bruno Strasser for making these and other letters from the Pauling Papers held at Oregon State University Library available to me.

[17] L. Pauling to R. Corey, 18 February 1948, and L. Pauling to J. Kendrew, 21 December 1955, Pauling Papers, individual correspondence, files C and K, Oregon State University Library. The project Pauling suggested taking up at Pasadena was the same as that pursued by Hodgkin in collaboration with Cambridge biochemists: the crystallisation and X-ray analysis of quarter molecules of insulin. After her first foray into protein structure in the mid-1930s, Hodgkin also proceeded from the analysis of smaller to that of more complex molecules, but Pauling's approach to protein structure was more systematic in that it built on the analysis of amino acids as the protein building blocks.

[18] Somewhat in contrast to this view, Kendrew, in a contemporary report of the conference written for *Nature*, noted above all the large area of agreement between various 'schools of thought' that the conference had revealed. According to his report, the most important

however, have always testified that it was Pauling's model building approach they had attentively studied and followed. And Bernal, in a later appreciation, also leaves no doubt that Pauling had a decisive impact on protein crystallography in Britain (Crick 1990, 60; Bernal 1968).[19]

What then gave structural research new impetus after the war? And how did computers enter crystallographic practice? Answering the second question will provide clues to the first.

The Cambridge Mathematical Laboratory

In 1949, not long after Hugh Huxley, the first research student, had joined the MRC Unit for the Study of the Molecular Structure of Biological Systems, Bragg, who closely followed the work on protein crystallography, suggested to him that it would be 'good for [his] soul' to calculate a two-dimensional Patterson projection of one of the haemoglobin forms. Huxley remembers: 'This was a terrible chore with Beevers–Lipson strips and an adding machine cranked by hand. It took me about two weeks of solid work with much loud complaining' (Huxley 1990, 133). The experience added to Huxley's disillusionment with protein crystallography which he soon left behind to dedicate himself to the study of muscle. But the exercise was not completely wasted.

One of the people Huxley complained to was John Bennett, his friend and fellow research student at Christ's College. Bennett, an engineer-mathematician from Australia, was part of Maurice Wilkes' team at the Mathematical Laboratory, where a pioneering electronic computer with a stored memory program had just started operating. He 'immediately' realised that the calculations Huxley was complaining about were programmable and managed to produce a program which could do Huxley's two weeks' work in about half an hour. Another half an hour was needed to print out the results (Huxley 1990, 134). Already determined to switch research field, Huxley handed the results over to Kendrew, his Ph.D. supervisor. With Bennett's help, Kendrew learnt programming himself and, together, they perfected the program. In 1952, they published a joint paper on their work. It was the first paper written on crystallographic computations on electronic digital computers (Bennett and

discussions concerned the α-helix suggested by Pauling, while Perutz's 'fresh attack on the [phase] problem using the classical "heavy atom" method' was presented as an 'important step forward' but one which still had to prove its promise. The double helical model of DNA commanded much less attention at the Pasadena conference dedicated mainly to proteins than at the just preceding Cold Spring Harbor meeting (Kendrew 1954).

[19] On Cambridge and Caltech see also Olby 1970a. On modelling and models see below (chapter 5).

Kendrew 1952).[20] This turned out to be only the beginning of a close and fruitful collaboration between Cambridge crystallographers and the Mathematical Laboratory.

The Cambridge Mathematical Laboratory was officially created in 1937 to provide computing facilities and advice to research workers in the university. The driving force behind this plan was John Lennard-Jones, Plummer Professor of Theoretical Chemistry at Cambridge (Croarken 1992). His group urgently required more advanced computational aids. For some time they successfully used a model differential analyser built of Meccano pieces (a toy construction set) on the model of a similar machine operating at Manchester University. The machine demonstrated the value of a full-size device. At that point Lennard-Jones launched the idea of a computing support centre which would house the full-size machine and be at the disposal of all those interested. For this plan, which was unusual at the time, he gained widespread support and formal approval from the university. The new Mathematical Laboratory was to be equipped with the new differential analyser, the model differential analyser, the 'Mallock Machine' (an electrical analogue calculating machine built in 1933 for the solution of linear simultaneous equations; Croarken 1990, 49–50) and several desk calculators (Croarken 1992, 11). Lennard-Jones was appointed part-time director and Wilkes, then a physics graduate, his assistant. However, before the centre came into operation, World War II broke out and the Ministry of Supply took over equipment and premises at Cambridge. Lennard-Jones was recruited by the Ministry of Supply and set up a ballistic research team in the Mathematical Laboratory, using the new differential analyser for some of his calculations. Wilkes became involved in radar and operational research. The mathematical and engineering skills he acquired during his war service were to become crucial for his later career.

The lease of the Cambridge Mathematical Laboratory by the military during World War II is only one example for the use and development of advanced computational devices for the exigencies of the war. Britain's 'best kept secret'[21] was the development of 'Colossus', a large electronic valve computer, built to break German military codes. War technologies and experiences as well as military funding continued to be crucial for postwar developments in computing.

After the war, the importance of mechanising and centralising computation was more widely recognised, in governmental as well as in academic

---

[20] Also interview with J. Kendrew, Cambridge, 14 July 1992.

[21] Publicity flier for Bletchley Park by the Bletchley Park Trust, Milton Keynes, UK [1997]. On code breaking at Bletchley Park and Turing's role in building the Enigma cipher machine see Hinsley and Stripp (1994) and Hodges (1992).

circles (Croarken 1990). Before the war had drawn to an end, Lennard-Jones, who had risen to the position of Chief Superintendent of Armament Research in the Ministry of Supply and was aware of new developments in computing technologies, advised Cambridge University that the Mathematical Laboratory should contribute to this field of research, while at the same time providing the central computing service as had been planned (Croarken 1992, 12). At the end of the war, Lennard-Jones himself chose to remain with the Ministry of Supply. Thus the task of reverting the Mathematical Laboratory to peacetime use fell to Wilkes.

The laboratory was soon able to offer central computing facilities to users in the university. But Wilkes' main interest lay in applying his electronic skills to developing new computing machines. In a report to the university he wrote: 'There is a big field here, especially in the application of electronic methods, which have made great progress during the war, and I think Cambridge should take part in trying to catch up some of the lead the Americans have in this subject.'[22]

The Cambridge EDSAC (Electronic Delay Storage Automatic Calculator) followed the design of the stored program computer, the EDVAC (Electronic Discrete Variable Computer), described by John von Neumann in an influential paper of June 1945 (von Neumann 1986). Wilkes had access to this paper before the summer of 1946, when he was among the participants of the lecture series on 'Theory and techniques for design of electronic digital computers' given at the Moore School of Engineering in the University of Pennsylvania. He became convinced that the future lay with stored program electronic digital computers rather than with analogue computers which were then much more entrenched (Edwards 1996), and on his return was determined to build such a machine for use by Cambridge University.

Wilkes has described in detail the work involved in turning the plan into a working machine, including the choice of mercury tanks as storage system (Wilkes 1985) (Fig. 4.2). The experience gained with electronics as a result of working with radar was crucial. Wilkes recalls: 'We thought of ourselves as electronics men, not as electrical engineers. We were used to wide bands with short pulses and we saw the possibility of achieving very high speeds with elegant economy of equipment by these techniques.'[23]

---

[22] M. V. Wilkes, 'The functions of the Mathematical Laboratory' (1946), quoted in Croarken (1992, 12).

[23] M. Wilkes, talk given at the Pioneer Day celebration commemorating the contributions of Howard H. Aitken, National Computer Conference (AFIPS), Anaheim, CA, 1983; quoted in Cohen (1988, 122–3).

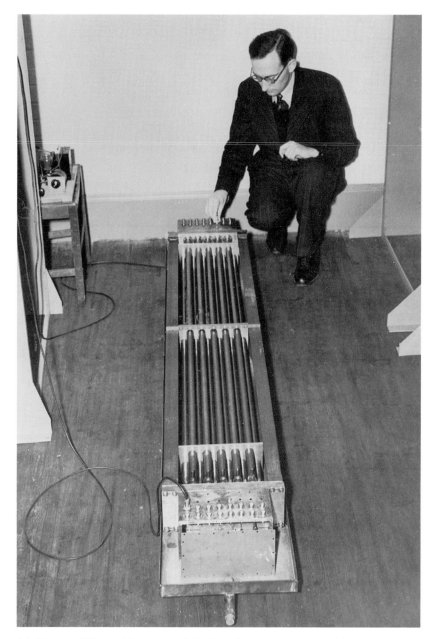

4.2 Maurice Wilkes with mercury delay lines for EDSAC 1. Copyright Computer Laboratory, University of Cambridge.

The EDSAC became the first fully operational stored program computer.[24] In contrast to Alan Turing, who designed the programming system of the computer built at Manchester University, Wilkes from the beginning chose to write programs by using mnemonic codes and decimal addresses which were easier to use. This was in line with the dedication of the machine for users outside the laboratory who would write their own programs.[25]

By the summer of 1950, a year after the computer had started to work, an operating service was set up and the EDSAC became available to researchers in the university. The racks with the vacuum tubes, the mercury storage units (each 5 feet in length), the control monitors and the input and output equipment filled a whole room on the top floor of the Mathematical Laboratory, while the tape punching and editing equipment was installed one floor below. As with all the early electronic computers, it was a 'thermodynamic monster', consuming a vast amount of power which produced a lot of heat (Dickerson 1992, 184). Douglas Hartree, Professor of Mathematical Physics at Cambridge since 1946 and a keen supporter of the EDSAC project, had long since started challenging researchers to think up problems that could be applied to the electronic calculating machines that were being developed (Croarken 1992, 13). In Cambridge, astronomers, nuclear physicists, economists and crystallographers were among the first users (J. M. Wheeler 1992) (Fig. 4.3).[26]

From punched card machines to electronic computers

Huxley was not the first one to complain about the tediousness of crystallographic calculations and in the past crystallographers had not shied away from using new calculation devices and services which had become available to alleviate their task. For the calculations involved in the structure determination of penicillin during the war, a project considered of highest importance and urgency, Dorothy Hodgkin sought the professional help of Leslie J. Comrie and his assistants at the Scientific Computing Service, a commercial firm, the first one of its type in Britain, operating since the

[24] The stress here is on *fully* operational. The very first stored program computer was the small experimental machine built at Manchester University. On the politics and perils of claiming 'first computers' see Agar (1998b). On the history of the EDSAC see also the fiftieth anniversary publication (Robinson and Spärck Jones 1999).

[25] On EDSAC programming see Campbell-Kelly (1992) and D. Wheeler (1992).

[26] Among the very first users of the EDSAC in Cambridge (and the very first geneticist to use a computer) was Ronald A. Fisher, who needed to calculate gene frequencies under special selective conditions, which led to complex equations (J. M. Wheeler 1992, 28). Other early users among the biologists were Huxley and Hodgkin, who calculated nerve action potentials on EDSAC 1 and 2 (Huxley 1972, 64–8).

4.3 Technician Rosemary Hill from the Mathematical Laboratory operating EDSAC 1, Cambridge's first electronic digital computer (*c.* 1950). Copyright Computer Laboratory, University of Cambridge.

mid-1930s (Hodgkin 1949; Croarken 1990, 42–6). Comrie performed the Fourier syntheses on punched card machines of the Hollerith type, adapting a method he had pioneered to calculate the positions of the moon, when still working as Superintendent at the Nautical Almanac Office at Greenwich in the late 1920s. Although it had been suggested at the time that punched card machines could be applied to crystallographic computations, the penicillin calculations were the first to be actually performed using that method (Croarken 1990, 31–2, 44). The costs of the service were covered by the MRC although the size of the bill caused considerable consternation and led to delays in payment.[27]

A few years later, Bragg and Perutz reverted to the same method and to the services of the same company for their three-dimensional Patterson

---

[27] By the time the structure determination of penicillin was completed, the plan to produce penicillin by chemical synthesis had been abandoned. But the fact that only crystallographers had been able to provide the conclusive proof for the structure of the molecule represented an important achievement of the field and raised the recognition of the power of X-ray crystallography among chemists.

analysis of haemoglobin. The calculations involved the summation of about 7000 Fourier terms for each of the 58,621 density points included in the analysis. In the method section of the paper in which the work was published, Perutz presented the 'photographing, indexing, measuring, correcting and correlating' of the data for the calculation as 'a task whose length and tediousness it will be better not to describe' (Perutz 1948–9, 475). The calculation by hand of that amount of data he declared as 'beyond the patience even of a crystallographer'. On the punched card calculating machine used by the computing service it 'did not take more than one person's working time for about four months' (Perutz 1949, 138). Besides the revision of *Chambers' Mathematical Tables*, this was the largest computation carried out by the Scientific Computing Service in the postwar years (Perutz 1948–9; Croarken 1990, 102–6).

Punched card methods, using machines of the International Business Machines Corporation, were developed independently by Pauling's group at Caltech (Shaffer, Schomaker and Pauling 1946). Indeed, before striking a deal with the Scientific Computing Service in Britain, Perutz, 'somewhat hesitatingly', contacted Pauling, enquiring if his haemoglobin calculations could be performed on the IBM machine at Pasadena. Perutz complained that 'although there exists a Hollerith machine in this country . . . insurmountable difficulties have so far prevented its use for our job, and there is little prospect that the situation is going to improve'.[28] Pauling refused the job, considering it unwise to perform a calculation in a 'routine way', but invited Perutz to come to Pasadena where their punched cards and machines would be at his disposal (Fig. 4.4).[29]

In 1946 this was not an easy offer to take up, but contrary to Perutz's prediction, the situation in Britain improved rapidly. Hollerith machines, marketed by the British Tabulating Machine Company, became more widely available and various researchers proposed simplified punched card methods which could be performed 'by any research worker in a reasonable time and without special training' (Cox, Gross and Jeffrey 1947; Hodgson, Clews and Cochran 1949). Beevers–Lipson strips and mechanical adding machines continued to be used well into the 1950s, but, especially for larger Fourier syntheses, punched card methods were seen as a

---

[28] M. Perutz to L. Pauling, 5 August 1946; individual correspondence, file P, Pauling Papers, Oregon State University Library. Unfortunately no records relating to the deal struck a little later by Perutz and the Scientific Computing Service seem to have survived.

[29] L. Pauling to M. Perutz, 29 August 1946, Pauling Papers, individual correspondence, file P, Oregon State University Library.

4.4 Punched card machine used for crystallographic calculations by Linus Pauling's group at the California Institute of Technology (1947). In the background crystal models. Copyright Rockefeller Archive Center, North Tarrytown, NY.

major improvement in terms of speed and accuracy and in terms of strain imposed on the 'computor', i.e. the person carrying out the calculations. At Cambridge the Mathematical Laboratory set up a computing service for X-ray crystallographers, performing Fourier syntheses with punched card equipment on loan from the British Tabulating Machine Company (Wilkes 1985, 141–2). This service, which was free, was used not only by the protein crystallographers, but also by the crystallography group under Taylor at the Cavendish.[30]

With this background in mind, the move from mechanical to electronic calculation aids on the side of crystallographers may seem a 'natural' one, bringing them in line with the newest computional technologies. But a closer look allows us to draw other connections.

[30] Kendrew, together with William Cochran from the Cavendish crystallography group, became involved in setting up the service.

True, once the Fourier program was in place, the crystallographic computations could be performed much faster on the EDSAC than with the more traditional punched card machines. Early users of stored program computers like the EDSAC were all impressed with the 'vast improvement in speed' (Sparks and Trueblood 1983, 228). Calculations took hours instead of months. But there were drawbacks. The EDSAC was an experimental machine and constant tests had to be run to achieve reliable results. To work on it required considerable technical skills. As Kendrew put it, 'if you did your programming correctly everything was checked, double checked, there couldn't be a mistake in the final result, but you could spend an immense time arriving at the final result because the machine would go wrong'.[31] The average time between machine errors on the EDSAC was about half an hour, but as one collaborator of Wilkes' original team noted, 'this is deceptive, as there were "good" and "bad" days and most faults were intermittent' (D. Wheeler 1992, 37).

Crystallographers, who ran the longest programs, usually worked on the machine at night or at weekends. 'If it was working', Kendrew recalled, 'people would bring you relays of sandwiches and beer to keep you going until it broke down, hopefully not before Monday morning'.[32] In the early years, there was no maintenance staff and users needed to know how to test and replace tubes, replace chassis and lift and test diodes (D. Wheeler 1992; Sparks and Trueblood 1983, 228). Input and output on the EDSAC was on punched tapes like those used by post offices for telegrams and this created further problems. Reading was not always reliable and the tapes tended to tear. A postdoctoral fellow working with Kendrew remembers: 'We developed the sloppy habit of editing minor tape glitches with patches of Scotch tape, but this gave the Computing Center fits. The patch eventually would be pushed off the computer tape, and the adhesive would gum up the tape reader' (Dickerson 1992, 184).[33]

When Kendrew started using the EDSAC, not only was the technology still in the experimental stage, there was also no strict necessity to turn to this new technology. The calculations he was doing could in principle still be done the traditional way. Huxley confirmed: 'You were not doing that many Pattersons. So, you know, you could live without it. Because most people did.'[34] As already mentioned, when Kendrew was already using

---

[31] Interview with J. Kendrew, Cambridge, 14 July 1992.
[32] Interview K. Holmes with J. Kendrew, Cambridge, 18 June 1997, LMB Archives.
[33] Dickerson's memories date from the late 1950s, when EDSAC 2 had replaced EDSAC 1. While a big improvement in many respects, the input/output system remained the same in the follow-up model (Wilkes 1992, 53).
[34] Interview with H. Huxley, Cambridge, 4 December 1997.

the EDSAC routinely, Perutz did not 'trust' the electronic computer and continued doing his calculations on the Hollerith machine.[35] Bragg also was not at once convinced of the necessity to move to electronic computing (Crowther 1974, 306). Perutz was, however, positively impressed with Ray Pepinsky's X-Ray Analogue Computer, a machine purposely built to calculate Patterson projections, which displayed the calculated electron density maps on the screen of a cathode ray oscilloscope (Pepinsky 1947) (Fig. 4.5). Perutz used the 'great machine', which moved with Pepinsky from the University of Alabama to Pennsylvania State University, on his first visit to the United States in 1950 in connection with a crystallographic computing conference held in Pepinsky's institution, and again in 1953, to run some preliminary calculations involved in the two-dimensional Patterson analysis of haemoglobin (Bragg and Perutz 1954).[36] Comrie from the Scientific Computing Service, possibly for professional reasons, held that, before building expensive new machines, it was more economical to exploit fully the potential of existing technologies. In his eyes, disillusion with the current technology most often depended on pure ignorance of their capabilities. However, his critique was directed in particular against expensive special-purpose machines that were periodically proposed also for crystallographic Fourier syntheses. He was much less critical of large general-purpose calculators and computers that began to appear after the war and to which the EDSAC also belonged (Coarken 1990, 104–5). In their paper on electronic computing of Fourier syntheses, Bennett and Kendrew reckoned it was too early to decide if general-purpose digital machines like the EDSAC or special-purpose analogue machines like Pepinsky's machine were of greater value in crystallography (Bennett and Kendrew 1952, 116). Digital machines were praised for their accuracy and versatility. But the amount of printing involved in crystallographic computations seemed to warrant at least a purposely designed output system, of the kind used in Pepinsky's machine. Another serious limitation of the

---

[35] Interview with J. Kendrew, Cambridge, 14 July 1992. According to Kendrew, Perutz's mistrust was mainly based on unfamiliarity with the new technologies. Perutz himself confessed to having been 'too busy mounting crystals, taking pictures and overcoming experimental difficulties' to share Kendrew's interest in computing; interview with M. Perutz, 1 July 1998.

[36] M. Perutz to G. R. Pomerat (Rockefeller Foundation), 18 May 1950, folder 566, box 44, series 401, record group 1.1, RAC. For the positive report of another British crystallographer visiting the Pepinsky machine on the same occasion see E. G. Cox to W. Weaver, 24 July 1950, folder 299, box 31, series 401, record group 1.2, RAC. Cox declared himself 'full of admiration' for Pepinsky's machine, but held that more experience on the machine needed to be collected, before British crystallographers should commit themselves to having such a machine. Meanwhile Cox continued to pin his 'faith to punched card methods, which although not perhaps so fast have the merits of flexibility and reliability' (*ibid.*).

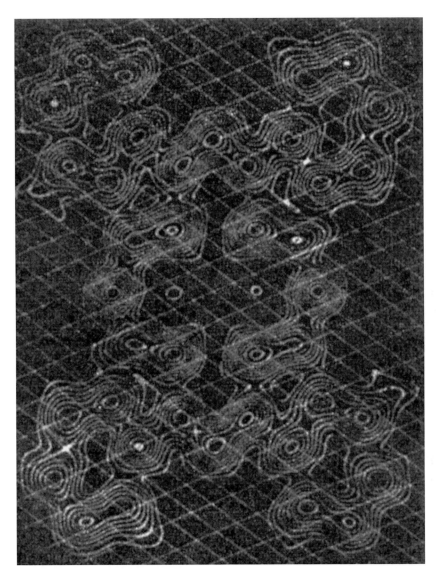

4.5 Display of electron density projection on Pepinsky's X-Ray Analogue Computer.
From J. M. Robertson, *Organic Crystals and Molecules* (1953), p. 107, fig. 49.
Reproduced by permission of the publisher Cornell University Press and of the
Syndics of Cambridge University Library.

EDSAC was its storage capacity. Even the planned doubling of the EDSAC
storage positions from 512 (little over 1 Kbyte) to 1024 was not sufficient
to calculate a three-dimensional synthesis without breaking it into parts,
since such an operation required about 8000 storage positions (Wilson
1951, 67). At a conference on Fourier methods in crystal structure analysis
held in London in 1950, computations on the EDSAC were discussed

alongside a wide range of other approaches and devices, all of which had their merits (Wilson 1951). This is a powerful reminder that in the early 1950s the capacities of digital electronic computers and their suitability for different tasks and different users had still to be established.

As ambitious users, protein crystallographers pushed to overcome the limitations of the machine. Wilkes, who headed the team which designed the EDSAC and its follow-up model, EDSAC 2, has made this point clear. In his memoirs he wrote:

X-ray crystallography illustrates very well the way in which new developments start under the umbrella of computer science, and later come to be viewed as computer applications pure and simple. At the time at which it was done, Kendrew's work was regarded as of great interest to all computer people. In the summer of 1959, when the British Computer Society held a conference in Cambridge, it seemed entirely natural for Kendrew to give an invited address with the title 'Why blood is red'. (Wilkes 1985, 192)[37]

Some of the programs designed by the crystallographers for the calculations of the protein diffraction patterns became so complex that they were used for the morning checks of the machine (Dickerson 1992, 184). Computer people also used the work of protein crystallographers in their grant applications to justify the demand for new machines and more powerful software.[38]

Managing data

What inclined Kendrew to spend so much of his time on a new calculating device which still had to prove its viability, at a point where his research did not really require it? And how can we explain Kendrew's and Perutz's different attitudes to the new 'gadget'?[39] The difference I see here is cultural rather than psychological, and concerns the specific way Kendrew perceived the crystallographic work and the computer. It refers back to his wartime experience in operational research. Kendrew, I would like to propose, from the beginning viewed protein structure analysis as a huge

---

[37] By that time Kendrew had proposed the first low resolution model of myoglobin based on extensive calculations on EDSAC 1 and was working towards a high resolution map of the molecule, making heavy use of EDSAC 2.

[38] Interview with D. Blow, London, 29 September 1992, and with D. Wheeler, Cambridge, September 2000.

[39] The term 'gadget' became common in the Cold War era, especially in the parlance of American physicists who adopted it as a code for the devices they created for the military (Forman 1996). The development of computers, even if not specifically of the EDSAC, was part of the same culture.

data handling problem and approached it in operational terms, that is, assessing the efficiency of the single steps and devices involved as well as of the operation as a whole. This general attitude would also determine his approach to the new electronic computer. In her informed guide to Kendrew's papers, Jeannine Alton confirms:

From the first, and long before the choice of a suitable material had been achieved, Kendrew realised how important a factor would be the rapid handling of very large data and information. It is interesting to see the fascination with note-keeping, filing and organisation present in the schoolboy and fostered by operational research in war finding a kind of bureaucratic apotheosis in the sustained effort of accuracy required for the long haul to the final successful three-dimensional picture. (Alton 1989, 65)[40]

She continues:

Kendrew was certainly ahead of many of his colleagues in seeing the necessity to harness the full range of automatic information handling techniques. This perception shows itself in two lines of enquiry. One, immediately after his return to Cambridge in 1946, is the organisation of his, and later his team's rapid access to all relevant literature via a system (then relatively new) of punched cards... The other is the investigation of computation and data processing, dating from 1949 and aimed in the first instance at the resources of EDSAC 1. (p. 65)

Kendrew was obsessed with effective methods of information storage and retrieval. From his schooldays and until late in his life, everyone who knew him was impressed by his filing systems and his expandable pocket notebook which, in a minute space, stored an extraordinary amount of information. A 'code-book' explained his classification system.[41] In operational research the storage and retrieval of information regarding the performance of devices or the effectiveness of operations was obviously a key issue, and one which very much occupied Kendrew, as clearly transpires from his different reports. As part of his efforts to build up operational research in the Middle East, for instance, he designed a punched card to be used by aircraft patrol crews to record operations in a comprehensive and uniform way such as to make it amenable to mechanical statistical

[40] Considering Kendrew's frequent references to Bernal with respect to organisational questions and information handling systems, especially in 1945/6, I would like to speculate that in their conversations in Ceylon the two operational researchers discussed the problem of protein crystallography in exactly these terms. On Bernal's interest in the indexing, storing and retrieval of information see Bernal (1939a); also de Chadarevian (1997b). On Kendrew's systematic search for suitable crystals, which took seven years to complete, see below.

[41] Addresses by A. Klug and H. Bondi at the Memorial Meeting in Honour of Sir John Kendrew, Cambridge, 5 November 1997, and interview with J. Kendrew, Cambridge, 15 March 1994.

analysis. Towards the end of the war, while still in government service, he proposed a 'Project for the compilation of a comprehensive manual of bombing data for the use of OR [operational research] personnel', which he sent to the responsible British and American divisions, pointing out its potential use in the 'Japanese War'. The publication he envisaged was to include a collection of all available data and a series of 'computers [i.e. statistical or mathematical devices] designed to solve all main types of problems usually encountered'.[42] Around the same time, he developed detailed suggestions regarding the national organisation of science, a centrepiece of which was the 'application of mechanical indexing systems to the Central Register with expanded information'. According to Kendrew, the effective use of scientific resources hinged on an 'effective Information Service', the organisation of which was 'a matter of mere mechanics and efficient filing systems, but to be effective it must be comprehensive'. The solution suggested involved the introduction of mechanical punched card systems (including Hollerith machines for the sorting of information) as employed in operational research. Kendrew explicitly stressed the 'close analogies' of the problems regarding the organisation of scientific manpower with those 'studied in OR sections'.[43]

Of course Kendrew was not alone in transposing operational research methods, developed for the analysis of weapon or defence systems and military operations, to other government concerns. However, when leaving government service behind, he applied the same methods to the organisation of his own scientific research. As Alton again puts it: 'The essential requirement of detailed and accurate data for the compilation of official OR reports...fostered, if it did not inspire, a continuing interest in the indexing and organisation of information' (Alton 1989, 50). Entering the field of protein crystallography, a 'comprehensive indexing system' appeared to him the only possible way to 'keep abreast of the vast literature of most of biochemistry and much of physiology, chemistry and physics' which pertained to his newly chosen field.[44] He corresponded widely to

[42] J. C. Kendrew, 'Notes on a project for the compilation of a comprehensive manual of bombing data for the use of operational research personnel', 1 August 1945, Kendrew Papers, MS Eng. c.2394, B.48, Bodl. Lib.

[43] J. C. Kendrew, 'Notes on the national organization of Science', August 1945, and 'On the application of mechanical indexing systems to the Central Register' [September 1945], Kendrew Papers, MS Eng. c.2394, B.50 and B.52, Bodl. Lib.

[44] J. Kendrew to Y. Chu-Pei (Unesco), 13 Januray 1947, Kendrew Papers, MS Eng. c.2395, C.2, Bodl. Lib. See also J. Kendrew to J. W. Perry (American Chemical Society), 24 August 1948, Kendrew Papers, MS Eng. c.2395, C.2, Bodl. Lib. Also in that file is a copy of a letter by J. von Neumann to E. Crowan (Cavendish Laboratory), 3 December 1946, giving details of Vannevar Bush's 'Memex' system for the electronic storage and retrieval of information and the development of electronic computers for calculations.

identify the best system, especially regarding the classification and coding of information, and finally devised his own punched cards and punch, settling for a needle sorting system. The system, which he later put at the disposal of his research team, remained in use for many years.

Kendrew's obsession with the organisation of data also explains his interest in computer programming and in devising ways in which data could be efficiently stored and displayed in the EDSAC. Indeed, Kendrew may well have viewed EDSAC not just as a fast calculating machine but as a system to handle and retrieve large amounts of data. The technology involved in EDSAC will have been familiar to Kendrew, who shared a brief spell in radar research with the new brand of electronic engineers engaged in constructing and improving the computer at Cambridge. The transposition of punched card methods for the calculation of Fourier syntheses to computer code was relatively straightforward. However, the problem which most occupied Kendrew and Bennett when writing their first Fourier synthesis program was the limited storage capacity of the EDSAC. The challenge lay in devising ways to economise storage space and to make effective use of it in the course of the computation (Wilson 1951; Bennett and Kendrew 1952). These questions corresponded well with Kendrew's interests. Another problem, to which Kendrew devoted much attention, was the selective retrieval and display of data. The EDSAC printed the calculated data (representing electron densities) in a single-digit code and arranged in a table that reflected the correct proportions of the crystal unit cell (Fig. 4.6). The computer print out could then be used directly to draw contours connecting points of equal densities and for model building (see chapter 5). The elaborated use of a magnetic 'backing store', a new feature of EDSAC 2, the machine following EDSAC 1, made it possible selectively to extract and display information from the overall calculation (J. M. Wheeler 1992, 32).

Assessing efficiency

Kendrew's approach to the EDSAC was aimed at assessing its efficiency for X-ray crystallography, very much as he had evaluated the efficiency of new gadgets or military operations during the war, even if the tools used for assessment were different and, especially, statistical methods were not involved. The efficiency of any computational aid in crystallography was generally expressed in the time it (or the person operating it) needed to perform certain tasks. In addition to speed, accuracy was an important criterion (though different crystallographic problems required a different level of accuracy). In evaluating the EDSAC in comparison to Pepinsky's

```
++942343210    00              0000    11
97301233221001221 0
100012100110001 10
111110                00      0
            00000100000          00   00
320   01110                00110 012
431   13321    00000       0000   023
11    1232100111 22210            00
331000110000000000
110       00110000
            00110000
    00000        00000000
00      0     00000001111 00
10    0110000000111111001110   000
11    01100 0000011110   010    00
0     0           00110

 000      0111000          00
1110      01111100     0   001100000
1110000000  000000110    011000112
000           000    00   022210 011
        00           01221000011
                        000100
                00      0111
                1110   01110000
110000   000000  011100112100011
.10000   0110000   00    1233210011
    0     010    000  01221100001
    00           110   011100
        0110   010     01110
        011    01100   01110011
    00  0000   00222111101222112 3
      0001110  013432112112343112 3
```

4.6 Patterson projection of electron density for whale myoglobin printed on EDSAC 1 (c. 1951). From J. M. Wheeler, *IEEE Annals of the History of Computing* 14, no. 4 (1992), 31, fig. 1. Copyright 1992 by the Institute of Electrical and Electronics Engineers, Inc.

machine its versatility played a role, while costs and availability were less of a concern for Kendrew, who had free access to the machine.

This interest in new devices and in more efficient (coupled to accurate) ways of doing things characterised Kendrew's approach to research more generally. A few examples illustrate the point.

At the time Kendrew entered X-ray crystallography, diffraction patterns were recorded photographically and the blackening of the film, a measure of the intensities of the diffraction spots, was estimated by comparison with a reference scale. A participant writes: 'This "eye estimation" needed enormous patience and experience; the tens of thousands of intensities used in the interpretation of the diffraction spots produced by crystals of haemoglobin, until the mid-fifties were all estimated by Max Perutz himself who would not trust any one else's measurements.'[45] When, on a visit to Maurice Wilkins at King's College London, Kendrew saw Peter Walker, a cell biologist, using a self-built densitometer to measure the density of cell sections, he realised that the same instrument, without any alteration, could be used to measure diffraction pictures. While the peaks of the trace which the instrument produced still had to be measured by hand, it none the less meant a big improvement, in terms of both speed and accuracy, to the traditional method of measuring reflections. Echoing the view just cited, Kendrew recalls: 'I started using that again as a matter of routine long before Max. I mean, Max again didn't trust it.'[46] Kendrew arranged to use the machine after working hours, and for some time travelled to London every evening to measure his pictures before taking the last train back to Cambridge. Later, the machine was commercially produced by Joyce Loebl of Newcastle, and the Cavendish group got its own instrument. It became the standard equipment to collect data until automated diffractometers came into use in the 1960s.

By the end of the 1950s, when both the myoglobin and the haemoglobin work were gearing up and tens of thousands of reflections had to be measured, women were recruited to do the work. An American visitor recalls: 'At one end of the long building was the densitometer room, where several young women spent their days feeding precession films through a Joyce-Loebl double-beam microdensitometer and measuring peak heights by hand with a millimeter scale' (Dickerson 1992, 183). Most of the women were untrained and worked on short-term contracts. They were hired straight away from schools or through personal networks. Kendrew introduced a routine to secure that the work was done accurately as well as efficiently.

At weekends, Kendrew took the pack of new data home and checked the accuracy of the measurements according to symmetry considerations which the pictures displayed. He explained: 'You really had to be rather tough with them [meaning "the girls"], because you had to set a limit.

---

[45] Arndt, 'The development of X-ray crystallographic techniques in molecular biology' [unpublished manuscript, *c.* 1987], pp. 2–3. Arndt, crystallographer-engineer first at the Royal Institution and then at the Laboratory of Molecular Biology at Cambridge, was later crucially involved in the development of a series of automated and computer-controlled devices for the measurement of diffracted X-ray intensities (see also below, pp. 281–2).

[46] Interview with J. Kendrew, Cambridge, 14 July 1992.

4.7 'Computer girls' doing the calculations following the densitometer readings (*c.* 1957). Courtesy of Medical Research Council Laboratory of Molecular Biology, Cambridge.

[If there was] more than 5 per cent error we had to throw the whole thing away and start again.' If the error repeated itself, the 'girl' was told she was 'no good' and was replaced.[47] One of the women, Mary Pinkerton, the sister of John Pinkerton who was in computing, took on the role of 'organising' the group doing the measurements (Fig. 4.7).

Kendrew's early interest in and assessment of protein sequencing as a rival approach to protein structure determination further illustrates Kendrew's way of going about research. As early as 1955 – that is, in the year Sanger first published the complete sequence of insulin – Kendrew 'searched the world' for someone who would undertake a sequence analysis of myoglobin. At that point in time, protein sequencing seemed to have won the upper hand over X-ray crystallography. Knowledge of the sequence was believed to be all that was needed to predict the three-dimensional structure of molecules. If this proved true, protein crystallography would become obsolete (Kendrew 1964, 696). Characteristically,

---

[47] Interviews with J. Kendrew, Cambridge, 14 July 1992, and Linton, 18 March 1993. Despite this rather strict regime and the dullness of the job, the 'computors' were seen as belonging to the 'inner sanctum' of the work and ranked higher than other assistant personnel in the laboratory; interview with Annette Lenton (née Snazle), Cambridge, 17 November 1999.

Kendrew took up the challenge and invited the collaboration of protein sequencers. Eventually, Allen Edmundson, a research student of William Stein and Stanford Moore at the Rockefeller Institute of New York, started work on myoglobin and joined the Cavendish group, followed by his automatic amino acid analyser which was shipped over from the United States. In the event, the sequence analysis presented unexpected difficulties, while the high-resolution map of myoglobin yielded more information than anticipated. But the sequence data still provided crucial confirmation for the atomic structure of myoglobin. Protein crystallographers became the biggest clients for protein sequences and all the first structure analyses relied on sequence information (Kendrew 1961; de Chadarevian 1996a).

It was Kendrew's constant aim for utmost efficiency regarding the tools, including especially those for data handling, and the organisation of the work that contributed in decisive ways to the successful completion of his pioneering work in the structure determination of myoglobin.

Setting the limits

EDSAC may not have been indispensable for the crystallographic calculations Kendrew was performing around 1950 when he first got interested in the new device, but increasingly it became so. As mentioned at the beginning of this chapter, Cambridge crystallographers always agreed that their later work crucially depended on the availability of digital computers. By the mid-1950s Perutz also came around to making use of the facilities at the Mathematical Laboratory, even if he never used the machine himself but rather left this part of the work to the younger people in his group. As he himself admitted, he was 'always hopeless at computing'.[48] Despite the increasing dependence on the electronic computer, however, the decisive breakthrough in their attempts to decipher protein diffraction pictures is generally seen to lie in Perutz's demonstration in 1953 that the isomorphous replacement method could indeed be applied to the structural analysis of crystalline proteins.

As Bragg has remarked, there was 'nothing original' in the idea of using the method for protein analysis. Bernal had pointed to this possibility many years before. Other groups had experimented with the method, providing

---

[48] Interview with M. Perutz, Cambridge, 1 July 1998. David Green, a graduate student, was the first of Perutz's people to make use of the EDSAC. Later Michael Rossmann and David Blow, two postdoctoral members of staff, both working with Perutz, became the 'computer specialists' in the laboratory (Dickerson 1992, 184). They helped with the calculations, run on the EDSAC, involved in the three-dimensional Fourier synthesis of haemoglobin which allowed the first structural model of the molecule to be built (Cullis *et al.* 1962).

important insights and tools (see above, and Hauptman 1998, 136–7). Perutz's achievement consisted in making the technique work, first by finding a heavy metal compound which could be attached to a specific site without altering the arrangement of the other atoms in the molecule, and second by estimating with sufficient accuracy the overall changes in intensity produced by the heavy atoms. According to Bragg, in this last respect Perutz's skill was 'at that time probably unique' (Bragg 1965, 12).

The isomorphous replacement method was eventually successfully applied to deducing the atomic structures first of myoglobin and then of an increasing number of other proteins, including haemoglobin. To this day, it is considered the key method to determine the crystal structure of proteins. At the time, however, it also led to disillusionment. The two-dimensional projection of the haemoglobin molecule, which Perutz calculated with the newly gained information from the replacement method, told 'disappointingly little' about the structure of the protein molecule (Bragg 1965, 14). 'The internal structure of the molecule', Perutz concluded the last of a series of six papers on his results, 'is still obscure' (Bragg and Perutz 1954, 326). Even moving to higher resolution proved not enough. It became clear that a three-dimensional analysis would be necessary. This, however, had drastic implications regarding the work involved: several heavy atoms replacements were necessary, many more measurements of intensities were required and the calculations became intractable. Without the availability of electronic computers and the experience acquired in running crystallographic programs on the EDSAC such an enterprise, the success of which remained uncertain, could not have been contemplated (Pepinsky's machine was not accurate enough for these calculations).

The first structure of a crystalline protein to be solved was myoglobin, to which Kendrew had shifted his attention in the early 1950s. At the beginning, the project seemed less promising, since it proved extremely difficult to grow crystals of sufficient size.[49] However, once Kendrew had found that the myoglobin of sperm whale meat, a stock of which was stored in the Low Temperature Research Station at Cambridge, grew beautiful crystals, the smaller size of the molecule, one quarter that of haemoglobin, provided a big advantage.[50] In addition, Kendrew and his collaborators,

---

[49] Kendrew's first attempts to crystallise myoglobin went back to 1946, when he had just joined the unit; J. Kendrew to L. Pauling, 10 December 1946, correspondence files, Pauling Papers, Oregon State University Library.

[50] During the war, whale meat was investigated as supplement to butchers' meat to meet the shortage of supply. Later batches of sperm whale meat were sent to Kendrew by air from Lima, where the meat could be bought on the market; interview with J. Kendrew, Linton, 18 March 1993. Before settling on whale myoglobin, Kendrew had grown myoglobin crystals from such diverse animals as horse, porpoise, seal, dolphin, penguin, tortoise and carp.

after much toil, managed to attach heavy atoms at five distinct sites of the molecule.

Before embarking on the calculation of the structure, the resolution to be aimed at had to be established. Kendrew decided to start with a resolution of 6 Ångstrøm which he expected to be sufficient to show the general layout of the molecule. From there the group jumped to a resolution of 2 and then 1.4 Ångstrøm. This meant moving from an analysis of around 400 reflections to one including 10,000 and then 25,000 reflections per heavy atom derivative.[51] Several reasons suggested such a stepwise approach. But it is certain that the calculations for the 2 Ångstrøm resolution required higher storage capacity than EDSAC 1 commanded. It would have been impossible to handle the calculations on that machine. EDSAC 2, with an expanded store, came into operation early in 1958, just in time for the new run of calculations (Fig. 4.8 a–e). In conversation, Kendrew put it in clear terms: 'Without that computer we would certainly have been simply unable, I mean, the amount of calculation by hand, it would have been impossible, even if you had money to hire twenty, fifty people.'[52] The problem was one not just of (wo)manpower (most if not all computors were women) and time, but also of accuracy. The work of data collection was shared between the Cambridge Laboratory and collaborators in Bragg's group at the Royal Institution which, until the early 1960s, was functioning as a kind of 'London office' of the Cambridge crystallography group.[53] Despite the introduction of an automated diffractometer, which decisively speeded up the process, data collection took three years to complete and was another time-limiting factor. The automated diffractometer was developed by Uli Arndt and David Phillips at the Royal Institution and was first applied to the myoglobin work.[54]

Computers became increasingly integrated into the various steps involved in crystal structure determination. By the mid-1960s, data were collected in a form that could be directly fed into a computer.[55] Computers

[51] By 1962, Kendrew and his collaborators had examined 110 myoglobin derivatives and measured the intensities of about 250,000 X-ray reflections (Hägg 1964, 651).

[52] Interview with J. Kendrew, Linton, 18 March 1993.

[53] U. Arndt, letter to the author, 15 October 1999. Both Kendrew and Perutz had adjunct positions as readers at the Royal Institution.

[54] The calculations for the further refinement of the electron density map required still larger computers. Kendrew arranged for these calculation to be performed on the most powerful machines available at the Atomic Weapons Research Establishment at Aldermaston (see below). On the method of successive refinement used to calculate the improved map see Kendrew (1964); for a partial publication of the data see Watson (1969).

[55] The new apparatus which automatically recorded the intensities of successive reflections on punched tape was designed by Arndt. Originally developed for measurements in nuclear physics, it was later adapted for use in X-ray crystallography (see p. 282, n. 30).

(a)

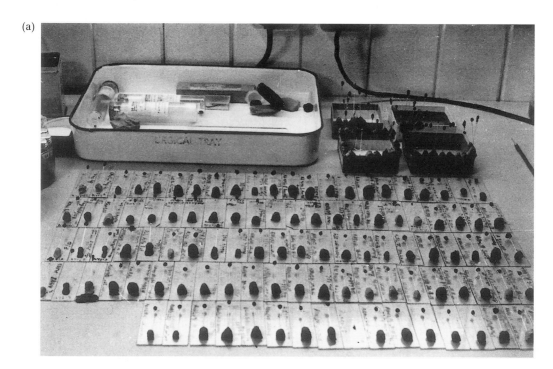

(b)

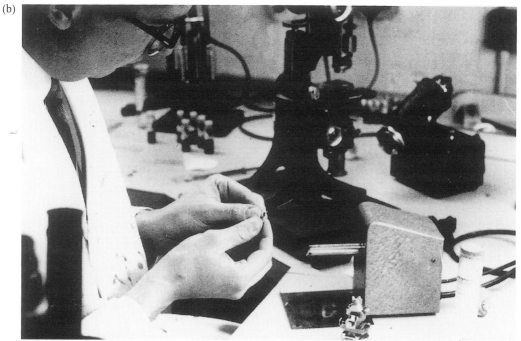

4.8 Data collection and data processing for the X-ray crystal analysis of myoglobin at 2 Ångstrøm resolution. (a) Some of the crystals of native myoglobin and of heavy metal derivatives used for the analysis. (b) Richard Dickerson preparing a myoglobin crystal for X-ray analysis by inserting it with its crystallisation liquid in a glass capillary.

(c)

(d)

4.8 *(Cont.)* (c) Danish postdoctoral researcher Bror Strandberg adjusting a crystal on one of the X-ray cameras. The photograph shows two rotating anode generators and four precession cameras. Basement, Cavendish Laboratory. (d) X-ray photograph of a myoglobin crystal (late 1950s).

4.8 *(Cont.)* (e) The hand-sorted data of the myoglobin calculations on EDSAC 2 are carried over from the Mathematical Laboratory to the MRC unit. Photographs (a)–(c) and (e) by Bror Strandberg; photograph (d) by John Kendrew. Courtesy of Bror Strandberg and Medical Research Council Laboratory of Molecular Biology, Cambridge.

also played an increasing role in the presentation of results (Fig. 4.9). By that time the Cambridge crystallographers, now housed in the new Laboratory of Molecular Biology (LMB) 2 miles away from the computing facilities in the centre of Cambridge, had their own in-house computer, while still using outside computing for large calculations (Gossling and Mallett 1968). With their needs soon outstripping the computing capacities of the Mathematical Laboratory, every day for many years tens of thousands of cards were put into shop trolleys and taken on the train to London to be processed, first at IBM and later at Imperial College.[56] It was

---

[56] Interview with R. Diamond, Cambridge, 1 July 1992, and D. Blow, London, 29 September 1992.

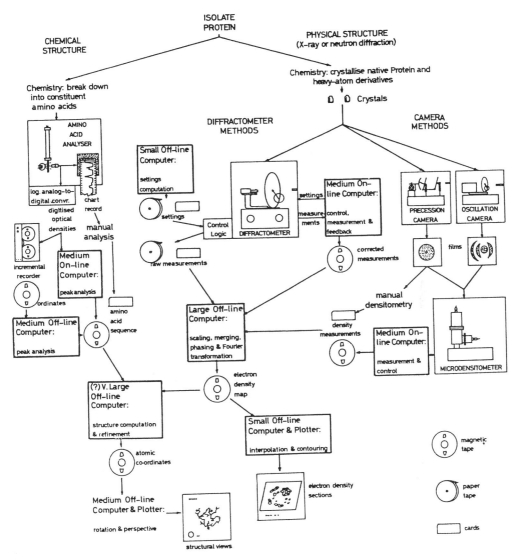

4.9 General diagram of the stages involved in solving a protein structure. Note the multiple ways computers are employed in the process. From T. H. Gossling and F. W. Mallett, *Fall Joint Computer Conference* (1968), p. 1090. Reproduced by permission of the authors.

not until 1980 that the LMB installed its own main frame computer, a VAX 11/780. Only then were the machines considered sufficiently manageable and reliable. And only then were the commercially available computers powerful enough to satisfy the needs of protein crystallographers for the foreseeable future. By that time the use of computers had also expanded from crystallography to other groups working in the laboratory, where they were employed for an increasing number of tasks ranging from picture storage and retrieval to sequence matching.[57]

Today, computer terminals are part and parcel of virtually every biological work bench. But even before computers became pervasive, computers and their programs were widely used to conceptualise the functions of cells, of genes in cells and finally life itself. This was part of a larger import of computational, informational, cryptological and cybernetical terms and technologies into biology, in many cases pioneered by the wartime inventors of these technologies themselves (Kay 1995; 2000; Fox Keller 1995b).[58] Yoxen has pointed to the connection between a biology which conceived of life in terms of a program and the managerial research system which became dominant, especially in the United States, in the decades following World War II (Yoxen 1981; 1982).

A new pace of research

This chapter has linked Kendrew's pioneering use of computers for crystallographic computations to his wartime experience with radar and operational research. These occupations made him familiar with technologies later deployed in electronic computers and fostered his interest in efficient and accurate methods of data handling and in the evaluation of new gadgets. Kendrew, I have argued, viewed the whole myoglobin project as a huge data operation problem. In the later stages the success of the project undoubtedly relied on effective methods of data collecting and handling and careful checks of accuracy. The EDSAC was crucial. Together with the powerful X-ray equipment built in the Cavendish workshop and the skills available in the lab, it contributed decisively to the making of Cambridge into a world centre for protein crystallography in the late 1950s.[59] An American postdoctoral fellow who spent time in Leeds

---

[57] On the use of computers and their programmes as technological and conceptual tools in Brenner's worm project, see de Chadarevian (1998b).

[58] On the wartime origins and the inscription of military logic into cybernetics see Galison (1994). On the connection of war and information and communication systems see also Kittler (1993).

[59] Of special importance for the work in the unit was a rotating anode X-ray tube, built by the in-house engineer Anthony Broad, improving on an earlier design by Abraham Taylor from Birmingham. The tube provided a beam ten times stronger than those commercially available at the time. Together with the precession cameras bought with Rockefeller Foundation funds in the United States, it made the unit better equipped than any other group in the field (Perutz 1996). Equipment and skills available in the laboratory attracted many visitors, especially from the mid- to late 1950s onwards. Among the first visitors were James Watson, who came as a postdoctoral researcher to learn X-ray crystallography; Peter Pauling from Caltech, joining as a Ph.D. student; Alexander Rich from Caltech, collaborating with Crick on the structure of collagen; and Anthony Allison from Oxford, who continued his studies on sickle cell haemoglobin.

before joining Kendrew's group in Cambridge in 1957 has given a graphic description of the advantages Cambridge offered at the time: 'Leeds University had a decent inorganic X-ray structure group...An electronic computer was available at Manchester University, just over the Pennine Hills, and one could always go there to calculate a three-dimensional Patterson map from a new data set. After that you were on your own at Leeds, with Beevers–Lipson strips and a desk calculator. Understandably we worked in projections...Cambridge, unlike Leeds, actually had a computer of its own: EDSAC 2, a marvel with 2000 words of fast access core storage, plus magnetic drum and tape' (Dickerson 1992, 182–3). Kendrew left no doubt: 'Supposing the Cambridge computer had not existed, until ten years later say. Well, we would not have got the result until ten years later. It would have been impossible without it.'[60]

Once computers became available and crystallographers started using them, computers (like accelerators in physics or electron microscopes in cell biology) increasingly informed the practice of the field. The tasks that could be tackled, as well as the pace of research, changed. Their use required a new set of skills and the interaction with a new brand of electronic engineers, computer operators and programmers. It also made protein crystallography dependent on expensive machines. Yet after the war, money was available for computer development as well as for biophysical research.

Postwar electronic computers built on war-related technological projects. Their further development remained bound to military interests. The construction of the EDSAC was not military funded, but the Ministry of Supply (the ministry responsible for military R&D) made no secret of its interest in the Cambridge developments and, for instance, took responsibility for printing the proceedings of a conference on electronic computers held at Cambridge in 1947 (Wilkes 1985, 144).[61] It also provided funds for the development of the first 'baby' version of the Manchester electronic computer into a full-size computer, engineered by Ferranti. In America the situation was similar. Pepinsky's X-ray Analogue Computer, for instance, was supported by the Office of Naval Research. Once the EDSAC was in operation, staff from the Atomic Weapons Research Establishment and the Telecommunications Research Establishment were sent to

[60] Interview with J. Kendrew, Linton, 18 March 1993.
[61] Funds for the development of EDSAC came from the research budget of the Mathematical Laboratory, provided by the University Grants Committee, from the Department of Scientific and Industrial Research and from J. Lyons and Company, which also provided an assistant for a year in exchange for advice on building its own computer for its catering business (see below). EDSAC 2 received funds from the Nuffield Foundation.

Cambridge to learn the know-how connected with the new electronic machines (Croarken 1990, 119). Soon the military commanded the most powerful computers in Britain, as in the United States. In 1959, Kendrew double-checked the 2 Ångstrøm Fourier synthesis of myoglobin calculated on EDSAC 2 on a 'defence computer' (Dickerson 1992, 185). Some of the calculations for the 1.4 Ångstrøm map of myoglobin were also done at the Atomic Weapons Research Establishment at Aldermaston.[62] Using these machines, protein crystallographers took advantage of the products of Cold War rearmament.[63] Computers also entered and radically changed many other sectors of society. The EDSAC, for instance, became the prototype for LEO (Lyons Electronic Office), a machine first built by the large London-based catering company to raise efficiency in their office work. The success was such that the machine was soon commercially produced and widely applied.

Of course there was no guarantee that the use of expensive and powerful machines would produce results. Nor was their use sufficient. Advances in crystallographic techniques were as decisive. But when results were achieved, as in the case of protein crystallography, the power of the new calculating devices added to the aura of what had been accomplished. The amount of data processed for the atomic resolution of the structure of myoglobin and haemoglobin was unprecedented, as Kendrew and Perutz never failed to underline. With conventional computing methods the task could not have been carried out. This made the achievements still more sensational.

Stressing the wartime and military connection of computer development, which was so crucial to protein crystallography, my aim has been not to redeem the military involvement of scientists and their role in the production of war-related technologies with the fruits of their peacetime applications, but rather to draw out the legacies of the scientific mobilisation for the postwar history of protein crystallography. Together with new political, economic and institutional opportunities for biophysical investigations and the influx of people with a new range of experiences and a new approach to research, computers decisively contributed to the postwar development of protein crystallography and the achievement of results

---

[62] The use of the Aldermaston computer, regulated by tight security and time restrictions, was not a long-term solution and Kendrew and his colleagues arranged to use other facilities (first at the IBM Data Center, then at Imperial College in London) as soon as these became available. For relevant correspondence see Kendrew Papers, MS. Eng. b. 2017, C.208 and C.209, Bodl. Lib.

[63] A striking example of this connection is represented by the interactive computer graphics systems, developed as straightforward defence project and early on adapted to displaying protein structures; see Langridge and Glusker (1987) and Francoeur and Segal (forthcoming).

which were later seen as landmarks of molecular biology. The same machines also created new dependencies and increasingly changed patterns and pace of research.

The most visible products of the work of protein crystallographers, those which colleagues came to inspect and all visitors were shown, which protein crystallographers carried along to lecture halls, which toured exhibitions, appeared on television and eventually ended up in museums, were the models, variously built of plasticine, meccano bits and steel clamps or colourful plastic balls. It is with these models and their role in the postwar advertisement of a new science of life that I shall conclude my discussion of postwar protein crystallography, the key technology of the Cambridge biophysics group well into the 1950s and beyond.

5

...

Televisual language

The double helical model of DNA built by Watson and Crick in the spring of 1953 is undoubtedly the best-known structure proposed by the Cambridge biophysicists. But the early protein models of myo- and haemo-globin are also familiar. While some of the original small-scale models are still in private hands, others, and especially the first atomic models, fragile and bulky constructions consisting of a maze of rods several feet long and brass building parts, are on show in the *Living Molecules* exhibition at the Science Museum in London (Fig. 5.1).[1] Replicas of the low resolu-tion model of haemoglobin, built of white and black plastic discs stacked on to each other, are part of many demonstration collections, and nearly every biological and biomedical textbook carries pictures of the model. The original of these models, protected by a perspex case and positioned in front of two gold-framed portraits of the benefactors of the building, until very recently decorated the otherwise unpretentious entrance hall of the Laboratory of Molecular Biology in Cambridge (Fig. 5.2). Its strategic position, forming an obligatory point of passage for every visitor entering the building, reflects the important role attributed to the model in the his-tory of the institution (see Part II). More models, among them a gigantic myoglobin model, a relic from the 1958 Brussels World Fair, are exhibited on the landing of the staircase in front of a large glass wall which renders them visible also from outside the building.

Despite their high visibility, models have been neglected by a historio-graphy strongly biased towards scientific theories and text. Even the more recent interest in scientific representations has focused on flat inscrip-tions and images (Lynch and Woolgar 1990). Protein crystallographers, however, have always insisted that the results of their work, embodied

---

[1] Since this chapter was written, the exhibition, set up in 1987 to celebrate forty years of molecular research in Cambridge, has been dismantled. Originally planned to stay on display for a few months, it remained open to the museum-visiting public for over twelve years. Most of the models, some of which still on loan, are now integrated in other displays in the museum.

5.1 Living Molecules exhibition at the Science Museum in London. Courtesy of Medical Research Council Laboratory of Molecular Biology, Cambridge.

in three-dimensional models, were hard to convey in words or pictures. Building and investigating physical models was as crucial to the experimental process of structure determination as to the appreciation of the proposed structures themselves. In addition, models played an important role outside the laboratory, in the lecture hall as well as in other public arenas, including exhibitions, World Fairs, museums and television, where three-dimensional visual displays not only were possible but had their unique place. Examining the models, then, allows us to combine a study of experimental practice with an analysis of the public presentation and image of that science. Applied more specifically to our concerns here, it offers a powerful avenue to understanding how protein crystallographers participated in the postwar scientific culture. If in the previous chapter I stressed the importance of technological developments built on the legacies of World War II, following the models will show how protein crystallographers used highly charged public events (like the Brussels

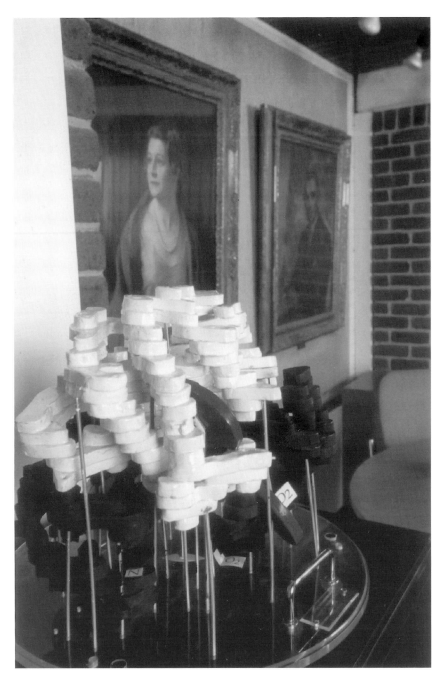

5.2 Max Perutz's first model of haemoglobin (normally protected by a perspex case) in front of the portraits of Lady Wadia and her husband, the benefactors of the laboratory, in the entrance hall of the LMB. The area is currently under reconstruction. Photograph by Lesley McKane (2000). Courtesy of Medical Research Council Laboratory of Molecular Biology, Cambridge.

World Fair) and new visual media (like television) to further their projects. Television emerged as a medium of mass communication in the postwar era. One of the early science series featured a programme on the protein models. With their models, crystallographers developed a 'televisual language' which was most effective in attracting interest and mobilising public support for their science (Yoxen 1978, 237).

I have mentioned before (see chapter 2) that crystal structure diagrams (including Patterson maps of myoglobin, haemoglobin and insulin), faithfully copied from laboratory notebooks and reproduced on textiles, wallpaper, tapestry, cutlery and tableware, were used by the Festival Pattern Group in 1951 to launch a new design for living. As I will argue in this chapter, the three-dimensional models of the protein molecules offered distinct and equally powerful ways to promote a peaceful image of science and a new understanding of life based on the knowledge of its fundamental structures. Erwin Chargaff, émigré biochemist trained before the war, who had scorned the biophysicists' reduction of life to a 'two dimensional pattern on wallpaper', has also ridiculed their practice of model building.[2] Taking the role of the 'old chemist' in a mock debate with a young biologist situated in 1961, he scarcastically remarked: 'We are then already in the middle of model building, the favorite occupation of modern biophysics. It is all done in front of mirrors, with wire and plastic, glue and papier-mâché; the knowledge of a child combined with the naïvité of the grown-up' (Chargaff 1963, 176–7). But in singling out model building as target for his pointed remarks, he himself acknowledged the impact of models in promoting a new understanding of life.

In this chapter my focus will be on the protein molecules (I will say more on the DNA model in a later chapter). I will first investigate the place of model building in the experimental practice of protein crystallographers and briefly discuss the ways in which researchers tried to overcome the limitations of the flat pages of scientific journals to publish their results. I will then deal with the uses and display of the protein models in public arenas and their role in the postwar advertisement of science, focusing especially (but not exclusively) on television. Finally, I will touch on the role of models in the making of molecular biology, a topic to be pursued in the following chapters.

Models as research tools

Model building has always been an integral part of crystal structure determination. In the trial and error method commonly used by X-ray

<hr />

[2] Quoted in Olby (undated, p. 6). On Chargaff see Abir-Am (1980).

crystallographers, probable structures were built and then tested against experimental data (Watson and Crick also proceeded this way). The protein models I will be discussing were built by 'direct' structure determination, that is by using exclusively crystallographic data (see chapter 4). Here the models were the last of a series of transformations involved in the interpretation of the highly complex diffraction pictures. Only by building models (first of the low, then of the high resolution analysis of the molecule) could the structure be viewed, the amino acids making out the structure be identified (at least in part) and the coordinates of the atoms be determined. The models could then be used for refinements of the structure. Even in this case, then, model building was part and parcel of the research process.

While crystallographers took over the practice of model building from chemists, who had used it since the 1860s, they had to be inventive in the way they applied it to their needs.[3] Modelling techniques from a whole range of disciplines besides chemistry, as well as from outside the sciences, provided clues, while building materials derived from most diverse sources.

The first step in model building consisted in calculating the distribution of electron density in the molecule. I have described the immense amount of measurement and calculation involved in this first step and the tools which made this work possible in the preceding chapter. The calculated densities were represented in electron density maps (also called contour maps) on which points of the same density were joined by a contour, much as altitudes are marked on topographical maps (Fig. 5.3). This technique was first introduced by Lawrence Bragg in his investigation of the structure of diopside, a complex mineral crystal, in 1929 (Bragg 1929). To capture the three-dimensional distribution of electron densities, sections were laid through the molecule and contour maps were plotted on perspex sheets which could be superimposed and illuminated. The series of sections through the molecule could be seen 'rather like a set of microtome sections through a tissue, only on a thousand times smaller scale' (Perutz 1964, 660) (Fig. 5.4).[4] But in three-dimensional contour maps the configuration of the protein molecules could not easily be made out. The structure needed to be 'pulled out'. Model building was intended to do exactly that. Again, the way to proceed had to be invented.

---

[3] On the introduction of models in nineteenth-century chemistry see Meinel (forthcoming).

[4] This is only one of many references by protein crystallographers to anatomical practices. On modelling techniques based on microtome sectioning in nineteenth-century embryology see Hopwood (1999). Generally, protein crystallographers did not refer to the three-dimensional contour maps as models, but reserved this term for the physical models based on these maps.

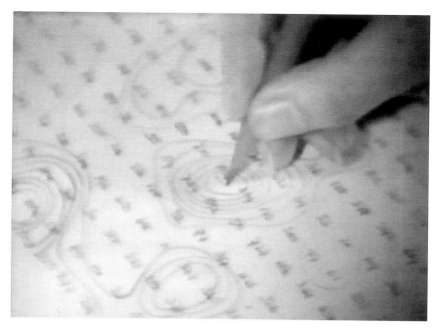

5.3 Drawing contours on electron density maps (*c.* 1957). Courtesy of Medical Research Council Laboratory of Molecular Biology, Cambridge.

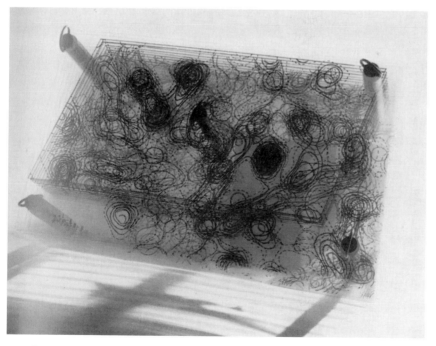

5.4 Three-dimensional low resolution electron density map of myoglobin. Courtesy of Medical Research Council Laboratory of Molecular Biology, Cambridge.

5.5 'Sausage model' of myoglobin (1957). *Source:* Science Museum/Science & Society Picture Library, slide no. SCM/PHY/C100369A.

Kendrew built his first model of myoglobin, also known as the 'sausage model', out of plasticine. Looking at his model, the first of a globular protein ever to be built, Kendrew expressed surprise at the unexpected twists the protein chain was performing (Kendrew *et al.* 1958) (Fig. 5.5).[5] To an MRC officer, who came to see the model, it looked much like an 'anatomical model of abdominal viscera'. The polypeptide chain was 'like intestine intricately coiled about a small red object representing the haem part of the molecule'.[6]

Perutz built the first model of haemoglobin at a similar resolution by reproducing the shapes of the high-density areas in the various sections of

[5] Kendrew presented the model shown in Fig. 5.5 made of black plasticine to the Science Museum, describing it as 'the first model of a protein' (J. Kendrew to F. Greenaway, 14 October 1975, Science Museum, file T/6762). How literally this needs to be understood remains open to interpretation. The model differs from the one which first appeared in print in a series of photographs and which has recently returned to the LMB from Kendrew's private collection (see Fig. 5.8). This (much better preserved) model is covered with a glossy white paint, except for the oxygen binding structure which is represented by a red disc. The structure is not sitting on pegs, but is supported by a few metal rods inserted between the bends. At the laboratory, priority is given to this model. The available record gives no definite answer. The 'white' model also exists in a few large-scale reproductions used for demonstrations and exhibitions and certainly had a wider circulation at the time. It was probably also the one described by a visitor to the MRC in the following passage.

[6] Dr Norton, 'Visit to the Molecular Biology Research Unit, Cambridge, 2nd December 1957', unpublished internal report file E243/29, MRC Archives.

the three-dimensional contour map in plasticine and assembling the discs in the right order. When this first model collapsed, Perutz resorted to more unfamilar material. The contour map was reproduced on sheets of thermo-setting plastic from which the shapes could be cut out, stacked together and baked ('like a cake') to set permanently. This resulted in the black and white disc model of haemoglobin mentioned before, which resisted the wear and tear of time (Perutz *et al.* 1960).

The interpretation of the 2 Ångstrøm resolution map of myoglobin with its 1260 atoms (excluding hydrogens) represented a new challenge. Kendrew invented a new modelling technique which took its inspiration from the toy construction kit Meccano. Steel rods, of the kind used in Meccano for the wheel axis of toy cars, but 6 feet tall, were positioned at the same intervals chosen for the mathematical analysis of the structure so as to form a vertical grid consisting of about 2500 rods. Coloured clips, used in Meccano to fix the wheels, indicated the electron density along the rods (the brighter the colour the higher the density). Skeleton-type atomic models giving the exact position for each atom could then be built between the rods, following the density indicated by the clips. Once the main chain was inserted, the density of the side chains could be seen emerging at the appropriate intervals. Careful observation of the lumps and model build-ing allowed Kendrew to identify many more side chains than anticipated. Again, model building was a crucial step in determining the structure. The atomic coordinates deduced from the model with a plumb-line were fur-ther used for mathematical methods of refining the structure (Kendrew *et al.* 1960; Kendrew 1961) (Fig. 5.6).

The scale of the model (5 cm = 1 Å) was chosen so as to allow a human hand to reach in and fix the clips and model bits. The model bits were designed by Kendrew, perfecting an idea first developed by Charles Bunn, research chemist and self-taught crystallographer at ICI. The 'Kendrew models' differed from conventionally used skeletal models in the connec-tor system, which consisted of a barrel with two screws fitted into grooves in the model rods. This tight fit was essential to build large structures like proteins. First built in the Cavendish workshop, the model parts were later commercially produced by Cambridge Repetition Engineers, a small precision-engineering firm started by John Rayner, a hobby model builder and car mechanic, and his business partner. They were widely used, and were only superseded by computer modelling in the 1980s (Fig. 5.7). For over two decades Cambridge Repetition Engineers employed three people full-time to cover the demand. The success of the business is an indica-tion of the impact Cambridge protein crystallographers had on the field. Most of the orders came from researchers who, at some point in their

5.6 'Forest of rods' model of myoglobin (1959). Courtesy of Medical Research Council Laboratory of Molecular Biology, Cambridge.

career, had been connected with the Cambridge laboratory.[7] The replication and marketing of the model parts also compensated for the difficulties of circulating the models in print, a point to which I will return.

To crystallographers, models provide ways to represent spatial relations of atoms. The large variety of available model parts underlines the conventional character of model building. This, however, did not preclude that crystallographers expected their models to reveal the 'real' structure of the molecules and that studying the models would give clues about how the molecules work. Their explanations made use of the mechanical properties

[7] See business correspondence, 1959–present, Cambridge Repetition Engineers, Green's Road, Cambridge. Around 80 per cent of the shipments went to America, where many researchers settled after spending their postdoctoral years at Cambridge. In the United States orders could also be placed with the Ealing Corporation which, however, marketed the models at much higher price. The parts were produced semi-automatically. Brass rods were cut with special machines (normally used to build parts for sewing machines) and soldered by hand using gauges. While the 'Kendrew models' found wide distribution, Kendrew's 'rod and clip' method was soon superseded by the invention of an ingenious optical device, known as 'Fred's folly', which projected an image of the electron density map onto the model while it was being built, greatly accelerating the task (Richards 1968). On the design and use of other molecular model kits, as well as on molecular modelling in the practice and historiography of the chemical sciences more generally, see Francoeur (1997).

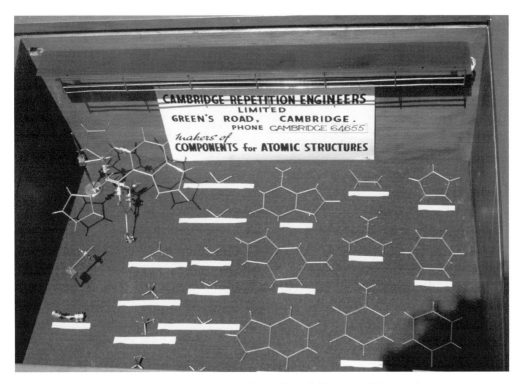

5.7 'Kendrew type' skeletal model parts. Display box of Cambridge Repetition Engineers. Photograph by the author.

of the models, which were manipulated and altered to predict changes in structure and functions. An example will illustrate the point. When Perutz, in his well-rehearsed discovery story of the structure and function of haemoglobin, produces his models of the molecule, he tells his audience: 'That is what it really is.'[8] Perutz used information from Kendrew's model of myoglobin, together with other evidence, to build a tentative atomic model of haemoglobin, composed of four subunits all closely resembling the myoglobin molecule (Perutz 1965). The predicted structure was later confirmed and refined by a full structure analysis. To understand the mechanism by which haemoglobin carries oxygen from the lungs to the tissues and helps to transport carbon dioxide back to the lungs, Perutz studied the atomic structure of both the oxygenated and deoxygenated forms of the molecule. The mechanism he proposed on the basis of his models he described as a 'clicking back and forth between two alternative structures' (Perutz 1978).[9]

[8] M. Perutz, *Science Is No Quiet Life*, Peterhouse Kelvin Club Video (1996).
[9] For further examples of how physical atomical models of proteins were used to describe functions see below (chapter 9 on chymotrypsin and chapter 11 on abnormal haemoglobins).

As this example underlines, models played a central role in the very research process of protein crystallographers. They shaped the way crystallographers talked and 'thought' about molecules and represented the very results of their investigations. This, however, posed particular problems regarding the publication of their work.

Publishing models

The results embodied in the models were not easily conveyed, either in words or in pictures, on the flat pages of scientific journals. Hardly any paper on the subject was published which did not refer to this problem. 'The whole question of publication is difficult', Bragg most emphatically summarised the problem. 'How can one convey all the information in a huge molecule... The "paper" sent to colleagues should be a model. There seems to be no simpler way of conveying the information.'[10] The outburst followed Bragg's repeated attempt to convince Kendrew finally to publish the results of the last step of his X-ray diffraction studies of myoglobin, the structure at 1.4 Ångstrøm resolution.'It is extraordinary that [the results] have never yet been written up', he confided to the Secretary of the Medical Research Council, whom he kept informed on the protein work. '[Kendrew] got his Nobel Prize in effect for unpublished work. The world ought to know about the structure, and many people who worked with him ought to get credit for what they did.'[11] As Bragg himself suggested, 'laziness' was not the (only) reason for procrastination. The positions of the atoms in the 'forest of rods' model were continually adjusted according to the refined data, but the full structure was not easily transferable into a text.[12]

Various solutions were found to convey at least some of the information the models revealed to the (trained) eye and hand. Photographing the models was the first, but not very satisfactory solution.[13] Kendrew's first (low resolution) model of myoglobin had already appeared in print in a series of photographs (Fig. 5.8), when Kendrew set out to find the help of an artist. '[Y]ou said that you know a draughtsman, experienced in anatomical drawing, who might be able to make a drawing of my myoglobin for me', Kendrew wrote to the Secretary of the Medical Research

---

[10] W. L. Bragg to H. Himsworth, 18 January 1968, file S18/1, MRC Archives.

[11] W. L. Bragg to H. Himsworth, 20 June 1967, file S18/1, MRC Archives. Kendrew shared with Perutz the 1962 Nobel Prize for Chemistry for his work on the structure of globular proteins.

[12] A full description of the structure, including the complete set of coordinates, was finally published by Kendrew's colleague, Herman Watson, in 1969 (Watson 1969).

[13] Photographs, however, remained central documents for artists to work from (see below).

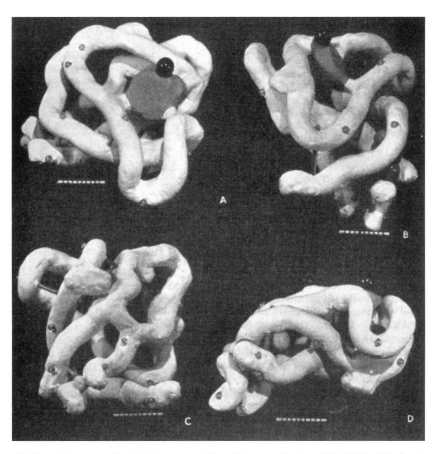

5.8 Myoglobin model as it first appeared in print. From *Nature* 181 (1958), 665, fig. 2. Copyright 1958 Macmillan Magazines Ltd.

Council. 'Since everyone is immediately reminded by my model of viscera, it seems a very promising way of going about it.'[14] The London-based artist Frank Price undertook to draw a 'front' and 'back' view of the model, stripping it of all constructive elements (and with these of its model character) and rendering only the course of the polypeptide chain (Bodo *et al.* 1959, 100). It soon became clear that, as in the original model, the haem group had the wrong inclination. Despite this error, precisely the drawing which showed the position of the haem group, and in which the problem was therefore more visible, was frequently reproduced, while the second one disappeared from circulation (see e.g. Kendrew *et al.* 1960, fig. 5; Kendrew 1964, fig. 3; 1966, plate 19). The inaccuracy of the drawing was pointed out in the 1960 *Nature* paper, but never thereafter. Clumsy verbal descriptions

[14] J. Kendrew to H. Himsworth, 17 December 1957, Kendrew Papers, MS Eng. c.2593, M.1, Bodl. Lib.

accompanied the drawing. Kendrew's best: 'embracing [the haem group], so to speak, are 2 segments of polypeptide chain arranged like the letter V'. Or: 'The polypeptide chain is shown as an irregular rod winding its way around the molecule' (Kendrew 1966, captions to figs. 18 and 19).[15]

Most artists worked only occasionally on molecules and it is as hard to find documentation about them and their work as about their counterparts in earlier centuries. One artist, however, has made his name and his career by drawing three-dimensional protein structures: the architect and illustrator Irving Geis. His career as 'molecular artist' started with the commission to draw Kendrew's molecular model of myoglobin for *Scientific American* in 1960. The drawing took six months to complete, cost the journal more than any illustration before, and met with the contempt of the editor, who stamped it 'a mass of crumpled chicken wire' (Gaber and Goodsell 1997) (Fig. 5.9). This devastating critique notwithstanding, Geis continued to depict molecular structures for the next forty years of his life, including several more for *Scientific American*. The production of each picture was a laborious process. Geis would travel to the model (even if this included a transatlantic passage), spend several weeks sketching and photographing it and then proceed to 'conceptualize . . . and finally paint it'.[16] Numerous conversations and much correspondence between the scientist and the artist accompanied the process. In his renderings of three-dimensional protein structures, drawn to exact coordinates, Geis practised what he called 'creative lying', by which he understood the introduction of small distortions and exaggerations to resolve overlaps and create an understandable image, in spite of the complexity. According to Geis it was this capacity which distinguished the human artist and gave his or her products an advantage over photographs or, later, the drawings of a 'mindless' computer.[17]

Geis's drawings deeply influenced the conventions for depicting and viewing intricate protein structures. The extent to which scientists came

---

[15] When Perutz's model of haemoglobin became available, he commissioned A. Kirkpatrick Maxwell to produce a drawing of the molecule which is still widely used in textbooks. Maxwell, then working on a freelance basis for the School of Anatomy at Cambridge, was well known for his illustrations of the many editions of *Gray's Anatomy* and for his work as Army medical illustrator in both World Wars (Archer 1997, 59–64). I thank Barbie Wells, former librarian at the School of Anatomy at Cambridge, for her time, enthusiasm and detective skills in reconstructing Maxwell's name from his initials. The constant association of protein crystallography and the structures it produced with (molecular) anatomy and the use of a corresponding set of representational techniques, including models, drawings and atlases of molecular structures, deserves further elaboration.

[16] Letter S. Geis to the author, 20 July 1998. I thank Sandy Geis for generously providing information on her father's activity and for making material available from the Geis Archives.

[17] On early uses of computer graphics programs for molecular modelling see Francoeur and Segal (forthcoming).

Myoglobin

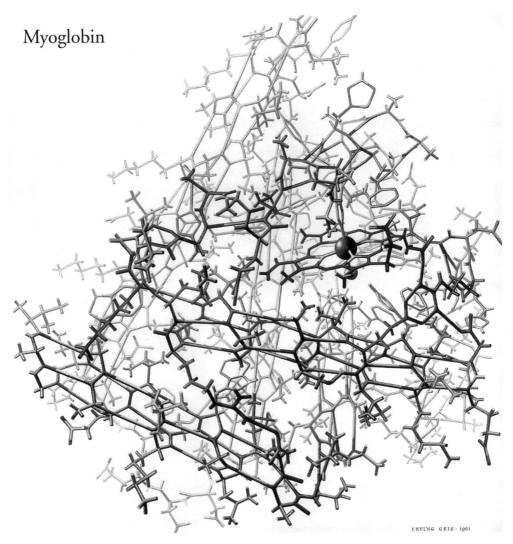

IRVING GEIS · 1961

5.9 Drawing of the atomic model of myoglobin by Irving Geis. From The Geis Archives [presentation booklet]. Copyright *Scientific American* (1961). Courtesy of Sandy Geis.

to depend on the artist's skills is documented by the fact that his name appeared as co-author on what has become a classic in the field: Dickerson and Geis's *The Structure and Action of Proteins*, a heavily illustrated introduction to proteins (1969). Geis's illustrations in this volume were reproduced in many biochemical textbooks, introducing a whole generation of biochemists to his style of representation. According to Richard Dickerson, who collaborated with the artist on two further book projects, Geis 'taught us all how to look, how to understand and how to show others what we saw' (Dickerson 1997b, 1249).[18]

---

[18] On Geis's molecular drawings see also Dickerson (1997a; 1997c).

In his later drawings Geis shifted from an atom-to-atom representation to an increasingly abstract and 'artistic', though always stereochemically correct, rendering of molecular structures or to what a critic called the production of an 'understandable metaphor for molecules' (Gaber and Goodsell 1997). This shift was an expression of his attempt to illustrate the *function* (besides the structure) of the molecules, but probably was also his response to the development of molecular computer graphics, which borrowed his ribbon-and-arrow style of representation. These later drawings also appeared in textbooks, copyrighted by Geis. Placing himself in a long and venerable line of tradition, he saw his role as that of a 'molecular Vesalius' who, like his Renaissance predecessor in the case of the new human anatomy, used art to teach the modern public the equally new field of 'molecular anatomy' (Dickerson 1997a, 2484).

Protein crystallographers resorted to stereoscopic drawings as an alternative way to convey spatial information in print (e.g. Perutz *et al.* 1968; Watson 1969). These were produced by re-tracing by hand photographs of models taken at different angles or using data of atomic positions generated by early mathematical model building programmes, a task requiring much patience and skill. In 1981, a whole *Atlas of Molecular Structures* with stereodiagrams of the myo- and haemoglobin molecules was published. To facilitate stereoscopic viewing, the diagrams were printed in red and green and a foldable pair of red–green spectacles was included in the back pocket of the publication (Fermi and Perutz 1981).

Recommendations and finally 'rules' approved by the International Union of Biochemists were set up to regulate the description and graphical representation of polypeptide conformation (Edsall *et al.* 1966; IUPAC-IUB Commission 1970). These allowed researchers to describe the position and orientation of every atom in a precise and standardised manner, greatly aiding communication and the comparison of different structures.

Yet all these efforts could not replace the importance of directly viewing, manipulating and actively building models to grasp the structure and function of the molecules. Parallel to publication, models were mobilised in other ways. Every new structure attracted many visitors who came to appreciate 'the body "in the flesh"'[19] and, to the extent to which the construction allowed it, models were carried about and showed around. With the complete set of coordinates, which were not always published but were generally available on request, the right model parts and some experience, models could also be replicated. In the case of the myoglobin model, this became easier once Cambridge Repetition Engineers started offering

---

[19] M. Perutz to H. Himsworth, 6 April 1953, inviting him to see the DNA model; file FD1/426, MRC Archives.

myoglobin component sets, complete with frames, clamps, angle gauges, and calipers, calibrated at the same scale as the model parts, allowing direct measurements in Ångstrøms.[20] The kits were mainly used to build models for demonstration or in tutorials for model building. Following growing demand in the mid-1960s, Kendrew also arranged for the commercial production of ball-and-spoke models of myoglobin, built to his specification by Alexander Barker of the Department of Engineering at Cambridge, who ran a small private business. Like other demonstration objects, these were sturdier and more colourful than the research models, but less accurate.

The shape of life

The same models which were so difficult to compress between the pages of a scientific article proved highly effective in all contexts and arenas in which display and manipulation were not only possible but called for. Protein crystallographers were keenly aware of this possibility and made effective use of their models, not only among scientific colleagues and in the lecture halls, but also in public arenas. Through these displays they created a public image and an institutional place for their science.

Television proved a particularly congenial medium to present protein crystallography to a wider public and promote the subject. The producers launching the first science series on television quickly seized the opportunity. The scope and background of these programmes is of interest here.

The BBC series *Eye on Research*, launched in 1957, proclaimed that its aim was to show what 'the backroom boys of science', employed in industry, government and university, did in terms of capital investment in knowledge for the future.[21] The programme, the most ambitious to that date, was supported at high political level and, as with most science information initiatives in the postwar era, was aimed at the appreciation rather than at a critical assessment of the scientific enterprise (the audience was

---

[20] Cambridge Repetition Engineers Price List [undated] and Ealing Catalogue 1975–6. Besides myoglobin component sets, Cambridge Repetition Engineers also offered model kits of other structures which were established, including the proteins ribonuclease, lysozyme and chymotrypsin, DNA and bacterial cell wall.

[21] The initiative for the series came from the BBC itself. The producer of the series was Aubrey Singer. The acquisition of a specialist scientific adviser to the production team was discussed, but finally dismissed on the ground that this would be a 'rather useless individual, especially if he had no knowledge of television' (A. Singer to O. B. O. Tel [date missing], file T14/1502/1, BBC-WAC). Most of the scripts for the series were written by Gordon Rattray Taylor, Cambridge science graduate and wartime reporter for the BBC. He later became editor of the science programme *Horizon*, broadcast on BBC2. and author of various books on science-related topics.

invited on 'a voyage of discovery') (Lewenstein 1992). Among the first to express his compliments on the series was the Director of Public Relations of the Atomic Energy Authority, commenting that if the public could be 'thoroughly impregnated with the idea that research is a fascinating but a long and painstaking business', life would be 'a lot easier for all of us in the science-information business'.[22] Congratulations also came from Alexander Todd, Cambridge Nobel Prize laureate and Chairman of the Advisory Council on Scientific Policy, who had been consulted on the series from the earliest stage. The series, which reached an audience of several million, appeared to Todd 'a very valuable means of educating the general public in the methods and development of science and in giving them a proper appreciation of their implications'.[23] It brought considerable prestige to BBC Television in academic and scientific circles and succeeded in 'selling' television as a means of 'putting across scientific information in a reasonably entertaining way'.[24] Even if British scientists did not directly depend on the public for funding, public interest in science was a legitimation for tax spending and was welcomed by scientists and science administrators alike.

In considering possible topics for a series, visual quality, next to intellectual content and treatment, was of overriding importance. Several academic subjects were *a priori* excluded as being too much of 'the pencil and paper kind' to provide material for television. From the beginning, a programme on the models of large molecules, even if certainly a complex subject for a programme targeted at a 'middle brow audience', figured among the subjects under consideration. Series producer and interviewed scientists agreed on this choice. In the round of discussions preceding the first series, for instance, Todd excluded chemistry, his own subject matter, as 'impossible to represent extensively in visual terms', but agreed that crystallography 'might well be made visual with the aid of models'.[25]

Examination of the production files of the first programme on the work of the Cambridge protein crystallographers, broadcast on May 1960,

---

[22] E. Underwood (Director of Public Relation, UK Atomic Energy Authority) to A. Singer (BBC), 13 February 1958, file T14/1502/1, BBC-WAC. Among those invited for the 'end of series party' (i.e. the end of the first series; many more series were to follow) were various scientists in leading administrative positions as well as high-level administrators from the Ministry of Supply, the Department of Scientific and Industrial Research, the Atomic Energy Research Establishment and the Industry Research Association; A. Singer [internal memo], 27 January 1958, file T14/1502/1, BBC-WAC.

[23] A. Todd to Sir Ian Jacob (BBC), 4 March 1958, file T14/1495/10, BBC-WAC.

[24] Miscellaneous correspondence in file T14/1502/1, BBC-WAC, relating to the first *Eye on Research* series in general.

[25] G. Rattray Taylor, TV Science Series, 20 November 1957 [Note on discussion with Sir Alexander Todd, Professor Meyer Fortes, Dr Robin Marris], file T14/1502/1, BBC-WAC. See also 'Note on a meeting with Lord Adrian [Vice-Chancellor of the University] at Cambridge, Oct. 29', same file.

confirms the central role attributed to the models and their 'performance' for a good production. Commenting on a draft of the script for the programme, the producer lamented the 'prosaic nature' of the series so far, and encouraged scriptwriter and cameraman to make maximum use of the models, which could produce 'exciting visuals' and bring up the reaction index. 'I hope', he intimated to his staff, 'you are planning to do that and not just show a lot of table top stuff. It is not sufficient to explain the subject in an interesting and exciting way.' After the production he deplored the lack of 'decent close-ups of the models'.[26]

While better effects might have been achieved with more 'imaginative camera work' and better use in terms of lenses, as the producer Aubrey Singer suggested, watching the programme today it is striking how many scenes revolved exclusively around the models.[27] Long sequences show only the models with hands turning them around, pointing to, taking apart and refitting parts of them – scenes and views only television, despite its 'flat' screen, could present. Time and effort was also spent on the choice of the most 'telegenic' models and on preparing them for the show. A 2 Ångstrøm myoglobin model, for instance, for better effect on the black and white screen, was equipped with fluorescing atoms in two nights' overtime by staff of the Royal Institution before being escorted to Cambridge to be filmed (Fig. 5.10).[28]

The audience (8 per cent of the television public, compared to 25 per cent for the 9.25 pm News and 32 per cent for the 'Play of the Week', both broadcast at the same time) reacted rather positively to the programme even if – models notwithstanding – most of the explanations went over the heads of many viewers.[29] Most complimentary, however, were the crystallographers themselves. They thanked the BBC for the 'magnificent effort' by the production team to put a difficult subject across to a lay audience and for the chance they were given to participate in the programme.[30]

For neither Perutz nor Kendrew was this the first experience with the media or lay audiences. They had presented their work in various science magazines, including *Endeavour* and the *New Scientist*, all of which had

[26] A. Singer to P. Daly [date missing] and 4 May 1960, file T14/1499/5, BBC-WAC. The general notes for production of the programme confirmed: 'As much as possible should...be made of the finished models. This is what the lay viewer wants to see: the "answer" to the problem.' G. Rattray Taylor, 'Perutz Program (Eye on Research 1960)', 27 December 1959, file T14/1499/5, BBC-WAC.

[27] *Eye on Research: Shapes of Life*. Script by Gordon Rattray; interview with Max Perutz and John Kendrew by Raymond Baxter, broadcast 3 May 1960.

[28] Miscellaneous notes and correspondence in file T14/1499/5, BBC-WAC.

[29] See 'Audience research report for *Eye on Research*, 5 – Shapes of Life', file T14/1499/5, BBC-WAC.

[30] M. Perutz to P. Daly, 6 May 1960, file T14/1499/5, BBC-WAC.

5.10 Preparing the models for the BBC programme *Shapes of Life* (1960). Model room, Cavendish Laboratory. Source: Medical Research Council Laboratory of Molecular Biology, Cambridge.

come on the market in the postwar years. They also had experience with radio broadcasting. Yet television proved much more congenial to their subject. Indeed, on at least one occasion the suggestion that Perutz should talk about his research on the structure of the haemoglobin molecule on a radio programme was turned down on the grounds that this was too dry and complicated a subject for transmission by radio.[31]

The programme on the Cavendish group was only one in a long series of TV programmes featuring the crystallographic models. The DNA model made its first appearance on TV in an earlier *Eye on Research* programme. Much discussion was spent on the choice of the most 'telegenic' model. In the end, a battery of models was used, with detailed planning involving ground-work men, studio designers, scene master, scriptwriter, producer and scientists going into the transport, safe-keeping and staging of the precious objects. One model was placed on a revolving table, where it kept turning in the background throughout the production for maximum

[31] L. Bragg to. V. Alford (BBC), 1 May 1944 and response; Talks, L. Bragg, file 1, 1938–62, BBC-WAC.

effect.[32] The programme was watched by 4.5 million people and was very well received.[33]

Models also played a central role in Kendrew's ten-week programme *The Thread of Life*, dealing with proteins, nucleic acids and viruses, shown on televison in 1964 (Kendrew 1966). Yet another *Horizon* programme, a talk with the title 'The Model Makers', centring on the work of the Cambridge crystallographers, was planned in 1965, but in the end did not materialise.[34]

Television programmes were important, but not the only occasions on which the models were presented in public arenas. In the late 1950s, Barker built a series of gigantic models, including a huge plaster replica of Kendrew's first myoglobin model. The models were shipped to Brussels to be displayed in the International Science Pavilion at the World Fair. In the hot phase of the Cold War and the atomic arms race, this exhibition, which carried the Atomium with its shining spheres as its symbol, was to foster a peaceful image of science, and of nuclear physics in particular (Lambilliotte 1959; 1961; Schroeder-Gudehus and Cloutier 1994) (Fig. 5.11).[35] In the six months it was open to the public, it attracted over 40 million visitors (Schroeder-Gudehus and Rasmussen 1992). Fifteen countries, including both the United States and the Soviet Union, contributed to the realisation of the Science Pavilion. Britain provided a sixth of the exhibits on show. The stress lay on the representation of fundamental or 'pure' science and the international character of scientific

[32] *The Thread of Life: DNA*, produced by A. Singer, broadcast 4 February 1958 and correspondence between N. Croll (BBC studio designer), C. MacIntyre (BBC ground man), A. Singer (producer), A. Todd, C. Waddington, M. Wilkins, A. Wurmser (model maker), file T14/1495/10, BBC-WAC.

[33] A. Singer to H. Ephrussi-Taylor, February 1958, file T14/1495/10, BBC-WAC. The choice of Waddington, the Edinburgh embryologist who opposed reductionist approaches in biology, as anchor man of the programme produced irritation and even open opposition in some of the groups active in the field. From the Cambridge group only Ingram was involved in the programme. Nor was the group keen to offer its models for the show. The Glasgow bacterial geneticists under Guido Pontecorvo cancelled their participation. Though DNA was declared as the 'first and main topic' of the programme, it was presented in a much broader biological and chemical framework than later programmes, which tended to focus much more exclusively on the Watson and Crick story.

[34] Kendrew cancelled the contract owing to 'unforeseen circumstances'. File TV ART 4 Kendrew, J. C., BBC-WAC. The file gives no clue as to what the actual reasons were.

[35] The 330-feet-high steel structure was itself a showpiece of technical achievement. The nine interconnected spheres were all accessible to visitors by a system of escalators, rolling carpets and lifts. Circulation was controlled by a man with a microphone and a series of television screens which allowed him to foresee and control bottlenecks. The lower spheres housed an international exhibition on the peaceful uses of atomic energy while the highest sphere contained a restaurant with a commanding view of the exhibition grounds.

5.11 The Atomium at the Brussels World Fair (1958). The structure representing the atomic lattice of iron crystals, enlarged 165 billion times, symbolised the peaceful image of the atom. The steel spheres, each 18 meters in diameter, were all accessible to the public. *Source*: Hulton Getty Picture Collection.

endeavour.[36] The organisers further emphasised the interconnection of all the sciences. The exhibits were organised under the four headings of atom, crystal, molecule and living cell, linking the physical and the biological sciences in a quest for the fundamental structures of inanimate and living matter (Bragg 1959). Three-dimensional models and animated films were regarded as most suitable media to represent scientific matters.[37] Bragg, who became responsible for the British contribution to the International Science Pavilion and himself coordinated the crystal section, corresponded widely with British colleagues to identify the most suitable exhibits. While some, including colleagues working in immunology and developmental genetics, retorted that their work did not lend itself to three-dimensional treatment, Perutz, Crick, Wilkins, Franklin (for the virus models) and Dorothy Hodgkin responded enthusiastically to Bragg's enquiries. Fred Sanger's pioneering work on the primary structure of insulin was a certain inclusion. Despite the fact that his work regarded the linear sequence and not the three-dimensional arrangement of the peptide chain, Sanger was convinced that the 'central feature of an exhibition on insulin' had to be an atomic model. 'This', he replied to Bragg, 'certainly gives people something to look at even if they cannot understand it' (Fig. 5.12).[38] Molecular models represented a substantial part of the British material sent to Brussels, featuring in all but the atomic section. Following the choice of the organisers, the DNA model, supplied by Wilkins, became a key exhibit of the living cell section (a second DNA model, built by Barker at Cambridge, featured in the crystal section). Photographs of the giant model appeared in media reports as well as in the official report of the exhibition (Lambilliotte 1959, 71 and photo 25). This is but one example in which the double helix was used to redeem the darker face of nuclear physics.

The science display at the international exhibition in Brussels very well encapsulated the appeal of the molecular models and the science that produced them in the postwar era. With nuclear physics, protein X-ray crystallography shared the aura of fundamental science. At the same time

[36] Exhibiting countries could still show off their national achievements in science and technology in their own pavilions. One major attraction was Sputnik, the first earth satellite, on show in the Russian pavilion.

[37] See for instance 'Exposition Universelle et Internationale de Bruxelles 1958. Palais International de la Science. Procès-verbal de la sèance de travail le samedi 23 juin 1956 à 10h. à la Fondation Universitaire avec Sir Lawrence Bragg'; RI MS WLB 83c/24 and annex, MS WLB 83c/25; also 'Design for presenting the nation', Britain in Brussels '58, Newsletter, Office of the Commissioner-General for the United Kingdom, Number 3, 19 July 1957, p. 2, RI MS WLB 83F/13.

[38] F. Sanger to L. Bragg, 4 February 1957, RI MS WLB 84A/4. Bragg was assisted by David Phillips, then at the Royal Institution, in the organisation of the British contribution to the International Science Pavilion. For the extensive correspondence relating to this task see RI MS WLB 82-6.

5.12 Ball-and-spoke model of insulin built by Alexander
Barker for display at the Brussels World Fair (1958). The
model is now on loan from the Science Museum to the
National Portrait Gallery in London, where it is on
display in the Science, Technology and Business Room
next to portraits of Fred Sanger and Dorothy Hodgkin.
The label of the exhibit specifies that the model, mainly
based on the amino acid sequence of the molecule, does
not reflect the actual arrangement of the atoms which
was established by Hodgkin's group on the basis of X-ray
analysis a decade later. *Source*: Science Museum/Science
& Society Picture Library, slide no. SCM/PHY/
C100106D.

its intricate models promised applications in the medical field, rather than in destructive weapon systems. Indeed, despite the dominant rhetoric, the notion that, once the structure of 'living molecules' was known, it would also be possible to reconstruct and 'repair' them, was never far away. In the *Living Molecules* exhibition in the Science Museum, mentioned at the beginning of this chapter, this connection is exemplified by a button. When pushed, a red light turns on in the haemoglobin model, indicating the position of the single amino acid which distinguishes sickle cell haemoglobin from 'normal' haemoglobin.[39] If the cause of a complex disease picture can be pinned down to a small chemical difference, this exhibit seems to suggest, a simple cure can be found. With its models and its investigations into the fundamental structures of life, protein crystallography then greatly contributed to the creation of a positive image of the sciences propagated through the new media and popular displays in the postwar and Cold War era.

A new science

The structural models of myo- and haemoglobin marked an important achievement. In 1962 both Kendrew and Perutz were honoured with the Nobel Prize for their pioneering work. Their achievements, I set out to show, crucially relied on opportunities and resources created for the sciences, and for biophysics more specifically, by postwar reconstruction. Taking advantage of these opportunities and applying for funds to the MRC, an institution whose own prestige had dramatically grown as a result of its active role in the medical mobilisation of World War II, Bragg managed to put protein crystallography on a safer institutional footing. Government funding via the MRC was particularly important, since protein crystallography remained at the borders of the traditional disciplinary divisions of the university. The attraction of the most visible products of protein crystallography, the models, relied on the same appeal carried by biophysics more generally, that is, on the expectations posed in the knowledge, acquired by powerful physical technologies, of the 'fundamental' processes of life.

However, protein crystallographers profited from the legacies of World War II not only in terms of funding and institutional support or ideologically. Instruments and technologies created during the war, or on the basis of wartime developments, also deeply transformed the scope and pace of research. The entrance of (both analogue and digital) electronic computers

---

[39] That normal and sickle haemoglobin differ in a single amino acid was first established by Ingram at the Cambridge biophysics unit in the mid-1950s. It was celebrated as a landmark result (see below, p. 192).

in crystallographic practice made this particularly clear. Skills and experiences collected in the scientific wartime projects shaped the approach of the researchers engaging in biophysical research to the new technical possibilities, as well as their approach to research more generally. Protein crystallography did hinge on increasingly big and expensive apparatus and on the collaboration with computer experts and engineers, but it was not 'big science' in the way nuclear physics became after the war (Galison and Hevly 1992). It did not itself depend on military funding as other fields of biophysics did. In its reliance on new biomedical funding, recruits from wartime scientific projects with particular skills, powerful computers and other electronic apparatus, and the postwar appeal of a physics of life, it none the less deeply depended on the legacies of the wartime scientific mobilisation. This becomes especially clear if, as attempted here, one focuses not only on the results but on the material practices of protein crystallographers, and on the circulation of their tools, in and outside the laboratory.

The myo- and haemoglobin models became part especially of popular histories of physics. A volume issued to celebrate the first hundred years of the Cavendish Laboratory dedicated ample space to the work of the protein crystallographers and also featured two large photographs of the protein models (Crowther 1974). In his book *The Physicists* C. P. Snow reported on the achievements of the protein crystallographers under the title 'A different harvest'. The chapter immediately followed one on nuclear fission with dramatic pictures of the victims and the physical destruction caused by the atomic explosions in Japan and a second one on the prospects of nuclear war (Snow 1981). If this presentation once more graphically underlines the role of protein crystallography as the 'good face' of physics, Cambridge protein crystallographers used their models for their own disciplinary aims. When, in 1958, Perutz went to London to present the case for a new laboratory to his patron, the Medical Research Council, he brought with him much the same set of models used in the BBC programme, making sure that a 'large table ... on which the exhibits could be arranged' was available'.[40] To their new project the MRC researchers, joined by Sanger and his group of protein chemists, gave the name Laboratory of Molecular Biology. The wider implication of this move and the place of the double helix in the history of the new field are the subjects of the next part.

[40] M. Perutz to R. C. Norton, 22 February 1958, file E243/109 I, MRC Archives.

PART II

Building molecular biology

In 1953 Watson and Crick presented their DNA model in the Cavendish unit. Important features of the model, such as the base pairing and the two helical chains running anti-parallel, stood the test of time. Few events in the history of science have attracted so much attention from scientists, historians and the popular media as the construction of this model.

In the history of molecular biology the presentation of the model is celebrated as the key event, if not as the proper 'origin' of the new science. This version of the history of the field was negotiated among participants in the mid-1960s. Taken up by historians, it has become an integral part of the standard account of the history of molecular biology – despite vociferous critique from many different directions. In the following chapters I will contrast this account with a local account of the construction of molecular biology at Cambridge.

In 1953 the unit still ran under the heading of biophysics. In 1957, it became the first institution world-wide to carry the term molecular biology in its name. The change of name occurred in conjunction with bigger plans for a new, independent Laboratory of Molecular Biology. A study of the events leading to this plan and the following negotiations will throw fresh light on the construction of the new science and on the role played by the double helical model of DNA in this history.

In the standard history of the field, places do play a role. Caltech and Cold Spring Harbor were the 'Mecca and Medina' of the phage group, and Crick and Watson met at Cambridge (Cairns, Stent and Watson 1966, ix). But these places are most often reduced to stages for meetings and discoveries. If, however, we take the proposition of the local production of scientific knowledge seriously, knowledge production and the construction of scientific fields become inscribed into the culture and history of places (Ophir and Shapin 1991).

My claim is not that molecular biology originated in Cambridge, even if Cambridge became an important centre for the practice of the new science. If we want to speak about 'origins', I would suggest that molecular biology has many origins. In this respect, as well as in their claim of local traditions, the concurring origin accounts proposed by participants are to be taken seriously. However, the Cambridge case offers the unique opportunity to explore to what extent and in what way the double helix reshaped existing research traditions in the laboratory where it was first proposed and how these changes were related to institutional developments. In this sense I would like to present the following account as an attempt to move beyond the division of intellectual and institutional histories.

I will argue that the move from biophysics to molecular biology in Cambridge was a political as well as a scientific move, based on a series of local contingencies regarding the people and institutions involved. Similarly, the way molecular biology became defined at Cambridge reflected local contingencies rather than describing a given empirical (ontologically defined) field of inquiry. The double helix, I will show, did not precipitate these events, but rather gained its role in the *course* of these events. That the move to molecular biology at Cambridge represented a solution to a local situation, however, does not exclude that the same move responded to and was aimed at setting trends for developments elsewhere. I will come back to this point in the last chapter of this part and deal with it in more depth in the final part of the book.

Stressing the retrospective construction of 1953 as the origin of molecular biology, I do not intend to belittle the scientific achievement of the proposed structure of DNA. However, scientific breakthroughs alone cannot explain the making of a new science. Once the explicatory power of the double helix has been suspended, a much richer history comes to light. This history also shows the work which was necessary to turn the double helix into the powerful icon it became.

After a brief review of the debates on the 'origins' and the role of the double helix in the construction of molecular biology, I will study investigative practices and institutional developments in the Cambridge unit, following Watson and Crick's presentation of the double helical model of DNA (chapters 6 and 7). The creation of molecular biology at Cambridge, I will argue, was the complex response to a protracted institutional crisis precipitated by Bragg's move to London in 1953. With Bragg's departure the unit lost its main patron in the Physics Department. In the ensuing negotiations, which led to the creation of the Laboratory of Molecular Biology, the double helix played only a subordinate role. The fate of the original model, which fell into neglect and only in the 1970s was reconstructed for public display, confirms that its pivotal role was established only in retrospect (chapter 8). A brief comparison with developments elsewhere, notably at the Biophysics Unit at King's and at the Pasteur Institute, highlights the specificities as well as the relations of the Cambridge story to events elsewhere.

6

...

Locating the double helix

James Watson's own account of the events leading to the presentation of the model in *The Double Helix* was a *succès de scandale* (Watson 1968).[1] Refused for publication by Harvard University Press, the publishing house connected to Watson's home institution, and vehemently opposed by two of the main protagonists, Crick and Wilkins, it was finally commercially published by Atheneum in America and by Weidenfeld and Nicolson in Britain in 1968 (Sullivan 1980). Two years later the book appeared in a paperback edition and, since then, it has been continuously reprinted. A critical edition prepared by Gunther Stent in 1980 included a selection of the innumerable reviews the book had attracted (Watson 1980).

Presented as 'a personal account of the discovery of DNA' (this was the book's subtitle), some took the story to be as close to the truth as it could be, while others vehemently attacked it for distorting events.[2] The same readers also opposed the new, irreverent way of writing about science and scientists. Historians pointed out that the competitive and ruthless behaviour of scientists celebrated in the book was more characteristic of the late 1960s than of Cambridge in the early 1950s (Yoxen 1985). But besides signalling this bold new way of going about research, the account aimed at legitimising Watson and Crick's own unconventional behaviour with respect to colleagues and superiors. This exculpation was not accepted by all, and in particular the way Watson had presented Rosalind Franklin and her contribution to the 'race' to the double helix was strongly criticised (Klug 1968; Lwoff 1968; Sayre 1975; Hubbard 1990, 48–66).[3]

---

[1] The manuscript first circulated under the title *Lucky Jim*. This was the title of Kingsley Amis's book, one of the 'Angry Young Men' of the 1950s (Amis 1954). The hero of the story, a grammar school educated ex-serviceman, now lecturing at a 'red-brick' university, enjoyed challenging the attitudes of the establishment.

[2] On the narrative structure of *The Double Helix* and its appeal to a certain scientific readership see Gross (1990, 54–65).

[3] In an appendix to the revised edition of his book on the history of molecular biology Judson launched an attempt to rescue Franklin from 'feminist appropriation' (Judson

Watson's book formed the basis of a television film which the BBC released in 1987.[4] *Life Story* turned Watson, Crick and the double helix into household names. In the production, an attempt was made to rebalance certain features of the book (for example we see Juliet Stevenson as Rosalind Franklin, whom Watson had described as a bluestocking, using a very red lipstick). However, the film ends in an apotheosis of the DNA model which, to triumphant music, is lifted up from its base and in a slow spiralling movement rises above the Greek temple near Naples, where Watson first heard Wilkins talking about X-ray diffraction patterns of DNA, into the blue sky.[5]

Life Story is not the only way in which the double helix has been turned into an object of consumption as well as into an aesthetic object with metaphysical meaning. The double helix is available as a DIY-kit; it decorates T-shirts and mugs, and comes as a perfume bottle. Beginning with Salvador Dali, it has been taken up by artists and sculptors and is a feature of the bell tower which dominates the campus of Cold Spring Harbor, home of the first university of DNA (Kamminga and de Chadarevian 1995).[6]

The structural model of DNA and the events leading to it were not only dramatised for the lay public and turned into consumer goods. They also moved centre stage in attempts to write the history of molecular biology and trace the 'origins' of the new science. The fact that we find some of the same people involved in writing disciplinary history and fashioning DNA as a cultural icon indicates that the two activities are related. In the following I will none the less focus on the role of the double helix in historical accounts.

I start with a brief analysis of the debate on the 'origins' of molecular biology and the role acquired by the double helix in this debate. I will then place the double helix back into its local context. Rather than giving my version of the events leading to the double helical model of DNA, however, I will focus on the role of the double helix in the subsequent

---

1994, 619–29), yet critique of Watson's treatment of her role did not come only from feminist quarters. On Franklin see also Glynn (1996) and Piper (1998). Two new biographies of Franklin by B. Maddox and L. Elkin are in preparation.

[4] It seems hardly a coincidence that the same year saw a first important allocation of funds by the American Congress to the Human Genome Project aimed at deciphering the complete sequence of human DNA. Watson was a main mover behind the project. From 1989 to 1992 he was director of the NIH-Human Genome Research Center. He resigned from this position in protest against US government policy regarding gene patenting.

[5] On the BBC production as well as on earlier film versions of the DNA story, see Crick (1990, 80–8). For a critical review of *Life Story* see Franklin (1988).

[6] On the architectural history of Cold Spring Harbor see E. L. Watson (1991). On DNA as cultural icon see also Myers (1989) and Nelkin and Lindee (1995).

development of the Cambridge unit.[7] The year 1953, we will see, marked as much a year of scientific achievements as the beginning of a serious institutional crisis for the unit. Research strategies changed much more gradually. Events in 1957, which coincided with Brenner's recruitment to the unit and the introduction of phage research in the Cavendish, as well as with Kendrew's presentation of the first structural model of a globular protein, marked a much more radical turning point in the history of the laboratory.

The debate on the 'origins'

The book which opened the debate on the 'origins' of molecular biology was a volume dedicated to Max Delbrück on his sixtieth anniversary by a group of friends and colleagues and published under the title *Phage and the Origins of Molecular Biology* (Cairns, Stent and Watson 1966). Joseph Fruton, biochemist turned historian and visiting scientist at Cambridge in 1962/3, has suggested that the publication of the Delbrück *Festschrift* was prompted by the public esteem gained by the MRC Laboratory of Molecular Biology in Cambridge (Fruton 1992, 211).[8] The volume, then, did not fail to call forth a response from one of the key members of that laboratory. In a review article published in *Scientific American*, Kendrew, Nobel laureate, deputy director of the Laboratory of Molecular Biology in Cambridge and editor-in-chief of the *Journal of Molecular Biology*, expressed his surprise that, in the collection of papers, molecular biology was supposed to have originated exclusively in the phage work of Delbrück and his group and that the central theme of the subject was biological information (Kendrew 1966). Instead, Kendrew suggested the existence of two schools: a British school, which was concerned with structure and stemmed from the X-ray studies of biological materials by Astbury, Bernal and their pupils, and the American or informational school with its emphasis on information. According to Kendrew, the first real link between the two schools was established when Watson migrated to Cambridge and started his collaboration with Crick. It was this encounter and the

---

[7] By this I do not mean to imply that the story of Watson and Crick's collaboration has been exhausted. As we continue to be reminded, several questions still await an answer; see for instance Abir-Am (1991, 332–3).

[8] In the preface to the new edition, John Cairns pointed out that the idea for the volume came from Watson. That Watson himself wanted to play out Caltech and Cold Spring Harbor against Cambridge is not necessarily plausible (Cairns, 1992, v). The suggestion that Stent, who became the chief spokesman for the history of the phage group, was the main engineer behind the volume is more convincing; see Abir-Am (1982b, 312, note 57). Watson's contribution to the volume, a personal piece on 'growing up in the phage group', formed the starting point for his fuller autobiographical account in *The Double Helix* (Watson 1966).

subsequent elucidation of the double helix which led to the extraordinary development of the field. Kendrew noted that communication between the conformational and informational school was still limited, but times were changing. The regulation and functioning of genes could not be elucidated without knowledge about the three-dimensional structure of both proteins and nucleic acids. Conversely, the study of protein structure profited from knowledge on genetic variability and the mechanisms of inheritance. Kendrew's main aim was certainly to defend the rights of the structuralist school in the history of molecular biology. However, in his argumentation Watson and Crick's collaboration on the structure of DNA acquired a key role.

Kendrew's version of the origin story of molecular biology was taken up and in principle accepted by Stent, one of the editors of the Delbrück *Festschrift*, even if he continued to attribute the more innovative role to the informational school (Stent 1968).[9] Thus, Stent elaborated the notion of the two schools and acknowledged the combination of structural and genetic reasoning in Watson and Crick's work on the structure of DNA. However, in Stent's account, the importance of Watson and Crick's work on the double helix lay not in the mixing of the two schools, but in the inauguration of a second, dogmatic phase, following the romantic one, in the history of the informational school of molecular biology.

Despite this marking of positions, there is no doubt about the central role accorded by the phage school to Watson and Crick's work in the making of molecular biology. As Stent, increasingly assuming the role of spokesperson of that school, put it in his introduction to the critical edition of Watson's book:

prior to April 25, 1953 [the day of publication of Watson and Crick's letter to *Nature* announcing the double-helical structure of DNA], no member of the Phage Group thought of or referred to himself as a 'molecular biologist'. But on that day, Delbrück's circle suddenly realized...that what it had been doing all along was molecular biology. (Stent 1980, xvi)

The 'compromise' on the history of molecular biology achieved by Kendrew and Stent did not pass uncontested. It rather precipitated the production of a whole series of books and articles on the 'origins' of molecular biology. In the audience at the Collège de France in Paris, where Stent

---

[9] To Kendrew Stent confided that he had been 'reading and re-reading' the advanced proof of the review he had received from the editor of *Scientific American* (the original instigator of the piece) and that he considered it 'an important essay in the historiography of the field' (G. Stent to J. Kendrew, 28 February 1967, Kendrew Papers, MS. Eng. c.2592, L.136, Bodl. Lib.). Incidentally, Kendrew was instrumental in attracting Watson to Cambridge in 1951 (Watson 1980, 27; Olby 1994, 308).

had first presented his revised origin account, were also the molecular bio-logists of the Pasteur Institute. They were in close contact with the phage group and had contributed to the Delbrück *Festschrift*. Yet a few years later, in a collected volume dedicated to André Lwoff's career, Jacques Monod underlined his teacher's original contributions to the phage school (Monod 1971, 7–9). In a celebratory volume for Monod himself, the Pas-teurians finally presented their own 'myth of origin' (Lwoff and Ullmann 1979; Abir-Am 1982b).[10] Significantly, in this account the double helix played only a marginal role.

Other scientists, notably biochemists, but also the group of structural chemists around Pauling, have claimed their place in the history of the field (e.g. Rich and Davidson 1968; [Krebs Report] 1969; Chargaff 1974; Cohen 1975; 1984; Kornberg *et al.* 1976; Fruton 1992, 195-214).[11] Confronted with the construction of a field of research which was attracting increasing recognition and prestige, these attempts were characterised by a dilemma. Those who felt excluded insisted on the originality of their own work, but sought wider recognition for it by inscribing themselves in the very story they contested. As in the case of the Pasteurians, these accounts have often taken the form of alternative 'origin' accounts of the field of molecular biology, which was accordingly redefined.

The only two book-length accounts dealing with the postwar history of molecular biology have, in different ways, tried to accommodate these alternative stories. They none the less confirm the central role of Watson and Crick's work on the structure of DNA. Both books, originally pub-lished in the 1970s, have recently been reprinted (Judson 1994; Olby 1994). Despite critique from fellow historians, they represent something like the textbooks or standard histories of the field. This status they have also achieved through their reliance on key participants as informants.[12]

Robert Olby's *The Path to the Double Helix* was written with the aim of providing a broad picture of the intellectual and institutional developments

---

[10] On 'myths of origin' and the role of anniversary rites in science for the construction of a collective memory see also Forman (1969–70), Abir-Am (1992c; 1998) and Abir-Am and Elliott (1999).

[11] On the practices, disciplinary strategies and disputes of molecular biologists and biochemists in the 1950s and 1960s see Abir-Am (1992a) and de Chadarevian and Gaudillière (1996), including the contributions of Creager, de Chadarevian, Rheinberger, Gaudillière, Kay and Burian to the same special issue of the *Journal of the History of Biology*. For a reflection on national differences in the response of biochemists to the rise of molecular biology see Abir-Am (1992a, 225–6) and Gaudillière (1993).

[12] All the scientists I have interviewed in the course of my research have unfailingly referred to Olby's and Judson's books, sometimes expressing preference for one or the other, some hinting that there was little left for me to do, others expressing the feeling that their part in the story was not adequately represented, but all testifying to the canonical status the two accounts had acquired.

which have yielded a physical and chemical description of the gene (Olby 1994).[13] Intended as a corrective to Watson's account in *The Double Helix*, it none the less celebrated the double helix as the climactic event, as indicated not only by the title of the book, but also by the space dedicated to the detailed reconstruction of the actual work on the DNA structure. The new subtitle of the reprint edition, 'The Discovery of DNA', even if suggested by the publisher, does nothing to dispel the impression that it is around this event that the account turns.[14] Despite focusing on at least four important research traditions on which Crick and Watson's work rested, Olby canonises Kendrew's interpretation of their work as the meeting of two schools.[15]

If Olby followed the path leading to the discovery of the double helix, Horace Judson, journalist turned historian, takes this same event as the starting point of his best-selling history of molecular biology. The 'undeniable drama of how the discovery was made' supplied just the kind of material and emotion Judson was eager to convey to his readers (Judson 1994, 24). But for Judson (as for some of his scientist informants) DNA is also 'primary', because 'it comes first conceptually' as it directs both its own duplication and the synthesis of proteins with their specific functions (Judson 1994, 23).

In an afterword Judson distances himself from what he calls the 'standard view' of the origins of molecular biology by claiming that the 'revolution' in biology consisted not in the clarification of the structure and function of genes, but in working out the idea of specificity. Ideas of specificity stood at the centre of all five disciplines which, according to Judson, constituted molecular biology. But Judson underlines that it was the structure of DNA described by Watson and Crick that made 'specificity comprehensible' (Judson 1994, 611). The new notion of specificity was spelled out by Crick in two basic tenets of molecular biology following the structure determination of DNA: the sequence hypothesis which postulated that the specificity of a piece of nucleic acid was expressed solely by the sequence of its bases which coded for the amino acid sequence of a particular protein, and the central dogma according to which information

[13] A parallel attempt at putting Watson and Crick's work in perspective, but more narrowly focused on DNA research, was undertaken by two scientists; see Portugal and Cohen (1977).

[14] In a note to the new edition, Olby has defended himself against the accusation that the notion of a 'path' indicated a 'teleological' or 'whiggish' approach to history (Olby 1994, xxv). Even if this defence is not completely convincing, the book remains a rich resource for historical work in the field.

[15] For a critical discussion of the concept of 'school' deflated of its socio-political meaning and reduced to its 'intellectual' content see Abir-Am (1985, 80–92). For a renewed interest in research schools as a subject for historical examination see Geison and Holmes (1993).

flows from nucleic acids to proteins, but never vice versa (Crick 1958). In attributing a central role to the elucidation of the structure of DNA and its implications, Judson's reinterpretation of the history remains compatible with the standard view.

Olby's and Judson's histories have been exposed as 'second-order legitimations' of scientists' accounts aimed at establishing authority structures in science (Abir-Am 1985). Revisionist work on the history of molecular biology has focused on prewar developments in molecular biology and on the role of the Rockefeller Foundation in promoting (and in some cases rejecting) physical and chemical approaches in the life sciences (Abir-Am 1982a; 1987; Kohler 1991; Zallen 1992; Kay 1993a;). Nevertheless, these authors who, implicitly or explicitly, place the origins of molecular biology in the 1930s – although the term was hardly current at the time – also concur in asserting that the successful institutionalisation of the new science took place in the 1950s and 1960s. Watson and Crick's work on the structure of DNA in 1953 remains a key event.[16] Are we to conclude that institutionalisation was subsidiary to intellectual achievement? And does the fact that Watson and Crick met at Cambridge explain the early establishment of molecular biology there? To answer these questions we need to take a closer look at events in the unit where Watson and Crick first presented their model. Participants' own accounts of the history of that institution offer a useful starting point.

Annus mirabilis at Cambridge

In an article first published in the *New Scientist* in 1980 under the title 'Origins of molecular biology' and reprinted seven years later as 'The birth of molecular biology' in a special issue of the same journal celebrating the fortieth anniversary of the Cambridge Laboratory of Molecular Biology, Perutz followed a familiar script, but gave it a personal interpretation. Emphatically he declared:

The year 1953 became the *annus mirabilis*. The Queen was crowned; Everest was climbed; DNA was solved; Huxley and the late Jean Hanson, then at MIT, discovered the sliding mechanism of muscular contraction; and I found a method of deciphering the X-ray diffraction patterns of crystalline proteins. (Perutz 1987)

---

[16] For Kay, for instance, the presentation of the double helical model of DNA and its reception marked the shift from the 'protein' to the 'nucleic acid paradigm' in molecular biology (Kay 1993a, 269–77). Note, however, that more recently Kay has provided evidence that, contrary to current assumptions, the information discourse in biology *preceded* Watson and Crick's work on the double helix, thus displacing the centrality of the double helix in this respect.

Perutz's statement is interesting on several accounts. Associating the coronation of the Queen and the conquest of the world's highest mountain peak with scientific breakthroughs implies a specific understanding of scientific work and of the role of the scientist in society.[17] In addition, there is a subtle link between the coronation, Everest and DNA, which Perutz may not even have intended. The 1953 coronation was 'television's coronation' (Briggs 1979, 458). With the cameras of the BBC positioned in Westminster Abbey, Elizabeth II became the first queen to be crowned 'in sight of all the people'. The ascent of Mount Everest, announced on television on coronation day, four days after the actual event, heightened the excitement of the media event. It marked a milestone in the history of television in England.[18] At the end of the decade, DNA made its first appearance on televison (see above, chapter 5). The structural model of the molecule lent itself to visual presentation, and the visual culture inaugurated by the new medium of mass communication has contributed in important ways to turning DNA into a cultural icon. As we have seen, Watson's dramatised account of the discovery of the structure of DNA offered further material for consumption by the viewing public.

Perutz's statement is also interesting because it maintains 1953 as a turning point in the history of molecular biology. For Perutz, however, this year is remarkable not only for the presentation of the DNA model, but also for breakthroughs in the structural interpretation of muscle fibres and globular proteins. These were all research projects pursued at Cambridge (Huxley, working on the mechanism of muscle contraction, later rejoined the unit). Perutz's view of history, therefore, like other origin accounts discussed above, reflects local research traditions and interests. As we shall see, molecular biology in Cambridge never was exclusively molecular genetics.

However, for the Cambridge unit, 1953 marked not only a year of achievements but also the beginning of an acute crisis regarding the future of the group working on the analysis of biological structures in the Physics Department. In the longer term, the crisis prompted the creation of molecular biology in Cambridge. This was not *just* a change of name or a political move. The negotiations ensuing from the institutional crisis of the unit created new research alliances and institutional structures. What molecular

[17] Intriguingly, the *Nature* issue carrying Watson and Crick's second article on the DNA structure also featured an editorial on the coronation and its meaning for science. One connection the anonymous writer mentioned was 'the astonishing miracle of science' which had made it possible for the coronation to be televised (Anonymous 1953a). Other salient political events of that year which Perutz could have mentioned included Stalin's death, the end of the Korean War and the test of the first Soviet hydrogen bomb.

[18] For a detailed account of the BBC coronation service see Briggs (1979, 457–73).

biology became in Cambridge depended on the course and length of the negotiations.[19]

A cuckoo's egg

In December 1953 Bragg left the Cavendish Chair of Physics in Cambridge to follow his father as Director of the Royal Institution in London. The newly designated Cavendish Professor, Nevill Mott, a solid state physicist from Bristol, let it be known that he did not intend to keep the MRC unit in his department. In the correspondence with the university which preceded his move to Cambridge, he made it clear that he needed the space occupied by Perutz's unit for his own research, for which there was no other place in the Cavendish.[20]

This crisis in the institutional legitimation of the unit was not completely unexpected. If anything, it underlined once more the importance of Bragg's patronage for the work of the unit. Bragg's move to London was not foreseen. But in view of his retirement, which was due in 1955, Bragg and the secretary of the MRC had already corresponded about the future of the unit and its researchers. A first informal exchange of opinion had occurred in 1950, when the five-year funding of the unit was only just half-way through. Bragg and Himsworth agreed that the main difficulty concerning the unit arose from the borderline character of its subject matter. Not really belonging to any department, it would be most liable to be cut down in times of financial crisis. Similar considerations arose in relation to Perutz's own future, a point which was particularly important to Bragg. Himsworth felt that since Perutz was 'only just in the medical field', the MRC might find it impossible to justify supporting him 'if circumstances became difficult'. He also saw the possibility that Perutz's line of research could move over or come to be included in one of the physical sciences, which again would make funding by a body concerned with medical research problematic. Therefore Himsworth suggested to Bragg that Perutz should be offered a university appointment, while continuing to act as honorary director of the MRC unit. In this way responsibilities would be shared and Perutz's future would be assured.[21] Kendrew was expected to be 'the type of man who almost certainly would have moved

---

[19] On the importance of the time dimension in decision-making processes for the design and function of technologies see Pestre (1992).

[20] On the reorganisation of research in the Cavendish under Mott see Crowther (1974, 339–57).

[21] H. Himsworth to W. L. Bragg, 16 January 1950, file FD1/426, MRC Archives.

away from Cambridge to another job' in a few years' time.[22] The two other members of the unit, Huxley and Crick, were also expected to leave before long. Following a conversation with Bragg in June 1952 on the future of the unit, Himsworth noted that Bragg's general concept seemed to be that Perutz would have a university post and people would come to work with him, while the unit would gradually dissolve.[23]

Despite these future uncertainties, under Bragg's protection the unit expanded. The group acquired a large new room in the Cavendish and applied to the MRC for the appointment of a biochemist to the staff and for a research assistant to Kendrew. Both requests were eventually met.[24] Vernon Ingram, an organic chemist with experience in protein chemistry, joined the unit in 1952 and built up a biochemical laboratory in the Cavendish. Support for the unit by the MRC was renewed for one and then two additional years till 1955. Perutz was offered a three-year university lectureship in biophysics which he took up in October 1953, just when Bragg was about to leave.[25] In fact, the position meant a cut in stipend and reduced research time, and Perutz was reluctant to accept it. What finally convinced him to resign his MRC staff position and take up the lectureship was the expectation that the university would feel more obliged to provide adequate laboratory space for the unit.[26]

Problems arose, however, as soon as Bragg left Cambridge. The Rockefeller Foundation, which continued to support the unit with important grants for equipment, required a guarantee that the university would provide accommodation before considering further funding. Mott,

---

[22] H. Himsworth, 'Interview with Sir Lawrence Bragg and Mr. Saunders at Cambridge, 15.3.50', file FD1/426, MRC Archives.

[23] H. Himsworth, 'Interview with Sir Lawrence Bragg, 13 June 1952', file FD1/427, MRC Archives. Yet a few months later, when the renewal of the MRC funding was up for consideration, Bragg assured Himsworth of the excellence of the work of the unit and of the very good prospects that it would find a home in the university even after his retirement, most likely in the Cavendish itself or otherwise in the Department of Colloid Science; W. L. Bragg to H. Himsworth, 11 October 1952, file FD1/426, MRC Archives.

[24] M. Perutz to A. L. Thomson, 9 October 1950, file FD1/427; W. L. Bragg to H. Himsworth, 1 March 1951, file FD1/427; W. L. Bragg to H. Himsworth, 18 June 1951, file FD1/426, MRC Archives. Kendrew's first research assistant was Watson. But Kendrew quickly realised that Watson was not patient enough to help him with his measurements; interview with J. Kendrew, Linton, 18 March 1993.

[25] General Board Paper 2900, University Archives, CUL. Perutz's lectureship was a joint lectureship in the Faculties of Physics and Chemistry, Biology 'A' and Biology 'B', but, according to the statutes which regulated these special lectureships, was assigned to one department, which Perutz chose to be the Department of Physics.

[26] M. Perutz to H. Himsworth, 26 June 1953, file Himsworth, LMB Archives. At the end of the academic year 1956/7, after a series of prolonged leaves for health reasons and when the application to turn the lectureship into a readership was not successful, Perutz resigned his university position and returned on the MRC staff.

who was to take up his appointment only in November 1954, promised to keep the unit till the summer of 1955, but not beyond. He reckoned that the range from nuclear physics to biophysics was much too large for one professor to look after, and that the biophysics unit contributed little, if anything, to the teaching and had few contacts with other groups in the department. He therefore deemed it 'most undesirable' to keep a large and expanding biophysics unit in the Cavendish.[27] The university administration convinced Mott not to take any further steps regarding the future of the unit until he had taken up his appointment. The search for alternative quarters became none the less imperative and soon proved to be anything but easy. However, despite the critical situation, Perutz and the other members of the unit were not ready to accept just any solution.

Bragg, upon leaving Cambridge, had tried to convince at least Kendrew to help him build up protein crystallography at the Royal Institution. But being split up and 'tucked away' in the Royal Institution, without the stimulus which Cambridge and the Cavendish offered, seemed unattractive.[28] From Bristol, Mott suggested that the canteen building near the linear accelerator on the Madingley site in Cambridge, just 2 miles out of town, should be adapted to laboratory use. Yet Perutz and the other members of the unit objected that the distance from the centre would undermine the possibility to engage in collaborations with other researchers and to participate in teaching. 'All colleagues', Perutz replied to Mott, 'prefer a cabin in the centre of Cambridge than a palace two miles away.' If they had to move out of town, Perutz predicted, the unit would shrink.[29] The Professor of Colloid Science, Francis Roughton, offered space to the unit in his Department in the centre of town, but not enough for the whole group to move together. The solution favoured by Perutz, even if removed in time, was the one proposed by the Professor of Chemistry, Alexander Todd, according to which the unit would be given accommodation in the old Chemistry block, once the Chemistry Department moved to its new quarters. 'This would give us enough space', Perutz reckoned, 'to keep the Research Unit together and would put us in a central position where we could continue our collaboration with other departments, such as Crystallography, Mathematics and Physiology.'[30]

[27] N. Mott to H. M. Taylor (Secretary General of the Faculties), 18 February 1954; General Board Note 11,630 for meeting of 10 March 1954, University Archives, CUL.
[28] M. Perutz to N. Mott, 19 February 1954, file Himsworth, LMB Archives.
[29] M. Perutz to N. Mott, 29 June 1954, file Himsworth, LMB Archives. This letter, however, was written only after the General Board had accepted the responsibility to provide adequate accommodation for the unit; see below.
[30] M. Perutz to H. M. Taylor, 22 May 1954, copied as General Board Paper 2982, University Archives, CUL.

Finding new accommodation clearly meant more than just finding a space for the single members of the unit to work. Space in the university was defined in terms of disciplines.[31] Finding a place meant redefining the work of the unit and positioning it in the disciplinary ecology of the university. It meant negotiating for resources, recruits and recognition and in this respect was an integral part of the research work. Perutz, as the director of the unit, knew this very well. In his correspondence with the Cavendish Professor designate and the university administration he unmistakably linked space to scientific considerations. He insisted that the work of the unit was at a critical stage. The 'Rosetta stone' to the solution of protein structure, Perutz explained, had been found in the unit, and unless they could exploit it, others would reap the benefits. To exploit it, the unit had to stay together. The members of the unit, by their different abilities and training, he argued, 'supplement each other in a hundred useful ways'.[32] By staying together, the chances of getting useful results were greater. Accommodation at least of the scale they were currently occupying and a 'reasonable security of tenure' were crucial conditions to pursue their investigations. 'Provided the university gives us accommodation', he summed up the situation, 'we are assured of generous financial support from both the M.R.C. and the Rockefeller Foundation, and I am sure that we shall be able to produce good work.'[33]

Perutz insisted on the importance of locating the unit in the centre of Cambridge, which meant in one of the two adjacent laboratory sites of the university, the Downing or the New Museum site, and on the importance of keeping up collaborations with other departments (Fig. 6.1). He did not, however, link the future of his group to a particular department. Teaching was not a central concern to Perutz and the other researchers, even if *some* teaching was seen as an important step in the career development of the individual members of the unit. Attracting research students was important in so far as they represented helpful hands, but training them was not seen as an aim in itself. The objectives of Perutz and his colleagues thus, in many respects, ignored the disciplinary tradition of the university and its mission as a research *and* teaching institution. Their asset, however, was that they could bring the university important research money.

After half a year of negotiation, the university administration, pressed by Perutz, finally accepted the responsibility to provide 'adequate'

---

[31] On 'place' as physical and cultural notion and on the relation of disciplinary distinctions and spatial arrangements in knowledge-making sites see Ophir and Shapin (1991).

[32] M. Perutz to N. Mott, 19 February 1954, file Himsworth, LMB Archives.

[33] M. Perutz to H. M. Taylor, 22 May 1954, copied as General Board Paper 2982, University Archives, CUL.

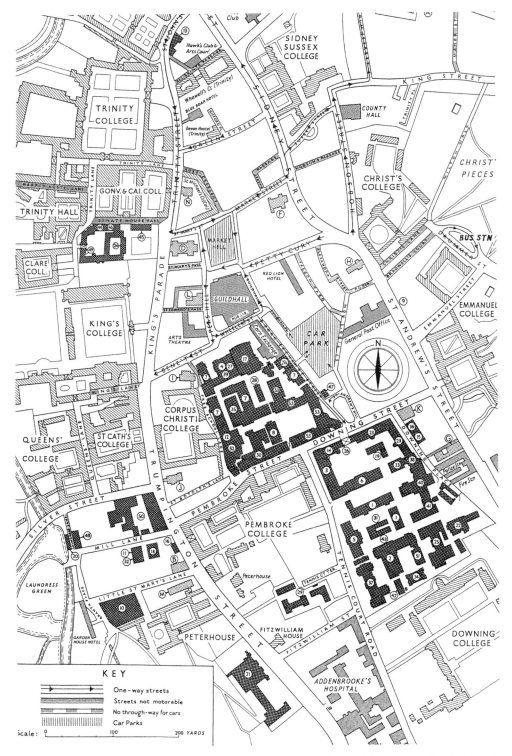

6.1 Map of laboratories at Cambridge (1960). Note the central position of the hut (number 28) which the MRC unit acquired in 1957. From *A Map of Cambridge*, W. Heffer & Sons Ltd (1960). Reprinted by permission of Heffers and the Syndics of Cambridge University Library.

KEY TO MAP OF LABORATORIES, ETC.

UNIVERSITY DEPARTMENTS, ETC.

1 Agriculture, Down. Site
2 Anatomy, Down. Site
3 Archaeology and Anthropology, Down. Site
4 Arts School, Bene't St.
5 Biochemistry, Down. Site
6 Botany, Down. Site
7 Cavendish Lab., New Mus. Site
8 Chemical Engineering, New Mus. Site
9 Churchill Coll. Office, St Andrew's St.
10 Classical Archaeology. Museum
11 Classics Faculty Lib., Mill Lane
12 Colloid Science, New Mus. Site
13 Divinity School, St John's St.
14 Economics Faculty, Down. Site
15 Egyptology Library, Downing Pl.
16 Estate Management, Trumpington St.
17 Examination School, New Mus. Site
18 Extra-Mural Studies, Board of
19 Faculty Rooms, Down. Site
20 Faculty Rooms, Laundress Lane
21 Fitzwilliam Museum, Trumpington St.
22 Geography, Down. Site
23 Geology, and Sidgwick Museum, Down. Site
24 Law Faculty, Old Schools
25 Low Temp. Res. Stn., Down. Site
26 Marshall Lib., Down. Site
27 Mathematics Faculty, Bene't St.
27a Mathematical Lab. New Mus. Site
28 Med. Res. Council, Molecular Biology, New Mus. Site
29 Medicine (Experimental) Lab., Tennis Ct. Rd.
30 Metallurgy, New Mus. Site
31 Microbiology Lab., Down. Site
32 Mill Lane Lecture Rooms
33 Mineralogy and Petrology, Down. Site
34 Molteno Inst. of Parasitology, Down. Site
35 Mond Lab. (Physics), New Mus. Site

36 Music School, Downing Pl.
37 Pathology, Down. Site
38 Pharmacology, Down. Site
39 Philosophical Lib., Arts School
40 Physiology, Down. Site
    Physics (see Cavendish and Mond Labs.)
41 Psychology, Down. Site
42 Public Health Lab., Down Site
43 Radiotherapeutics Lab., Down. Site
44 Seeley Historical Lib., Old Schools
45 Senate House
46 Squire Law Lib., Old Schools
47 Statistical Lab., St Andrew's Hill
    Stuart House (see Extra-Mural Board)
48 University Assistants' Club, Laundress Lane
49 University Offices, Old Schools
50 University Press, Trumpington St.
51 Veterinary Anatomy, Down. Site
52 Whipple Mus. of Hist. of Science, New Mus. Site
53 Zoology, New Mus. Site

PLACES OF WORSHIP

A All Saints
B Emmanuel Congregational
C Friends' Meeting House
D Great St Mary's (University Church)
E Holy Sepulchre (Round Church)
F Holy Trinity
G St Andrew's Baptist
H St Andrew's the Great
I St Bene't's
J St Botolph's
K St Columba's Presbyterian
L St Edward's
M St Mary's the Less
N St Michael's

6.1 (*Cont.*)

accommodation for the unit for as long as the MRC supported it and wished that it remained in Cambridge.[34] Informed of this decision, Himsworth assured the university that he felt under 'moral obligation' to leave the unit in Cambridge where it 'was born and has grown-up' and where it was producing good work. He was, therefore, very pleased that the university was committed to providing adequate accommodation.[35] The formulation of a more formal agreement between the university and the MRC regarding the accommodation of the unit was postponed until a decision was reached about the department to which it should be attached. However, the agreement in principle provided important assurance regarding the future of the unit. A major hurdle seemed to have been cleared, even if a site still had to be found.

[34] General Board Minute 12,102, meeting of 9 June 1954, and H. M. Taylor to H. Himsworth, 30 June 1954, ref. GB.546443, University Archives, CUL. The Professor of Chemistry, Alexander Todd, apparently played a decisive role in this decision. See M. Perutz to H. Himsworth, July 1954, file Himsworth, LMB Archives.

[35] H. Himsworth to H. M. Taylor, 7 July 1954, file Setting up and Constitution of Laboratory, LMB Archives.

The question of the accommodation of the biophysics unit was addressed as soon as Mott took up his position at Cambridge. A special committee was formed to discuss this and other matters regarding the Cavendish, notably the question of outside funding of nuclear physics and radio astronomy. The committee accepted that Mott could not provide as much space for the unit as it had occupied so far, and decided that the needs of the unit would be considered along with other needs for accommodation on the New Museum Site.[36] This, however, meant putting things on hold.

With the postwar expansion of research in the university and the rise of student numbers there was much pressure on central laboratory sites. Unlike other British universities, as for instance Oxford, Cambridge followed a politics of containment and did not build up a new science area outside the centre. The central laboratory area was to be remodelled, but planning as well as new construction projects proceeded slowly. Competing interests of various departments, all striving to increase their own influence, further slowed down decision processes. The plans for the new Chemistry laboratory also advanced much more slowly than expected. Thus, the hope that space in the old Chemistry block would be available in the near future proved an illusion. While the tenure of the unit in the Physics Department was tacitly prolonged, the space problem became ever more pressing. Brenner, who was going to join the unit by January 1957, was prepared to work 'in a cupboard' (Crick 1987, 66).[37] A few months later the unit successfully negotiated that a prefabricated hut in the court of the Cavendish, recently vacated by Metallurgy, would not be demolished, but be assigned to them. On first inspection an MRC officer noted that the hut 'was quite unsuitable – one could say – useless for any refined scientific work'. Perutz and his colleagues, however, considered it the 'most suitable accommodation available'. Sharing costs, the MRC and the university adapted it to laboratory use and a five-year tenure was agreed.[38] The unit also secured two derelict rooms in the old Anatomy building, belonging to the Department of Zoology, for Crick and Brenner's phage

[36] General Board Paper 3112, University Archives, CUL.

[37] See also F. Crick to S. Brenner, [20] October 1955, Sydney Brenner, private collection, box Archive 1. Crick and Brenner corresponded regularly after Brenner's return to South Africa (see below). Unfortunately, only one side of the correspondence, Crick's letters to Brenner, is preserved.

[38] M. Perutz to H. Himsworth, 7 March 1957, file Himsworth, LMB Archives. 'Report on visit to MRC unit in Cambridge on 13 June, 1957', and 'Accommodation for the Unit for Research on Molecular Structure of Biological Systems' [note by Himsworth dated 24 June 1957], file E243/109, vol. 1, MRC Archives. Note that, contrary to a Cambridge myth, Watson and Crick's work on the double helix happened not in the hut, which was not occupied by the unit before 1957, but in the Austin wing of the Cavendish.

6.2 The 'hut' on the central laboratory site in Cambridge (*c*. 1960). The MRC unit occupied the space from 1957 to 1962. Courtesy of Medical Research Council Laboratory of Molecular Biology, Cambridge.

work. These were temporary solutions, but allowed research to continue (Figs. 6.1–6.3).

The phage was newly introduced as a research object in the Cavendish group by Brenner, who joined the unit at Crick's suggestion in January 1957. The twenty-year-long close collaboration between Crick and Brenner has received much less attention than Crick's collaboration with Watson. Arguably, however, Brenner's move to Cambridge marked a much more radical shift in the research strategies pursued at the Cavendish unit than the collaboration between Watson and Crick. To understand this it is important to trace in more detail the impact of the DNA work on Watson's and Crick's later work and the place of the double helix in the development of the Cambridge unit.

After the double helix

In his account of the history of the laboratory, Perutz has acknowledged that Watson stirred the minds of the protein crystallographers in Cambridge by introducing the question of where proteins came from (Perutz 1987, 40). Watson, however, came to Cambridge to learn X-ray

6.3 Group picture of some of the people working in the hut (*c.* 1958). From left to right, back row: Larry Steinrauf, Richard Dickerson, Hilary Muirhead, Michael Rossmann, unidentified, Ann Cullis, Bror Strandberg, Wibeke, unidentified; front row: Leslie Barnett, Mary Pinkerton and Max Perutz. Courtesy of Medical Research Council Laboratory of Molecular Biology, Cambridge.

crystallography. Crick's expertise in the interpretation of X-ray diffraction pictures proved crucial for their common work on the structure of DNA. What distinguished their approach to structure determination in relation to their Cambridge colleagues was an earlier and stronger heuristic use of model building, inspired by Pauling.[39] A few months after submitting their two papers to *Nature*, both Watson and Crick left Cambridge to take up temporary positions in the United States, and the MRC unit reverted exclusively to structural work on proteins. Watson and Crick, in different ways, followed up their work on DNA structure. Both researchers brought back their work to Cambridge, Crick on a more long-term basis, Watson as a frequent visitor to the unit. But their projects following their collaboration consisted to a large extent in the prolongation of approaches and techniques already applied in the unit. The helix itself did not become an

[39] For a comparison of institutional set-ups and experimental approaches in X-ray crystallography at Caltech and Cambridge see Olby (1970a).

object of inquiry at Cambridge. In the longer run it did change research directions in the laboratory, but in indirect ways.

In the summer of 1953, Watson took up a two-year fellowship with Delbrück at Caltech. The pairing mechanism he and Crick had proposed for the two strands of the DNA molecule suggested new avenues of research for the problem of replication which was of particular interest to the phage people. However, Watson did not pick up this line of investigation. Rather, together with Alexander Rich, a postdoctoral fellow in Pauling's adjacent lab, he set out to determine the structure of RNA by X-ray crystallography.

Watson has described this work in his Nobel lecture. Here he claims that the problems of gene replication and of the genetic control of protein synthesis (in which RNA was implicated) could be 'logically attacked' only when the structure of DNA became known (Watson 1964, 785). However, Watson started work on RNA while in Cambridge, after his and Crick's first shameful attempt to present a structure of the DNA molecule (Olby 1994, 357–63). At that stage it was agreed that DNA studies should be left to Randall's group, while Crick was to focus on his thesis, which concerned the structure of proteins. Watson, who was formally attached to the plant virologist Roy Markham and was expected to work on the structure of nucleoproteins and viruses by the Merck Fellowship Board, embarked on the crystal structure analysis of tobacco mosaic virus (Olby 1994, 308, 354, 364). An important component of the virus was RNA, and Watson hoped that studying viral RNA could provide clues on the structure of DNA, while not blatantly impinging on the field of research claimed by the King's College group. But the structure and function of RNA *per se* represented as intriguing a problem as DNA. Crick and Watson were convinced that in RNA viruses (most plant and some animal viruses) RNA was the carrier of genetic information (Watson 1980, 68). RNA was also implicated in protein synthesis since Jean Brachet and Torbjörn Caspersson's pioneering experiments had shown that, in eukaryotes, protein synthesis occured not in the cell nucleus but in the cytoplasm, and that activity was correlated to RNA content. A scheme jotted down in his notebook in November 1952 confirms that, well before his second attack on the structure of DNA, Watson attributed a central role to RNA in the functional relationships between DNA (genetic information) and proteins (Watson 1980, 90) (Fig. 6.4). The RNA pictures he took at Cambridge, however, were 'very diffuse' and therefore not interpretable (Watson 1964, 787).

Working on RNA *after* DNA made a difference, but it is important to see in what way. The complementary structure of DNA suggested that genetic specificity was transferred from DNA to RNA through base pairing along a

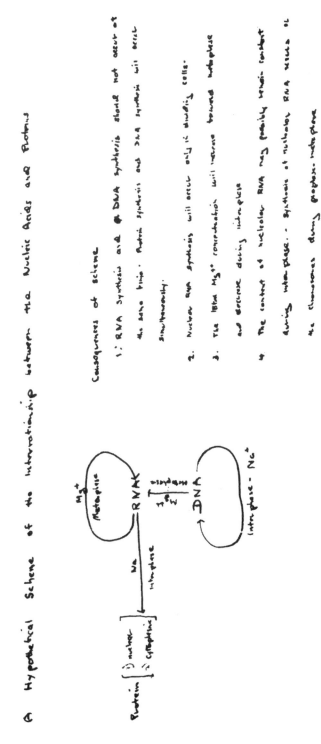

6.4 Page of James Watson's laboratory notebook showing a hypothetical scheme of the interaction between nucleic acids and proteins (1952). From J. Watson, *The Double Helix* (1980). p. 90. Reprinted with permission of Cold Spring Harbor Laboratory.

template.[40] The relation of DNA and RNA remained none the less a highly controversial issue (Rich and Watson 1954b; Rich 1995, 107). But we have already seen that knowing how DNA made RNA was not a prerequisite for studying RNA structure. What Watson took over from his work on DNA structure was experience with a set of techniques he had acquired in his work at the Cavendish and the expectation that the crystallographic and model building approach which had proved so successful in the case of DNA would lead to similar results in the case of RNA. In a striking manner, the double-stranded structure of DNA had suggested a possible mechanism for the function of the molecule. This, however, was only true for the mechanism of replication of the genetic material. Moving on to RNA was carried by the hope that the structure of this molecule would suggest the mechanism of protein synthesis which the structure of DNA had not given away.

The DNA work guided Watson and Rich's attempt to establish the structure of RNA from the methods of fibre making to the interpretation of the RNA diffraction pictures (Rich and Watson 1954a; 1954b). The RNA work, however, presented unexpected technical as well as interpretative problems. As in Watson's first attempt, the X-ray diffraction photographs of RNA did not show the same resolution as Franklin's DNA pictures, whether from lack of skill or because RNA fibres did not pack as neatly as DNA remained unclear. Analytical data indicated the existence of at least two different classes of RNA, one with a random distribution of bases, the other with complementary base pairing. But these data were at odds with crystallographic evidence which indicated that all RNAs produced the same X-ray pattern. Model building proved equally unsatisfactory (Rich and Watson 1954b).

In 1955–6 both Watson and Rich, the latter meanwhile working at the National Institutes of Health in Bethesda, visited the Cambridge unit to pursue different crystallographic projects (see below).[41] Later Watson moved on to study the structure of the ribonucleoproteins known to be the site of protein synthesis in the cell. Again the hope was that the establishment of the structure would reveal the mode of function of the molecules. But the approach which had proved so fruitful in the case of DNA structure did not lead a second time to quick success. The mechanism of protein

---

[40] That RNA originated on a DNA template had been suggested by Alexander Dounce on theoretical grounds before Watson and Crick presented their model of DNA in 1953. Crick was aware of Dounce's work (Dounce 1953; Judson 1994, 247, 296). Watson and Crick's contribution was to suggest an original mechanism according to which the template would exercise its function.

[41] Rich's departure may well have been a reason for Watson to abandon the RNA studies. Rich continued working on the problem for many years.

synthesis turned out to be more complicated than expected and other avenues of research had to be pursued.

Crick submitted his Ph.D. thesis in July 1953. Briefly thereafter he left
Cambridge to take up a temporary position at Brooklyn Polytechnical
Institute in the United States. Crick's thesis comprised theoretical and experimental studies on the structure of proteins and polypeptides, especially
horse and ox haemoglobin. Crick considered incorporating the DNA studies, but decided against it, because it was a collaborative piece of work and
because it would have meant recasting the whole plan of the thesis (Crick
1953, [4]). The thesis ended with a new hypothesis on the structure of
globular proteins and suggestions on how to test it which Crick hoped to
pursue in the future. All this clearly indicates that Crick situated his work
on DNA in the context of his crystallographic studies which he proposed
to continue.

In Brooklyn Crick was expected to transmit crystallographic skills and
experience acquired in Britain to the people working under David Harker,
head of the Protein Structure Project at the Polytechnical Institute.[42] The
possibility of expanding his visit to spend six to twelve further months with
Pauling at Caltech was also discussed. Perutz was keen for Crick to return
to the unit after his visit to the United States, and he pursued the question
before Crick's departure. In view of the uncertainties surrounding the
unit's future and unsure if Crick met the prerequisites, the MRC did not
accept Perutz's suggestion of unlimited tenure, but offered a seven-year position. Crick, writing from the States, keenly accepted.[43] His first months
abroad had evidently confirmed the view he had expressed before, that
the MRC unit was 'the obvious choice for anyone wishing to work on the
structure of biologically important molecules', both in terms of facilities
and for its 'intellectual atmosphere'.[44]

On his return to the Cavendish in autumn 1954, Crick continued to
work in protein crystallography. His main collaborations also continued
to be in this field. When Watson came to visit, they published a couple
of joint papers on the structure of viruses, based on Watson's crystallographic analysis of tobacco mosaic virus. They suggested that viruses were
constructed from a number of identical small sub-units arranged in a regular manner. They supported their claim arguing that the information the
virus RNA could carry was limited (this was assuming that the RNA did

---

[42] On Harker and his contributions to X-ray crystallography see Hauptman (1998).

[43] F. Crick to H. Himsworth, 15 June 1953; M. Perutz to J. G. Duncan, 14 July 1953; J. G.
Duncan to Perutz, 25 July 1953; F. Crick to J. G. Duncan, 4 February 1954; file P3/125,
vol. 1, MRC Archives.

[44] F. Crick to H. Himsworth, 15 June 1953, file P3/125, vol. 1, MRC Archives.

determine the genetic properties of the virus) and that the virus would find it easier to control the production of a large number of identical sub-units rather than that of one or two very large ones (Crick and Watson 1956). When briefing Monod on his DNA work for the Nobel nomination, Crick expressed surprise that people did not more often refer to his and Watson's work on virus structure which, he reckoned, was 'quite original' at the time and influenced 'all the modern X-ray work on viruses'.[45] With Rich, also visiting from the United States, Crick worked on collagen, proposing a triple helical structure for the molecule (Rich and Crick 1955).[46]

Work on the double helix, however, led Crick also into new directions. A letter addressed to Watson and Crick by the theoretical physicist, cosmologist and cryptologist George Gamow in the summer of 1953 proved decisive. Gamow had been intrigued by the DNA structure and its interpretation presented by Watson and Crick in their second paper in *Nature*. 'Your article', he wrote exuberantly, 'brings Biology over into the group of "exact" sciences'. It opened the 'very exciting possibility' of studying the genetic characteristics of organisms 'by theoretical research based on combinatorix [*sic*] and the theory of numbers'.[47] A little later he suggested a way in which the double helix could serve as a direct template for protein synthesis. According to Gamow's scheme, which he published in a brief note to *Nature* in October 1953 and later elaborated in a full-length publication, the helix could form twenty different cavities, depending on the sequence of the bases (Gamow 1954a; 1954b). Boldly cutting down the large number of known amino acids to a basic set of about twenty, Gamow concluded that there was a cavity for each amino acid. Gamow further assumed that each amino acid was coded for by a triplet of bases and that the triplets standing for successive amino acids overlapped (this was necessary for stereochemical reasons).

Crick, together with Watson, quickly realised that Gamow's suggestion was untenable. But Gamow's approach made Crick realise 'that solving the code could be viewed as an abstract problem, divorced from the actual biochemical details' (Crick 1990, 93). This made it attractive to him, and in

[45] F. Crick to J. Monod, 31 December 1961, fonds Monod, Mon. Cor. 04, AIP. The genetic argument was evidently based on assumptions regarding the function of virus RNA and the mechanism of virus replication which were still hypothetical. On tobacco mosaic virus as a model system in biomedical research see Creager (in press).

[46] At Cambridge Rich also pursued his studies of the structure of RNA, using the powerful X-ray tubes in the unit to take diffraction pictures of artificial RNAs. The aim of these experiments remained that of understanding the function of RNA in protein synthesis; interview with A. Rich, Cambridge, MA, 20 July 1993.

[47] G. Gamow to J. Watson and F. Crick, 8 July 1953, Sydney Brenner, private collection, box Archive 1. On the decisive role of Gamow, his cryptological interests and his military connections in the history of the code see Kay (2000).

the following years he became deeply involved in the problem. However, again it is important to see exactly how his work on the code, and that of other people in the unit he recruited to work on the problem more generally, built on the helix.[48]

Codes, maps and sequences

The double helix had surely precipitated Crick's interest in the question of how genes function on the molecular level, but in his contributions to the 'code problem', which were mainly on a theoretical level, the actual structure of the DNA molecule played only a subordinate role. Experimental work on the genetic control of protein synthesis in the Cambridge unit started with Ingram. A more large-scale project combining biochemical and genetic techniques was set in place with Brenner's move to Cambridge.

While at Brooklyn, Crick set out to disprove Gamow's first suggestion of a genetic code. Structural arguments spoke against Gamow's scheme, but this was not Crick's main line of attack. Using the few protein sequencing data so far available, he dismissed the possibility of an overlapping code which formed a crucial part of Gamow's argument (Crick 1990, 94).[49] In a talk presented in Paris on the occasion of one of the numerous fortieth-birthday celebrations of the DNA model, Crick, commenting on the importance of the model for later research, declared that the fact that it was a helix represented 'a nuisance', since it posed a series of theoretical problems regarding DNA function, but that base pairing proved important.[50] In Crick's first approach to the code problem, not even the latter was relevant.

The same is true for another of Crick's most celebrated contributions, his paper on the 'adaptor hypothesis' first circulated in the RNA Tie Club. This club, initiated by Gamow, never actually met, but, according to participants, became an important forum for the circulation of bold ideas and experiments regarding the nature of the genetic code. The name, as well as the embroidered insignium on the ties of its members, makes clear that RNA rather than DNA represented the main focus of interest (Fig. 6.5). In his paper Crick postulated the existence of a series of small molecules able to bind a particular amino acid and to recognise a specific sequence on

---

[48] The crystallographic work aimed at the refinement of the DNA structure was left to Wilkins, honouring the former agreement between Cambridge and London and following Franklin's move to Bernal's group at Birkbeck College. But there is no indication that Crick himself would have been interested in pursuing this line of research.

[49] Here Crick implicitly already used what he later formulated as the 'sequence hypothesis'. For further discussion see below.

[50] F. H. C. Crick, 'Looking backwards', talk presented at the Symposium From the Double Helix to the Human Genome. 40 Years of Molecular Genetics, Paris, UNESCO Headquarters, April 1993 (quote taken from notes written during the lecture).

6.5 Meeting of some members of the RNA Tie Club in Crick's house at Cambridge (autumn 1955). Left to right: Francis Crick, Alexander Rich, Leslie Orgel and James Watson. Courtesy of Alexander Rich.

the RNA template. Contrary to what is sometimes believed, Crick did not postulate base pairing as the mechanism by which the adaptor would bind to the template, but spoke much more generally of a 'specific hydrogen-bonding surface' of the adaptor molecule.[51]

It should also be noted that information language and the notion of a 'genetic code' had entered genetics even before DNA had replaced

---

[51] The relevant passage of the unpublished manuscript is quoted in Judson (1994, 290). The 'soluble RNA', later called transfer-RNA, identified by Mahlon Hoagland in Paul Zamecnik's laboratory at Massachusetts General Hospital in Boston, fulfilled in principal the functions postulated by Crick for the 'adaptor molecules'. Hoagland, however, has stressed that his work developed independently of Crick's suggestion (Hoagland 1990, 96). For a discussion of these events see Olby (1970b, 966–7) and Rheinberger (1994). In 1957, Hoagland visited Cambridge for a year to collaborate with Crick in isolating and characterising the soluble RNAs (see below). On the RNA Tie Club see Kay (2000).

proteins as the hereditary material. As has been convincingly shown, these terms were imported into biology from cybernetics, information theory, electronic computing, and control and communication systems in the aftermath of World War II. People like Norbert Wiener and John von Neumann, who had been involved in the wartime creation of the new technologies themselves became instrumental for their application in biology. Their work was soon picked up by some biologists, among them the British geneticist John Burdon Sanderson Haldane, his colleague at University College London, Hans Kalmus, and Henry Quastler, Austrian émigré radiobiologist at the University of Illinois (Kay 1995; 2000). When Watson and Crick, in their second paper on the genetic implications of the DNA structure, concluded that the sequence of the bases most probably represented 'the code which carries the genetical information', they used these notions without much ado (Watson and Crick 1953b, 965). This confirms that by 1953 the concept of a genetic code was not new and that the information language was commonly used. Discussing the application of information theory to the problem of protein synthesis, American biophysicist Hubert Yockey, organiser of a 1956 symposium on information theory in biology at Oak Ridge National laboratory, Tennessee (a laboratory taken over from the Manhattan Project by the Atomic Energy Commission and supporting a large biophysics section), stressed that even if using Watson and Crick's suggestion concerning the role of DNA for developing the formalism, the central ideas of his paper were independent of much of the detail embodied in the DNA model (Yockey 1958a, 51). In addition, genetic transfer was only one of a whole series of biological problems in relation to which the application of information theory was discussed (Yockey 1958b). Watson and Crick were not participants in these discussions. They used the notion of information in a colloquial, rather than a technical sense. The picture Crick had in mind was apparently the transmission of information in Morse code by the telegraph. Only later did he realise that the term 'genetic cipher' would have been more appropriate, since the Morse code was also a cipher (Crick 1990, 90).[52]

[52] The image of the Morse code was also used by Erwin Schrödinger to illustrate the way genes acted in development (Schrödinger 1992, 61). Interestingly, Crick corresponded with Schrödinger, but in relation to his notion of an aperiodic crystal and not in relation to his concept of a code-script (I thank Hans-Jörg Rheinberger for pointing this out). This again indicates that Crick used and understood the term in an unproblematic way. Note, however, that just before his communication with Crick on the structure of DNA, Watson co-authored a letter to *Nature* which suggested replacing a battery of confusing terms regarding genetic changes in bacteria with the term 'inter-bacterial information'. This change, the authors noted, 'recognizes the possible future importance of cybernetics at the bacterial level' (Ephrussi *et al.* 1953). On the suggestion that Watson and Crick consciously adopted the information discourse, although using it in a metaphorical way,

The plan for Brenner to join the Cavendish was devised in the summer of 1954, when Crick and Brenner spent some time together at Woods Hole and later both attended the phage meeting at Cold Spring Harbor. Crick and Brenner had met for the first time in the spring of 1953, when Brenner, together with the theoretical chemist Leslie Orgel and the crystallographer Jack Dunitz, had come from Oxford to Cambridge to see the DNA model. But the decisive discussions between Crick and Brenner took place only in the following year.

Brenner, who had a medical degree from the University of Witwatersrand in Johannesburg, had just finished his Ph.D. on bacteriophage resistance in Hinshelwood's chemical laboratory at Oxford (Brenner 1954).[53] Hinshelwood, a Nobel Prize winning chemist, was interested in the kinetics of bacterial growth and had written an authoritative volume on this subject (Hinshelwood 1946). He held the view that resistance in bacteria was an adaptational rather than a mutational phenomenon and, with his authority, was to a large extent responsible for the late reception of bacterial genetics in Britain.[54] Brenner had started working on phages at Hinshelwood's suggestion. He threw himself into the work, reading widely in the literature. He became convinced that only crossing experiments could decide between the adaptational and the mutational hypothesis and taught himself the necessary techniques. His conclusions contradicted those of his teacher. Through his work Brenner became well versed in bacterial and phage genetics, but he was self-taught. Milislav Demerec's invitation to come to Cold Spring Harbor that summer was his first direct contact with the phage school. Since 1945, when Delbrück gave his first phage course at Cold Spring Harbor, phage geneticists convened every summer at the Long Island laboratory, originally the seat of intense eugenic research, both to work and to relax (Kevles 1985).

While in Oxford, Brenner had heard Sanger giving a lecture on his protein sequencing work which had much impressed him. Learning about Seymour Benzer's fine-scale mutation map of a phage gene, he saw the

---

see Kay (2000, 58–9). By not signalling the metaphorical use they made of the terms code and information, Kay argues, Watson and Crick helped to invest DNA with the cultural power of a new semiotics of communication and control (Kay 2000, 59). On the molecular biologists' use of information language see also Fox Keller (1995b).

[53] For a first-hand and lively account of Brenner's upbringing in South Africa and his scientific career see Brenner (1997).

[54] Interview with Elie Wollmann, Pasteur Institute, Paris, 26 June 1996. There were none the less a few people who worked on bacterial genetics in the 1950s. William Hayes of the London Postgraduate Medical School in Hammersmith became the leader of the later expansion of the field in Britain. On his work see his contribution to the Delbrück *Festschrift* (Hayes 1966). For a brief but useful overview of the development of genetics in postwar Britain see Fincham (1993). On the development of ecological (and later medical) genetics in Britain see also Zallen (1999).

possibility of studying the genetic control of protein synthesis by combining Benzer's mapping techniques with Sanger's biochemical techniques. Matching the place of mutations on the genetic map with changes in the amino acid sequence of the corresponding protein would allow some inferences on the nature of the genetic code. Crick, who was thinking along parallel lines (remember that he had disproved Gamow's genetic code using protein sequencing data), was keen that Brenner should come to Cambridge to pursue this project. Thus, it was not phage work generally but Brenner and his particular project that Crick was eager to attract to the Cavendish. Later asked for the reasons which made him work so hard for Brenner to come, his reply was: 'He was just the kind of chap we wanted.'[55]

The plan for Brenner to come to Cambridge, however, had to be postponed, since his scholarship (an Exhibition Overseas Science Scholarship) committed him to return to South Africa for two years, while the acute space problem at Cambridge made it impossible to appoint a new researcher.

From South Africa, Brenner tried to beat isolation by engaging in vigorous correpondence with numerous scientists he had met during his stay in Europe and America. Stent from the Virus Laboratory in Berkeley and Crick wrote back most assiduously, supplying Brenner with long and detailed reports of recent scientific developments, discussions in meetings, news of common friends. Meanwhile, Brenner energetically set out to organise pipettes, tubes, water baths and incubators and start phage experiments with the cultures which he had brought with him or asked his friends to supply. As he wrote to Stent: 'At the present moment, I do not think of going away again, but feel I must settle down and do a solid piece of phage work and establish myself. Perhaps in a year or so I shall think again.'[56]

By summer 1955, Crick felt the time was approaching 'when we should give serious thought to you coming back to England'.[57] By that time, Brenner had already received offers to consider positions at the Department of Zoology in Edinburgh and at Glaxo Laboratories. Also Stent was suggesting ways for Brenner to join the Berkeley laboratory. Crick continued: 'You would, I think you will find, have no difficulty in going to Edinburgh, but naturally I should much prefer to see you here. If only this lysozyme business would get started, I could put up a case to the MRC for "molecular genetics" '.[58] It was the first time Crick used the term 'molecular genetics' to describe their research project dealing with the

---

[55] Interview with F. Crick, Cambridge, 25 May 1993.
[56] S. Brenner to G. Stent, 27 December 1954, Sydney Brenner, private collection, box Archive 3.
[57] F. Crick to S. Brenner, 6 July 1955, Sydney Brenner, private collection, box Archive 3.
[58] *Ibid.*

relations of genes and proteins. Since his return from America, Crick himself had started doing some experiments in this direction while continuing his crystallographic work. He decided that the first problem was to prove that a mutation inherited in a Mendelian way (and therefore most likely coded for in a nuclear gene) produced an amino acid change in the corresponding protein. He enrolled Ingram, the protein chemist in the laboratory, to collaborate on the problem. Ingram was engaged in adding heavy atoms to haemoglobin and myoglobin, but he was familiar with techniques to isolate and characterise proteins. Together they settled on the protein lysozyme which was easy to obtain from tears (Crick's own tears) and egg white. Running hundreds of samples over ion exchange columns and testing the lysozyme fractions, they set out to find mutants of the protein (Crick 1990, 103–5). However, success, so far, had eluded them.

Phage was another obvious material to try the experiment on. But as Crick confessed to Brenner, 'I feel a little out of my depth here, at any rate for the moment.'[59] This was one reason why he was so keen to attract Brenner to the laboratory.

Since Crick had never asked for a collaborator to join him, he felt he stood a good chance with the MRC. But the situation was none the less delicate. He explained: 'Of course my own position here is anomalous (I still do theoretical crystallography!) so I would rather let things drift than call attention to it before I need to.'[60]

A little later all his cautiousness was set aside. Two main events had occurred: first, the Rockefeller Foundation had approved a new grant of £40,000, to be spent over four years. This changed 'a position of acute financial stringency into one of comparative plenty'.[61] The additional money could pay for new apparatus necesssary to build up phage work in the unit. Second, a senior researcher (Huxley) had decided to leave, making it possible for Crick to make a case for someone to take his place. Brenner preferred Cambridge to any other solution and was prepared to adjust to crowded conditions in the laboratory.

The research proposal Crick submitted was much wider than originally suggested. In a half-page statement which accompanied Perutz's short note to the MRC, Crick explained that Brenner's current work concerned the growth of phage in protoplasts. According to Crick, the future of this work was not easy to predict, but after a year Brenner might not wish to pursue it any further. The general type of work Brenner was interested in, Crick

---

[59] F. Crick to S. Brenner, 4 November 1955, Sydney Brenner, private collection, box Archive 3.

[60] *Ibid.*

[61] F. Crick to S. Brenner, 20 October 1955, Sydney Brenner, private collection, box Archive 3.

explained, could be called 'molecular genetics', i.e. the connections of protein synthesis and genetics. Crick listed three approaches to this problem
which Brenner suggested be pursued: the study of the genetic control of
the protein of bacteriophage; the study of 'mutant proteins' in higher organisms, as for instance the abnormal haemoglobins of man; and the study
of the occurrence of errors in protein synthesis. Crick also mentioned that
Brenner had started preliminary work on the last subject.[62] Brenner joined
the unit in 1957 on a three-year research fellowship.

Meanwhile, Ingram, moving from lysozyme to sickle cell haemoglobin,
had brilliantly proved Crick's point. Sickle cell anaemia, a genetic defect,
could indeed be traced down to the exchange of a single amino acid in the
protein molecule.[63]

Ingram studied sickle cell haemoglobin following Crick's suggestion,
but also Perutz was interested to settle the chemical difference between
normal and sickle cell haemoglobin after his failed attempts to find a structural difference between the two molecules. To compare the amino acid
composition of sickle cell and normal haemoglobin without resorting to
a complete chemical analysis of the molecules, Ingram applied a two-
dimensional 'fingerprinting' technique which he developed following an
approach first devised by Fred Sanger in the Biochemistry Department in
Cambridge. Proteins were selectively digested and the resulting peptides
subjected to electrophoretic and chromatographic separation on paper.
Perfecting this technique, Ingram was able to pin down the minimal difference between the two haemoglobin molecules which had eluded earlier
researchers (Ingram 1956; 1957). This was the first evidence that an inherited defect was in fact laid down in the protein sequence of a protein. The
work applied biochemical techniques to genetic questions. The phage introduced by Brenner would offer a more powerful tool to combine genetic
with structural biochemical (or protein sequencing) techniques.[64]

Both Crick and Brenner saw their programme in molecular genetics
as directly following from the structure determination of DNA. This was

---

[62] M. Perutz [to MRC], 23 November 1955, 'Crick's collaborator', and attached note by
F. Crick, 'Tentative research programme', file P2/326, vol. 1, MRC Archives.

[63] Since Pauling had shown that sickle cell haemoglobin differed electrophoretically from
normal haemoglobin, this molecule was attracting much attention from researchers (de
Chadarevian 1998a). Supply, however, especially in Britain, was very limited. Ingram
used sickle cell haemoglobin that Anthony Allison, the Oxford geneticist who promoted
the theory that carriers of sickle anaemia were more resistant to malaria, had left behind
when visiting the laboratory. Brenner sent further samples from South Africa. Note that
Brenner also planned to work on abnormal haemoglobins.

[64] For Ingram the biochemical and genetic study of abnormal haemoglobins became a
lifelong interest. He left Cambridge in 1958 to take up a position in the newly formed
Department of Biology at the Massachusetts Institute of Technology in Cambridge, MA.

also the way Perutz presented the project to the MRC when applying for Brenner.[65] But to what extent and in which respect was this really the case?

Curiously enough, Crick later recalled that even before he joined the Cavendish he was convinced that the amino acid sequence held the key to the three-dimensional structure of proteins and that genes would have to determine this sequence.[66] The simple structure of DNA supported the suggestion that the genetic information was linearly arranged. However, Benzer's fine-scale mutation map of a chromosomal region of bacteriophage T4, based on functional data, provided much more direct evidence.[67] It should also be remembered that in 1951 Alexander Todd and Daniel Brown of the Chemistry Department at Cambridge had already established that nucleic acids were unbranched linear polynucleotides which differed from one another in molecular size and in the composition of their nucleotides.[68] From the chemists' point of view this meant that the problem of nucleic acid structure was 'solved' (Todd 1983, 87, 124).[69] With the knowledge of the structure, the synthesis of nucleic acids could be attempted, a programme pursued by several of Todd's students, among them

---

[65] Crick (1990, 108); interview with S. Brenner, Cambridge, 30 June 1993; M. F. Perutz, 'Crick's collaborator', file P2/326, vol. 1, MRC Archives.

[66] Crick (1990, 22) and interview with F. Crick, 25 May 1993; see also Judson (1994, 245). Crick's memories relate to a talk he gave at the Strangeways Laboratory while working there from 1947 to 1949. Unfortunately no notes of this talk are preserved.

[67] Benzer himself leant strongly on Watson and Crick's work to formulate aims and results of his investigations, but Delbrück for one was critical of this side of Benzer's work and advised Benzer to discuss his results on their own terms (Holmes 2000). On Benzer see also Weiner (1999).

[68] Important data on the changing chemical composition of nucleotides in nucleic acids were also produced by Erwin Chargaff's group at Columbia University, following Oswald T. Avery's work on the transforming principle of DNA. The results of the analyses made it conceivable that DNA represented the genes and thus lent strong support to Avery's hypothesis (Olby 1994, 211–15).

[69] Despite a certain polemic in this statement, Todd was adamant that his group did not contribute to the development of the Watson and Crick model. For their lack of interest in the physical structure of DNA, Todd blamed the disciplinary divisions between physics and chemistry. In reality, a fair amount of interest and exchange existed betweeen Todd's group and the crystallographers in the Cavendish. Because of Todd's interest in nucleotides, X-ray analyses of the compounds were performed in the Cavendish. The detailed structural data of the nucleotides were available and of great use to Watson and Crick while they were working on the structure of DNA. Bragg consulted with Todd how Pauling could have chosen the alpha-helix from among three structures all equally possible on the basis of X-ray evidence. When Todd was able to give chemical reasons for this choice, Bragg resolved that no nucleic acid structure should leave the Cavendish without being approved by Todd, as indeed happened for the DNA model (Todd 1983, 88–90). Later Todd became involved in the negotiations for the future of Perutz's unit. On this occasion he made a series of constructive suggestions, but in the end did not put enough force behind the plan to provide space for the group in his own new department (see below).

Gobind Khorana who, by producing synthetic oligonucleotides of a definite sequence, contributed in decisive manner to the 'cracking' of the code.

If in the case of proteins Crick and others boldly assumed that the 'folding problem' could be ignored and that what mattered was the amino acid sequence, a similar assumption could surely have been made regarding the nucleic acids. To understand how bold the assumption was regarding proteins, one should remember that insulin, the first protein whose chemical structure was known, did not consist of a simple linear polypeptide chain, but of two chains linked by three disulphide bridges.

The double helix certainly helped to organise some ideas and direct research. It may well have stimulated both Crick and Brenner to formulate their programme in molecular genetics. This, however, is not the same as (logically) deducing everything which followed from the double helix. The double helix found its place and importance in a complex web of experimental data and hypothesis, much of which was acquired quite independently and sometimes itself provided the very evidence for the proposed structure. Arthur Kornberg, whose work on DNA polymerase has been seen as following on from the Watson and Crick model, has made this point most forcefully. He has stressed that his search for DNA polymerase was not spurred by the double helix but was based on his interest in metabolic pathways and his fascination with enzymes (Kornberg 1989, 121).[70]

Considering Crick's (slow, but then definite) shift from X-ray crystallography to molecular genetics more generally, we also need to remember that important changes had occurred in protein research since the late 1940s, when Crick had first entered the field. At this time, proteins were seen as the key of all life processes. Their specific functions seemed to depend on their exact three-dimensional structure and X-ray crystallography promised to yield some insight into their architecture. Sanger's complete determination of the structure of insulin in 1955 represented a serious threat to the future of protein crystallography, since it was generally expected that it would be possible to deduce the three-dimensional structure of proteins from their sequence (evidence that genes coded only for the amino acid sequence of proteins strengthened this view). Indeed, the sequence of insulin was used to build (hypothetical) models of the molecule.[71] In his Nobel Lecture in 1962 Kendrew predicted that

[70] Kornberg is much less precise on when and how the model did make an impact on his work. Interestingly, in his Nobel Lecture which occurred before the double helix gained prominence in molecular biologists' claims to authority, Kornberg paid ample attention to Watson and Crick's DNA model, so much so that Crick, in his Nobel Lecture three years later, could refer to it regarding the description of the proposed structure (Kornberg 1964; Crick 1964).

[71] One such person-sized model was exhibited in the entrance of the Biochemistry Department in Cambridge. Eventually it was dismantled and removed, since it was

protein crystallographers could soon go out of business, 'perhaps with a certain sense of relief' (Kendrew 1964, 696; de Chadarevian 1996a, 372–3). If these sentiments were predominant among protein crystallographers, there were strong independent reasons for Crick to diversify his interests. However, proteins and their specific structure remained a key issue also for his work in genetics. In Cambridge the central question was protein synthesis, not as for instance in the phage group gene replication.

In his celebrated lecture at the Symposium of the Society of Experimental Biology in September 1957, in which Crick first formulated the sequence hypothesis and the central dogma, he still stressed 'the central and unique role of proteins' (Crick 1958, 139). Indeed, according to Crick, it was the importance of proteins that provided the 'psychological drive' to believe that genes controlled the synthesis of proteins and that there was 'little point in genes doing anything else' (Crick 1958, 138–9).

Phages in the Cavendish

On the instrumental level, the new project in molecular genetics introduced major changes in the Cambridge unit. Apparatus and materials hardly commonly in stock in the stores of a Physics Department like bacteria and phage stocks, growth media, incubators, columns and fraction collectors, centrifuges, paper chromatography basins, electrophoretic apparatus, an amino acid analyser, enzymes, mutagens and radioactive markers entered the Cavendish.

Brenner's first task at Cambridge was to organise equipment and facilities and to train the new assistant, Leslie Barnett. He then threw himself into experimental investigations. He soon became the focus of the new experimental programme in molecular genetics. The phage work introduced by Brenner built on techniques and results of the phage school, but combined these with structural approaches. His aim was to study the genetic control of protein synthesis. On the way he made important contributions regarding the structure of phages, new electron microscopy preparative techniques and the effect of mutagens. His work developed in collaboration with researchers in other departments and with a constant flow of researchers from abroad whom the new project attracted to the Cambridge unit.

considered not right to show students a wrong structure. Parts of the model have recently resurfaced and attempts are underway to restore it to full size. A much smaller transportable ball-and-spoke model of beef insulin, built for display at the Brussels World Fair in 1958, has found its way to the National Portrait Gallery in London where it stands next to a commissioned portrait of Sanger by Paula McArther (1991) and Maggi Hambling's portrait of Hodgkin (1985) (see Fig. 5.12).

Brenner's first step was to devise a method to break up phages and study their components. Here he could build on preliminary work done in Johannesburg. Used to improvising, he developed a method to grow large quantities of phages in a Hoover washing machine, delivered for free from the manufacturer and adapted to the particular task (Brenner 1997, 116). For the characterisation of the break up parts of the phages by electron microscopy Brenner collaborated with Robert W. Horne from the Electron Microscope Group in the Cavendish Laboratory. Together they developed a new negative staining method, the phosphotungstate method, which allowed an unprecedented resolution of structural details. The next step was to fractionate the protein components of the phages and to characterise them chemically and functionally. Brenner's aim was to find a component part for which a genetic map existed or could be constructed and which was amenable to sequence analysis.

In the autumn of 1957, a few months after Brenner had joined the unit, several researchers assembled at Cambridge to collaborate with Crick and Brenner. Mahlon Hoagland and Crick, working at the Molteno Institute, tried to isolate the soluble RNAs identified by Hoagland, while Benzer, Sewell Champe (a research student of Benzer's) and George Streisinger from Cold Spring Harbor joined forces with Brenner in the Cavendish.[72] The protein coded for by the chromosome region of bacteriophage T4 mapped by Benzer could not be isolated. The group thus decided to use Streisinger's mutants which affected the tail fibres of bacteriophage T2 at the tips. The experiments did not show the expected results, probably because the isolated protein was not the right one, and the work was never published. Brenner and Barnett, however, continued to pursue the project, when the visitors had long left (Brenner and Barnett 1959; Brenner *et al.* 1959).

Another important trail followed by Brenner was to study the effect of different mutagens. This work led to the identification of two classes of mutagens, the base analogue class which affected particular base pairs, and the acridine class which caused the insertion or deletion of a base pair (Brenner, Benzer and Barnett 1958; Brenner *et al.* 1961). Brenner and his collaborators exploited these differences to probe the chemical nature of the genetic fine structure. Crossing experiments with acridine-induced mutants allowed Crick, Brenner and their collaborators to settle that the genetic code consisted of three-letter codons (Crick *et al.* 1961).

The new programme also led to first contacts with François Jacob and Jacques Monod's group at the Pasteur Institute in Paris (see below, p. 257).

---

[72] For a reminiscence of that year in Cambridge see Hoagland (1990, 97–116). From Johannesburg Brenner had already corresponded regularly with Benzer and Streisinger about their work; see Sydney Brenner, private collection, box Archive 3.

One important result was Jacob and Brenner's collaborative work with Matthew Meselson at Caltech which provided evidence for the existence of an unstable or messenger RNA.

It is worth remembering that this line of investigations developed parallel to the crystallographic work in the unit. Crick became increasingly involved in the new programme, leaving behind protein crystallography. But in the same year in which Brenner joined the unit, Kendrew presented a low resolution model of myoglobin. This was the first structure of a globular protein ever to be solved. It was celebrated as a major achievement.

What is surprising is how unproblematic the expansion of research interest seems to have been. Explanations given retrospectively range from 'the MRC did not really know what we were doing' to 'by then the MRC had taken the view that it was fine what we were doing'.[73] With Brenner joining Crick and fully implementing his new project in molecular genetics, a divergence of research interests in the unit was none the less clear. While the work on DNA structure in 1953 had not made it necessary to change the general description of the unit's work, the entry in the MRC Annual Report of 1956/7 noted that 'the scope of the unit's work has been much enlarged during the past year; it now includes the mechanism and genetic control of protein synthesis and the effect of genetic mutations on protein structure' (Fig. 6.6).[74]

Some of the options the researchers considered in order to solve the ever more pressing space problem confirm the diverging interests of the group. In spring 1957, Crick, ready to change his disciplinary affiliation, applied for the Chair of Genetics which Ronald A. Fisher had left vacant. Fisher, who was friendly with Crick, supported his candidature. If appointed, Brenner would have moved with him. But apparently the Appointing Committee decided that Crick did not understand anything of genetics and did not offer him the chair – a powerful reminder that molecular genetics was not of general interest to geneticists at the time. Crick, who was never keen on teaching, later commented, 'luckily so'.[75] But at the time, even if involving a split of the unit, a successful application must have seemed a possible solution to an extremely precarious situation. By that time, however, a much more ambitious plan involving a radical redefinition of the work of the unit and its place in the university was taking shape.

[73] Interview with F. Crick, Cambridge, 25 May 1993.
[74] *Report of the Medical Research Council for the Year 1956–1957* (London: Her Majesty's Stationery Office, 1957), p. 91.
[75] Interview with F. Crick, Cambridge, 25 May 1993; also interview with S. Brenner, Cambridge, 17 July 1992. The chair was occupied only two years later and according to statute, after two candidates nominated by the Committee had declined the offer. I am indebted to Elizabeth Leedham-Green for this information.

6.6 Francis Crick (centre) with Sydney Brenner (right) and Alexander Rich (left) in front of the Cavendish Laboratory (1957). Courtesy of Alexander Rich.

Watson and Crick's double helical model of DNA, we have witnessed, had only a slow and indirect impact on research directions in Cambridge. A programme in 'molecular genetics' took root only several years later, when Brenner joined the unit. Throughout the 1950s, however, protein crystallography remained the dominant research strategy. As we will see in the next chapter, not the double helix, but Kendrew's achievement in his myoglobin work, together with a new alliance with Sanger from the Biochemistry Department, led to a turning point in the negotiations for the future of the unit.

7

...

Disciplinary moves

In June 1957 Harold Himsworth, Secretary of the Medical Research Council, invited Sanger from the Biochemistry Department in Cambridge, and since 1950 on the external staff of the MRC, to discuss 'some proposals' he had heard Sanger was considering. He also solicited a written statement on the subject beforehand.[1] A few days later Himsworth received a substantial memorandum. Sanger kept the actual proposal he had to make to the very end of his statement.

The document started with a general outline of what Sanger saw as the present position of protein chemistry, including possible future developments and applications to other fields. Only in the last section did Sanger turn to the work of his own group. The present concern of the group, Sanger explained, was to develop more efficient sequencing methods for proteins. He anticipated that in the near future standard methods for amino acid sequencing would become available which could be applied to a vast array of problems. Much of this work would be routine structure analysis that could not easily be carried out in a university department. Slightly changing his line of argument, he added that there was not much opportunity for the expansion of his group in the Biochemistry Department. This was not a problem at present, but could become one in the near future. Seen in this light, Sanger explained, finally coming to the decisive point of his argument, the proposal of a new laboratory to be shared with Perutz's unit seemed most attractive. Such a set-up, he reckoned, would allow his group to expand, creating an establishment which was big enough to justify central facilities like a library, a workshop and certain apparatus and to guarantee a stimulating workplace. Perutz's group was also working on proteins, though from a different and in many ways complementary point of view. Collaborations between the two groups, Sanger pointed out, already existed, especially

---

[1] H. Himsworth to F. Sanger, 21 June 1957, file E243/109, vol. 1, MRC Archives.

regarding the use of sequencing techniques for studying the molecular basis of genetics.[2]

A letter from Perutz to Himsworth followed suit. Doubtless Perutz had been informed about the Secretary's enquiry. In this letter the new plan got a name. After recapitulating the ongoing uncertainties regarding the future of the unit, Perutz presented to Himsworth the proposal to build a new Institute of Molecular Biology in Cambridge for his and Sanger's group. The institute, funded by the MRC, would be outside the central laboratory area, but be linked as closely as possible to the university. Perutz estimated that the sum to build and equip the new laboratory would amount to between £150,000 and £200,000, but suggested that these funds could be raised from private and public sources.[3]

As soon as he was informed of these plans, Himsworth initiated a series of consultations to sound out the views of all relevant parties, including the Secretary General of the Faculties who presided over the General Board, one of the main self-governing bodies of the university, and the professors most involved in the discussions about the future of the unit, namely the Professors of Physics and of Biochemistry, as well as the Professor of Organic Chemistry, who was interested especially in the crystallographic work of the unit. As chairs of large departments and with strong college affiliations, these three people also wielded considerable power in university politics.[4] Among the direct participants of the project, only Kendrew harboured doubts. He was concerned about losing links with the university and also foresaw difficulties of organisation and personalities.[5] Among the consulted heads of department, Mott from Physics and Todd from Chemistry were on the whole supportive, but Frank Young, the Professor of Biochemistry and a former member of the MRC council, strongly opposed the plan. His main reservation regarded the creation of a research institute in the university whose members would not take over their full share of teaching. The Secretary General seemed to share Young's preoccupation.[6] Despite this opposition, it was agreed that, as a next step, Perutz, together with Sanger, would draw up a memorandum setting out 'the

[2] 'Memorandum to the Medical Research Council by F. Sanger' (undated; attached to a letter of 26 June 1957), file E243/109, vol. 1, MRC Archives.

[3] M. Perutz to H. Himsworth, 27 June 1957, file Setting up Lab, LMB Archives.

[4] Alexander Todd from Chemistry, who was also chairman of the Advisory Council of Scientific Policy and had received a knighthood in 1954, was the most established among the three. In the autumn of 1957 he was awarded the Nobel Prize. He became a Lord in 1962, taking his place in the upper house as an independent.

[5] 'Proposed Institute of Molecular Biology at Cambridge', note of 26 August 1957 by H. Himsworth, file E243/109, vol. 1, MRC Archives.

[6] 'Proposed Institute of Molecular Biology at Cambridge' (notes of 7 and 8 August 1957 by H. Himsworth), file E243/109, vol. 1, MRC Archives.

scientific case for bringing together the various interested parties under one roof where they would also have more room'.[7] Scientific reasons seemed to be given precedence over institutional powers and policy matters.

Suggesting the construction of a new laboratory for themselves and Sanger's group, Perutz and his colleagues followed a completely new strategy for dealing with the need for a new affiliation after the loss of Bragg's protection in the Cavendish, as well as to confront the ever more pressing space problems and the increasingly diverging research interests of the different group members. Instead of splitting up crystallographers and molecular geneticists, trying to find separate accommodation for them, they now recruited Sanger to increase the weight of the whole group. They no longer speculated on the hospitality of other departments, be it Physics, Colloid Sciences, Chemistry or Genetics, but redefined their work such that it represented a distinct and independent programme for which they created a new name. Building above all on the support of the MRC, they moved out of the narrow spaces and disciplinary politics of the university. But hoping to get the best of two worlds, they none the less wished to keep tight links with it.

What had provoked the change in strategy? How was Sanger drawn into the plan? And what was behind the new name?

Forging links

It was the opportunity to draw up a common proposal with Sanger that precipitated the change in strategy. This opportunity was based on strategic research as well as political motives, with local contingencies playing a decisive role. Analysing the decision behind this move offers the possibility to show the complex way in which collaborations at the bench level and institutional and disciplinary negotiations are interconnected and played out in a local context.

Neither from documents nor from interviews was it possible to establish who first launched the idea of a common laboratory. While Sanger's formulation in his letter to Himsworth seemed to imply that the scheme was developed by Perutz's group, Perutz, in his first letter to Himsworth, indicated that the suggestion came from Sanger.[8] Both Sanger and Perutz were probably implicitly presenting an excuse for not having informed Himsworth of their plans before. Later Perutz insisted that they alone were the originators of the scheme.[9] But this move too was strategic, since the MRC was

[7] H. Himsworth to M. Perutz, 9 August 1957, file E243/109, vol. 1, MRC Archives.
[8] M. Perutz to H. Himsworth, 27 June 1957, file Setting up Lab, LMB Archives.
[9] M. Perutz to H. Himsworth, 5 July 1957, file Himsworth, LMB Archives.

already committed to finding new accommodation for his group. Perutz remained Himsworth's main interlocutor throughout the negotiations.

Later interviews with the participants did not provide a clear picture either. In one version, the proposal to move together with Sanger appeared as the response to a more grandiose plan pursued by the MRC. According to this scheme the MRC, in an attempt to remove all space restrictions, was to construct one new building in Cambridge into which all Cambridge researchers funded by the Council would move together. Perutz and his colleagues regarded this 'administrators' plan' as unacceptable. But it encouraged the alternative plan of asking the MRC for a new laboratory just for the Cavendish group and for Sanger, who was also on the MRC staff and was therefore drawn into the discussions.[10]

Most people, however, regarded the decision to move together with Sanger as more or less obvious. This may well be a retrospective view, produced by twenty or more years of life in the same institution. However, as already pointed out, contacts between the Cavendish group and Sanger existed beforehand. These were based on the interest of several researchers in the Cavendish group in Sanger's protein sequencing techniques.

The first contacts dated back to around 1950. The Cavendish group was looking for a protein biochemist and informally approached Sanger with the suggestion that he move from the Biochemistry Department to the MRC unit. At the time, the main channel of communication with Sanger was through Herbert Gutfreund, who was working in the Department of Colloid Science, estimating the molecular weight of insulin by measuring osmotic pressure. He was friendly with both Sanger and Crick, later also with Watson. But Sanger turned down the suggestion, thinking that he would not feel at home in a Physics Department.[11] But the contacts between the two groups did not end here.

In the mid-1950s, Kendrew, still working on the first low resolution model of myoglobin, tried to convince someone from Sanger's group to take over the sequencing of his protein. As we have already seen, around this time Sanger's success in determining the complete amino acid sequence of insulin, together with the first promising-looking attempts to use this information to build a three-dimensional model of the molecule, posed a serious challenge to protein crystallography. Sequencing and model building could well make protein crystallography obsolete. But Kendrew

[10] Interview with S. Brenner, Cambridge, 19 July 1992. No documents regarding this plan could be retrieved. Possibly it was only informally discussed.

[11] Interview with F. Crick, Cambridge, 25 May 1993, and letter F. Crick to the author, 30 January 1996. The unit appointed Ingram from the Rockefeller Institute instead. In this case the contact was also established through Gutfreund; interview with V. Ingram, Cambridge, MA, 20 July 1993.

strived to combine crystallographic and sequencing data. Only when no one in Sanger's group was available for the job did Kendrew look further afield and he finally recruited Edmundson, a doctoral student of Stein and Moore at the Rockefeller Institute in New York (see above, pp. 124–5). This episode indicates the interest of protein crystallographers in sequencing data. But it also shows that proper collaborations between the protein crystallographers and the protein chemists in Cambridge were difficult to achieve.

More direct contacts existed between Crick, Ingram, Brenner and Sanger. Sanger's sequencing work, together with evidence on the linearity of genetic information, had allowed the formulation of the 'sequence hypothesis' which, as we have seen, was central to Crick and Brenner's research project (above, chapter 6). More important still, protein sequencing also provided the tools to test the hypothesis and to attack the problem of the code. Refining Sanger's fingerprinting techniques, Ingram had just provided the first experimental proof that inherited defects affected the amino acid sequence of proteins. According to Crick these experiments first made Sanger (and biochemists more generally) appreciate the importance of his sequencing work for genetics (Crick 1990, 106). Crick and Brenner themselves were mainly interested in Sanger's radioactive techniques for fingerprinting proteins. Radioactive labelling was much more sensitive than other chromatographic methods and the two Cavendish researchers were keen to adopt the new technique for their phage experiments. Sanger himself was working with slices of oviducts which he incubated to get labelled ovalbumin. This was quite a lengthy and laborious procedure. Already in a letter of October 1956 Crick was writing to Brenner who was still in Johannesburg: 'I stressed to Fred [Sanger] how extremely favorable the phage system might be for this method…He seemed very interested'.[12] After moving to Cambridge in January 1957, Brenner regularly visited Sanger in his lab to learn his techniques and talk to him (Figs. 7.1 and 7.2).

Crick and Brenner were interested not only in Sanger's techniques, but also in 're-educating' him in genetics. Apparently on Sanger's own suggestion, in the autumn of 1957, a series of lectures on genetics was organised for Sanger's group in Crick's house (Crick 1990, 106).[13]

---

[12] F. Crick to S. Brenner, 26 October 1956, quoted in Judson (1994, 331).

[13] Brenner's recollections of these lectures differ quite markedly from Crick's. According to him, the lectures were set up not specifically for Sanger's group, but for the large group of visitors working on genes and proteins who had gathered in Cambridge that year. Sanger attended some of them. As for Sanger, he has altogether forgotten about these meetings; interviews with F. Crick, Cambridge, 25 May 1993, with S. Brenner, Cambridge, 30 June 1993, and with F. Sanger, Cambridge, 25 March 1993. According to both Brenner and Crick, they also tried to interact with Ernest Gale, who was working on protein

7.1 Fred Sanger with protein chromatogram in his laboratory in the Biochemistry Department (1958). Courtesy of Fred Sanger.

Seen in this light, the decision to move together seems to follow directly from benchwork connections. It is, indeed, quite likely that Crick

biosynthesis in bacteria in the same department as Sanger. Gale, however, apparently showed no interest and the attempts failed; interviews with S. Brenner, Cambridge, 30 June 1993 and with F. Crick, Cambridge, 25 May 1993. On Gale's work see Rheinberger (1996).

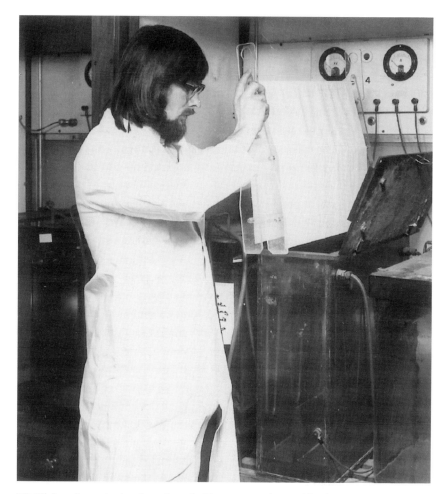

7.2 High-voltage electrophoresis tank. The same tanks used in the 1950s for the separation of peptides in protein sequencing were later (as seen here) used for the separation of radio-labelled nucleic acid fragments. Courtesy of Bart Barrell and Medical Research Council Laboratory of Molecular Biology, Cambridge.

and Brenner – who most regularly interacted with Sanger and had a vital interest in his techniques – first brought up the idea of a common laboratory. But Perutz also stressed that Sanger would decisively strengthen the chemical aspect of the crystallographic work.[14]

Sanger's interest in benchside collaborations with the Cavendish researchers and especially with the X-ray crystallographers was much less pronounced. Regarding the 'scientific case' for bringing him into association with the Cavendish group, Himsworth noted: 'My impression

[14] M. Perutz to H. Himsworth, 27 June 1957, file Setting up Lab, LMB Archives and Perutz (1987, 41).

was that he felt that they could get a good deal from him but he was not quite clear as to what he could get from the others.'[15] The work connections were none the less strong enough to make a scientific case. To discard this as 'mere' rhetoric would be too easy. As we will see, the fact that Sanger and his group were part of the new project decisively shaped the way the new field of molecular biology was defined and the laboratory planned. It also shaped the work practices which developed in the new laboratory, even if in other ways than projected in the original proposals. Benchwork and institutional negotiations were tightly connected, even if the relation was not a causal one.

In the choice of the name for the new enterprise the same mixture of scientific and local political motives prevailed, with the participants offering distinct readings of the events.

A new name

Names matter. They are not only labels or reference terms for historical accounts, but strategic tools. Why then was the term molecular biology chosen for the new enterprise?

A new name was needed. As Perutz put it: 'I could not well ask the Medical Research Council for a new Laboratory for the Study of Molecular Structure of Biological Systems.'[16] Sanger on his side reckoned that a new name had to be invented, because there already was a laboratory of biochemistry in Cambridge.[17] According to Crick and Brenner, the new term best described the work of the unit, which had expanded to comprise more biological questions.[18] As we have seen, they had labelled their new, more biologically orientated research project 'molecular genetics'. Today molecular genetics and molecular biology are often used interchangeably. But what was needed in Cambridge was a name which comprised three distinct enterprises: molecular genetics in the sense Crick and Brenner had defined it, protein crystallography and protein sequencing. 'Biophysics', the term so often used to refer to the unit's work, did not seem appropriate anymore.

---

[15] H. Himsworth, 'Proposed Institute of Molecular Biology at Cambridge', notes of 7 August 1957, p. 2, file E243/109, vol. 1, MRC Archives.

[16] Interview with M. Perutz, Cambridge, 3 July 1992.

[17] Interview with F. Sanger, Cambridge, 25 March 1993.

[18] Interviews with S. Brenner, Cambridge, 17 July 1992, and with F. Crick, Cambridge, 25 May 1993. Famously, Crick later claimed that he had started to call himself a molecular biologist as a shortform, 'because when inquiring clergymen asked me what I did, I got tired of explaining that I was a mixture of crystallographer, biophysicist, biochemist, and geneticist' (Crick 1965, 184). He also defined molecular biology as 'anything that interests molecular biologists' (Crick 1970, 613n). For an early use of the term see Crick (1953, 1).

The term 'molecular biology' was not new. Warren Weaver, director for the natural sciences in the Rockefeller Foundation, first used it in print in the Annual Report of the Foundation for 1938 (but not regularly thereafter) (Weaver 1970; Kohler 1991, 299–300).[19] It may well have been suggested to him by the British crystallographer and Rockefeller grantee William Astbury, who used it in print as early as 1939 to describe his structural approach to biological problems (Astbury 1939). His 1950 Harvey Lecture 'Adventures in molecular biology' helped to propagate the term, especially in Britain (Astbury 1950–1).[20] However, it did not really catch on and world-wide there was no institution which carried that name.[21] Perutz, like Astbury and other British scientists, was a grantee of the Rockefeller Foundation programme of molecular biology, or 'vital processes' as it was more frequently referred to, from 1938 to 1945, but at this time certainly did not think of himself as a 'molecular biologist'. He tended to describe himself as 'a chemist who worked in the Physics Department on what was in fact a biological problem' (Perutz 1987, 40).

The fact that the term 'molecular biology' was around, but not really current, may well explain the decision to use it to name the new enterprise. In addition, it worked well for Crick and Brenner, who wanted to stress the more biological side of their work, as well as for the crystallographers, who could relate to it through Astbury's use of the term. As for Sanger, he never drew a sharp distinction between molecular biology and biochemistry, at least not concerning his own work (de Chadarevian 1996a). There was, however, a significant coincidence, which, if nothing else, corroborated the choice. It also indicates that the term, even if not institutionalised, was becoming more current in America in the late 1950s.

[19] Weaver's statement in *Science* followed a widespread debate on the origin of the term and the science it named (see chapter 6). Weaver's early use of the term was not widely known, even among those who were grantees of the programme, indicating that it did not make an impact at the time (see below). An exchange of letters in the late 1960s between Cambridge biochemist Edward F. Hartree, who, with a letter to *Science*, had intervened in the debate on the origins of molecular biology, pointing to Keilin's decisive role in promoting the new science, and Hans Krebs, the well-known Oxford biochemist and himself a grantee of the Rockefeller Foundation in the 1930s, confirms this point. In their attempt to determine the origin of the term 'molecular biology', Krebs could not trace its use further back than to Astbury's Harvey Lecture in 1950, while Hartree reported that Astbury described himself as a molecular biologist as early as 1939. Weaver's name was not mentioned; Hartree (1968) and letters H. Krebs to E. F. Hartree, 20 May 1968; E. F. Hartree to H. Krebs, 13 June 1968; and H. Krebs to E. F. Hartree, 17 June 1968; Krebs Papers, Sheffield University Library. I thank Lawrence Aspden from Sheffield University Library for pointing me to this exchange of letters and making it available to me.

[20] In this lecture Astbury described molecular biology as 'predominantly three-dimensional and structural', but also concerned with 'genesis and function' (Astbury 1950–1, 3).

[21] In 1945 Astbury had suggested the name for his chair in Leeds, but the biologists vetoed it on the grounds that his programme was not biological enough. He had to content himself with 'Biomolecular Structure' (Astbury 1961). In 1952, the National Science Foundation started a fellowship programme in molecular biology; see Appel (2000).

In the spring of 1957, while the plan for a new laboratory was taking shape, Kendrew was in negotiation with Kurt Jacobi of Academic Press to take over the editorship of a new journal called provisionally *Journal of Molecular Biophysics*. This term corresponded to a generally acknowledged division of the British biophysics community in a molecular, radiation and muscle–nerve group (see above, chapter 3). Solicited by Kendrew to express his views regarding the new editorial project, Paul Doty, physical chemist at Harvard, reckoned that he saw the need for a *Journal of Molecular Biology*. '[M]olecular biology and biophysics', Doty reflected, 'are turning out to be the natural subdivisions of the application of physical and physico-chemical principles of biology.' If this was true, the introduction of a new title such as 'molecular biophysics' would be 'extraordinarily confusing'. 'Biophysics as such is', Doty thought, 'beginning to bear the connotation of not being involved with the molecular and molecular structural side of biology, but more with the direct application of the five major subdivisions of classical physics to the study of biological systems as a whole. Moreover, much of biophysics is pretty bad as you know.'[22] In a letter of a few months later, Doty increased the dose: 'The more I hear and think about the problem I believe that Biophysics will take shape as the field of work of frustrated physicists who turn to biology and who do not know or learn anything about structure. In this sense molecular biology will take its place midway between biochemistry and biophysics and I think have the best of both worlds.'[23]

Kendrew agreed at once that *Journal of Molecular Biology* was a better title and was indeed the one he had thought of himself.[24] This was just a few weeks after the proposal for an Institute of Molecular Biology was presented to the Medical Research Council. Asking Himsworth for permission to take over the editorship of the new journal, Kendrew did not fail to point out the 'precise correspondence' between the titles proposed for the journal and the new laboratory.[25] It is generally acknowledged that the journal, the first issue of which appeared in May 1959 and for which Kendrew assembled a prestigious Anglo-American editorial board, was decisive in establishing the term molecular biology.[26]

[22] P. Doty to J. Kendrew, 24 July 1957, Kendrew Papers, MS Eng. c.2593, M.1, Bodl. Lib.

[23] P. Doty to J. Kendrew, 14 October 1957, Kendrew Papers, MS Eng. c.2593, M.1, Bodl. Lib.

[24] J. Kendrew to P. Doty, 5 August 1957, Kendrew Papers, MS Eng. c.2593, M.1, Bodl. Lib.

[25] J. Kendrew to H. Himsworth, 22 October 1957, Kendrew Papers, MS Eng. c.2593, M.1, Bodl. Lib. While Doty's initial response dated from three days before the first letter with the proposal for a new Institute of Molecular Biology was sent to Himsworth, it is unlikely that Kendrew received it before that date. Discussions regarding the new name will also have preceded the formulation of the letter to the MRC.

[26] The first members of the editorial board were Doty and Watson from Harvard, Andrew Huxley from Cambridge, Robert Sinsheimer from the Californian Institute of Technology and Wilkins from London. Later more members were coopted. For many years Kendrew

In Cambridge the new name took hold even more quickly. While still working on the proposal for the new institute, Perutz suggested to Himsworth a change of the unit's name. The MRC had repeatedly urged Perutz to think of a more concise name, suggesting Biomolecular Research Unit as an alternative. But Perutz had expressed his 'strong dislike' of the word 'biomolecular' because it mixed words of Latin and Greek origins and was therefore 'rather ugly'.[27] 'The old name "Molecular Structure of Biological Systems"', Perutz now explained, 'no longer covers all our activities and the name "Molecular Biology" has grown up in recent years for a science which looks at biological events from the point of view of the structure of the large molecules involved in them.'[28] In fact, this definition of the term did not represent much more than a reformulation of the old name. The MRC, however, at once approved the proposed change, not least because 'it will certainly save a lot of ink'.[29] The adoption of the new name was celebrated in connection with the move of the unit into the hut which also coincided with the tenth anniversary of the unit's foundation in 1947 (Fig. 7.3).

While creating a name, founding a journal and building an institute or laboratory were classic moves in creating academic disciplines, in this case the relationship of the proposed institute to the university had still to be negotiated. Support for the plan was first sought from the outside grant-giving body.[30]

The case for a laboratory of molecular biology

Writing the memorandum on the proposal for a new institute of molecular biology proved much more arduous than expected. After promising it for September, in October Perutz asked for a further delay of two or three weeks. The reason was not that other activities took the upper hand. 'I find that I do not really know enough about the new activities which have sprung up in this unit during the past year and that I do need to do some reading before putting it on paper', Perutz excused himself. 'I have been

edited the journal from his college office in Peterhouse. Interestingly, Perutz at first was against the idea of a new journal, which has long been a classic step in the creation of a new scientific field; see J. Kendrew to H. Himsworth, 20 October 1957, Kendrew Papers, MS Eng. c.2593, M.1, Bodl. Lib.

27 M. Perutz to H. Himsworth, 15 August 1951, and M. Perutz to A. Landsborough Thomson, 20 August 1953, file FD1/426, MRC Archives.

28 M. Perutz to H. Himsworth, 27 September 1957, file Himsworth, LMB Archives.

29 H. Himsworth to M. Perutz, 4 October 1957, file Himsworth, LMB Archives.

30 In fact, Himsworth himself was most apprehensive that the matter should not be officially raised with the university, before the Council had expressed its view on the proposal; H. Himsworth to M. Perutz, 15 August 1957, file E243/109, vol. 1, MRC Archives.

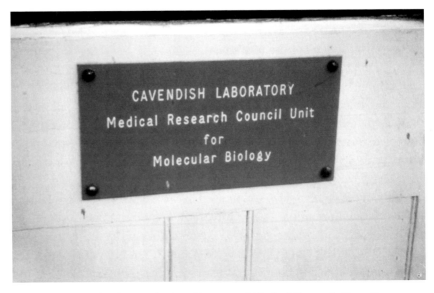

7.3 Name tag on the door to the hut. Courtesy of Medical Research Council Laboratory of Molecular Biology, Cambridge.

doing it for the past month or so and I am still at it.'[31] The estimates for next year's expenditure of the MRC were done in November. But Himsworth was anyway very doubtful that in the 'present frigid financial climate associated with a bank-rate of 7 per cent' much could be done for a big scheme of the kind envisaged.[32]

At the end of December Himsworth finally received two documents in draft form, a shorter, programmatic one succinctly presenting the scientific case for a laboratory of molecular biology and a much longer one reviewing the present state and possible developments of the field. The first document was signed by Crick and Perutz, while the second one carried only Perutz's name. Both documents, however, were approved by Sanger

[31] M. Perutz to H. Himsworth, 14 October 1957, file Himsworth, LMB Archives.

[32] In his statement Himsworth was referring to measures put in place to strengthen the pound. The launch of Sputnik, the first space satellite, by the Soviet Union in that same autumn did not have the same dramatic impact on science spending in Britain as in America, although in Britain also the science budget grew at a faster rate in the late 1950s than at the beginning of that decade. Britain did, however, profit from the increased number of American fellowship programmes which brought American researchers to Europe and vice versa as part of a post-Sputnik measure aimed at mobilising science in the whole of the Western alliance. On the role of the Ford Foundation in this politics see Krige (1999). On the importance of the high influx of American postdoctoral researchers in the economy of the LMB see below (chapter 9).

and all members of the Cavendish unit. In fact, they had all probably contributed in important ways to the drafting.[33]

The documents were primarily political. They were designed to convince a particular funding agency to invest substantial sums in the building of a new laboratory. They were, however, much more than that. They represented the attempt to subsume the work of all participants into a coherent project. In this sense the documents fulfilled a function in some ways comparable to that of a textbook. As has been repeatedly argued, textbooks not only display acquired knowledge to a student audience, but themselves create a body of knowledge and provide the tools for disciplining the body of future researchers. Similarly, the signatories of the documents created a new field of knowledge and drew up standards of recruitment and training.[34] Doing this, they also claimed authority over a particular field of knowledge. This is a second sense in which the documents were political. How did the signatories argue for their case?

Let us look at the shorter, programmatic document first. Its aim was to convince the relevant parties that there was a case for a new laboratory, rather than to present a detailed plan. Comparing this document with Bragg's first application to the MRC for Perutz and Kendrew's work in 1947 makes clear how much the research interests of the group had shifted and how decisive Crick and Brenner's inputs in the formulation of the programme were, regarding both content and style. The text started with the bold assumption that the 'molecular basis of *simple* living systems' (emphasis in the original) would be discovered in the next fifteen or twenty years.[35] By then the basic phenomena of living matter such as reproduction, metabolism and growth would be understood in terms of the interaction between molecules of known structure. All biological functions, the authors explained, were reducible to the functions of genes and proteins which were very similar in all living systems, from bacteria to man. This, they concluded, 'encourages us to seek for general explanations in terms of molecular structure and function, rather than in biological concepts'. This formulation clearly distinguished the enterprise of molecular biology

---

[33] References in the following are to the official versions of the two documents. Of the shorter document an undated draft version exists. Handwritten corrections include the terms 'pathological' and 'medicine' at all conceivable places throughout the text. Besides some minor corrections regarding style, this was the only, but revealing, change introduced in the official version; see 'A Laboratory for Molecular Biology', file Setting up Lab, LMB Archives.

[34] Significantly enough, work on the background document became the preparation for a first textbook of molecular biology written by Perutz some years later (see below).

[35] Here and in the following, F. H. C. Crick and M. F. Perutz, 'The case for a laboratory of molecular biology', Council Paper MRC.58/307, MRC Archives.

from 'classical' biology. Less clear was the distinction from other fields, notably from biochemistry. 'In biochemical terms', the authors reckoned, 'our problem is the structure, synthesis and action of the macromolecules DNA, RNA and protein.' This formulation, while justifying the collaboration of biochemists in the project, was also poised to raise authority disputes with biochemists.

According to the authors, research in the field was characterised by 'a very rapid and striking progress'. Good reasons could be given why this progress was likely to continue. Referring for details to the accompanying document which summarised present knowledge and 'future lines of advances' in the field of molecular biology (note here and elsewhere the use of military terminology) the authors directly moved on to ask: 'What can we deduce from this about the general nature of the work and about the people and the organisation required to carry it out?'

Regarding work practices, the authors noted that the biological materials used were very wide ranging. They extended from viruses to man, including whales, rats, horses and salmonella. The reasons for this wide choice, the authors explained, lay in the unity of the processes they were studying. Similarly, the methods employed on any one problem derived from many different scientific disciplines. Typically one would combine techniques from chemistry, physical chemistry and X-ray crystallography with techniques from biochemistry, cytology and genetics. People specialising in one small area or in one technology were working in the field, but in the eyes of the authors this was not very satisfactory. 'It leads to protein chemists who do not understand genetics, to geneticists who know little of molecular structure, etc.' However, in recent years a small group of workers had grown up 'who understand the subject as a whole, and are prepared, irrespective of their original training, to learn and to produce new techniques to solve the problem before them'. Interestingly, Max Delbrück, 'a physicist at the California Institute of Technology, who essentially founded modern work on bacteriophage', was singled out as 'most outstanding example' of that kind of person. Others were obviously the proponents of the project.

The document presented at least three main reasons for the creation of a large new laboratory of molecular biology, even if these were not analytically separated. First, only a big laboratory would allow for the collaboration between people trained in many different disciplines, for the presence of different research materials and a good proportion of different techniques. Second, such a laboratory would foster the formation of a common viewpoint. Finally, it would provide a focus for this kind of work in Britain. Only thus, the authors reckoned, could Britain keep its lead in the field of structural analysis of proteins and nucleic acids, and counter the dominance of American researchers in other fields, notably in molecular genetics and protein biosynthesis.[36]

The laboratory was to provide postgraduate and postdoctoral teaching to 'train people in the subject as a whole' and in techniques not touched by their previous experience. Undergraduates should be made familiar with the results of molecular biology, but according to the authors this could best be achieved by integration of the new knowledge into existing teaching courses rather than by the establishment of a separate course based at the laboratory. Putting up their own careers as paradigmatic, the authors remarked that the best researchers in the field had studied 'either two complete subjects in succession, or learnt at least one subject thoroughly before acquiring knowledge of a second or third through informal studies and futher research work'. An undergraduate course in biophysics (here we find a slip into the older denomination) 'would leave people with a superficial acquaintance of several subjects, but without a thorough knowledge of any'.

Always shifting between more general statements about the formation of the field and their particular case, the authors noted that in Cambridge there already existed a number of people, mostly on the Council's staff, but presently housed in temporary and inadequate quarters, who could form the nucleus of such a laboratory. What would happen if a laboratory of molecular biology were not to be built? The development of a common viewpoint, as well as the recruitment of new researchers, would be much more difficult. Most likely, however, those groups which were now in Cambridge would disperse. If, on the other hand, the laboratory were built, it would attract into the biological field 'first-class men' from physics, chemistry and medicine. It would allow 'a full-scale assault' on the problems now under discussion, but also a move on to new problems, like embryology and differentiation.

The actual organisation of the proposed laboratory was only touched upon in a closing paragraph. What little was said was clearly tailored to give space to those people who had been drawn into the plan. There was no reference to a similar existing institution. The laboratory would embrace sections of X-ray crystallography, electron microscopy, protein chemistry, molecular genetics and physical chemistry. It would have physicists, chemists and biologists on its staff. It should be close to the central laboratory area in Cambridge and could be associated with one or several university departments.

---

[36] On the 'brain drain' argument in British science politics and its uses by molecular biologists see below (chapter 10).

This programmatic document was accompanied by a second, much more voluminous one which represented the ambitious attempt to review recent developments in the field of molecular biology. Even if Perutz appeared as sole author, the text represented a palimpsest, integrating text passages prepared by other group members, but adapted by Perutz to fit the larger argument.[37] The document did not deal exclusively with the work of the unit members. Rather it created the context in which the work done at Cambridge by Perutz, Kendrew, Huxley, Crick, Watson, Brenner, Ingram, Sanger and others would find its place and mark important breakthroughs. An important issue then becomes what was marginalised from the newly defined field.

In an introductory paragraph the field was given its ancestor. Frederick Gowland Hopkins, first Professor of Biochemistry at Cambridge, was credited with having created the 'scientific climate' in which research on the chemical structure and spatial architecture of proteins was begun. According to Perutz, this work was based on the 'creed' that all living systems were governed by the same biochemical processes, that these were catalysed by enzymes and that all enzymes were protein molecules.[38] More recently, biochemists had started asking themselves what controlled the synthesis of enzymes. It was found that genes controlled protein synthesis. Genes therefore possessed a dual function: they could replicate themselves and determine the structure of proteins. This posed the question of the chemical nature of the gene.[39]

The main problems to tackle therefore concerned the structure and replication of the genetic material, the genetic control of protein synthesis and the structure and function of protein molecules. According to Perutz these questions defined the field of molecular biology. They also structured the following review.

Today we are used to the definition of molecular biology as dealing with the study of proteins and nucleic acids. To my knowledge, however, it was the first time that research on these two molecules (excluding other biological macromolecules like lipids and polysaccharides) was separated out

---

[37] When Perutz later published the document as an article under his name only, this raised some misgivings.

[38] M. F. Perutz, 'Recent advances in molecular biology', Council Paper MRC.58/303, MRC Archives.

[39] In view of the problematic relationship of Cambridge molecular biologists and biochemists, this account of the biochemical 'beginnings' of molecular biology may come as a surprise. In this context, however, it served the function of creating a 'local' tradition and to give legitimacy to the new enterprise. Another strategic reason to turn Hopkins into a founding father might have been that the MRC had generously funded the beginnings of biochemistry at Cambridge (Kohler 1978). On Hopkins and biochemistry at Cambridge see also Kamminga and Weatherall (1996) and Weatherall and Kamminga (1996).

and subsumed under one heading. Equally new was the presentation of structural and genetic work as belonging to one and the same enterprise. The definition was also less coherent than it may appear to a present-day reader. Molecular biology as presented in the review was not, or not only, molecular biology in the common present sense of molecular genetics. It rather represented the conglomeration of various lines of research. Considerations of molecular structure were found to be the common denominator. The definition of the new field and the selection of work reviewed both closely mirrored the interests of the proposers.

Structural studies represented the main preoccupation of the Cambridge unit. The term 'structure' could further conveniently be used to cover both the study of the three-dimensional structure of proteins by X-ray crystallography and Sanger's work on the primary structure of protein molecules. Structure was seen as the key to an understanding of function. Also genetic questions were approached through the study of the structure of the genetic material and the gene products, that is, through structural studies of nucleic acids and proteins. Watson and Crick's double helix exemplified how structure displayed function. The very expectation that function would be laid down in the structure had guided the investigation and the mechanism of replication suggested by the structure gave it credibility.

The selection of work discussed showed the same local biases. Structural descriptions of genes and proteins preceded considerations regarding their function. The question of protein synthesis was discussed from the point of view of its 'structural aspects'. Here the work by Benzer and Hoagland, both current visitors at the unit, was mentioned. Relevance was given to attempts at establishing the structure of RNA, while biochemical studies of protein synthesis were hardly mentioned. Ample space was dedicated to the crystallographic study of proteins. These were discussed in their own right, quite independently of genetic considerations. But almost no attention was given to the structural studies of fibrous proteins. Also excluded was Huxley's work on the muscle proteins actin and myosin. Muscular contraction, Perutz reckoned, 'is a large problem of its own which belongs to quite a separate field of molecular biology, the field of structure and movement, and should really be dealt with in a separate review'.[40] Huxley, co-proposer of the sliding filament mechanism of muscle contraction, had left the unit. He would join the new project only at a later stage. The inclusion of numerous figures, all depicting structural relationships,

---

[40] M. F. Perutz, 'Recent advances in molecular biology', Council Paper MRC.58/303, MRC Archives, p. 10. Note that here 'molecular biology' was clearly used synonymously with 'structural studies'.

further underlined the importance attributed to structural studies while at the same time using the rhetorical power of the figures to argue for the value of the structural approach and the results achieved.

Summing up his review of a field which the same text had only just created, Perutz remarked that in the past six years molecular biology had changed 'from a subject of speculation and uncertainty to an exact science'.[41] Research which had run under different headings could thus retrospectively be reappropriated as belonging to the (pre-)history of the newly created field. If we count back, we find that 1953, Perutz's *annus mirabilis* marked the turning point. The document was only written for limited circulation. However, it represented the first formulation of a history with its turning points and heroes which would often be repeated as well as contested.

Council decision

The case for a new laboratory of molecular biology was discussed by the MRC in April 1958. In addition to the two documents already mentioned the dossier also included a report of the unit's activities in the last three years.[42] The report was geared to providing further arguments for the laboratory project. In contrast to the two other documents, it dealt specifically with the history, achievements and plans of the unit. Quite unusually, however, Sanger's papers were included in the publication list. A brief history pointed to the increasing lines of investigation pursued in the unit, to the large number of important achievements and to the overcrowded conditions in which research was pursued. Thus it built on excellence and productivity as arguments for the provision of new quarters.

Perutz was invited to London to present the case for a new laboratory of molecular biology before the Council members. He equipped himself with models, photographs and paper chromatograms which he intended to exhibit on a table to illustrate the unit's work and to help him convince the audience of its attractiveness. His task proved easier than expected. As he himself has often told, even before he started talking, he was congratulated on the most interesting project in years.[43] The Council agreed that there was 'a strong case, on scientific grounds', for bringing together the various groups working in the field of 'molecular structure', particularly in relation

[41] *Ibid.*, p. 13.
[42] 'Progress Report 1956–1958 of the Molecular Biology Research Unit, Cavendish Laboratory, University of Cambridge', Council Paper MRC.58/306, MRC Archives.
[43] E.g. Perutz (1987, 41) and Perutz in the programme *Eureka*, BBC Channel 4, 21 February 1996, 9.30 pm. On Perutz's plan to bring a variety of exhibits see M. Perutz to Dr Norton, 22 February 1958, file E243/109 I, MRC Archives.

to the development of the biological aspects of the work, and that this possibility should be further explored.[44]

It is difficult to assess on what grounds the Medical Research Council based its agreement.[45] While in Perutz's reconstruction of events the Council decision clearly marked a turning point, it may well be that for Council members it represented more the confirmation of ongoing obligations. All applicants were already funded by the Council. In particular the protein crystallographers had been receiving funds for ten years. The recent results, especially Kendrew's first model of a globular protein molecule, were a confirmation of the Council's politics and of the trust it had put in these researchers. It was the practice of the Council to expand successful research groups. The necessity to provide new quarters for the groups in question had already been recognised.[46] The recourse to the old denomination of 'field of molecular structure' would seem to confirm that for Council members not much new was at stake. It should also be noted that expenditures were not yet debated, and that the decision was in fact only an encouragement to pursue the matter further. The complete separation of 'scientific' and 'financial' considerations in this case does indeed appear as a tactical move on the side of Himsworth who, from the beginning, was a strong supporter of the plan (Fig. 7.4).[47] The positive statement

[44] Minutes of Council, 18 April 1959, minute 95, Council Minutes MRC.58/408, MRC Archives.

[45] In his later correspondence with the Secretary General of Cambridge University, Himsworth mentioned that the Council had 'the advantage of several independent scientific opinions' to consider the scientific case for a Laboratory of Molecular Biology' (H. Himsworth to H. M. Taylor, 16 October 1958, file E243/109, vol. 1, MRC Archives). However, no documents relating to the Council's decisions could be retrieved. Besides Perutz, Charles Harrington, the Director of the National Institute of Medical Research, and Walter L. M. Perry, the Director of the Division of Biological Standards at the National Institute of Medical Research, were invited to the Council meeting. Among the Council members present at the meeting, no one was in any particular way connected with the Cambridge researchers. The one most committed to the scheme was certainly the Secretary himself. According to all testimony, he had much power to influence decisions. In his strong support of the Cambridge group he built in important ways on Bragg's judgement, from whom he regularly received enthusiastic and detailed progress reports on the work of the unit. Bragg kept sending his yearly newsletter when he had long left Cambridge for London.

[46] See for instance 'Visit to the Molecular Biology Research Unit, Cambridge by Dr. Norton on 2nd December, 1957', file E243/29, MRC Archives.

[47] Himsworth's own background was in clinical medicine. He took up office as Secretary of the MRC (1949–68) when the National Health Service was taking shape and became instrumental in using the new opportunities opened up by the reform to promote clinical research. This was seen as his major achievement (Landsborough Thomson 1975, 30). In later years, he himself took particular pride in his support of molecular biology. His engagement in this latter field should be seen in conjunction with rather than in contradiction to his support of clinical medicine. On the development of clinical research as strongly based on basic research in Britain see Booth (1989).

7.4 Harold Himsworth, Secretary of the MRC (1949–68).
From A. Landsborough Thomson, *Half a Century of
Medical Research,* vol. 1 (1973), plate 3. Reprinted with the
permission of the Medical Research Council, 2000.

of the MRC, however, certainly emboldened Perutz and his colleagues and
gave new impulse to the negotiations at Cambridge.

Using the media

The founding document 'Recent advances in molecular biology' had hardly
been discussed and approved by the MRC when Perutz submitted it for
publication to the magazine *Endeavour*. 'This should help to interest people in the subject', Perutz reckoned.[48] *Endeavour* was a widely distributed
quarterly science review. It was one of a number of new scientific publications, intended for a wider readership, started during or in the years
following World War II (chapters 2 and 5 above). Perutz's article was

[48] M. Perutz to H. Himsworth, 15 October 1958, file Himsworth, LMB Archives.

reprinted without major changes and accompanied by a series of glossy pictures, predominantly of models, in the autumn issue (Perutz 1958).

In the same autumn, an abstract of the document also appeared in the Medical Research Council Annual Report presented to the Privy Council.[49] It finally also served as blueprint for what Perutz referred to as the 'first textbook of molecular biology'.[50] The book, based on a lecture series held at the Weizmann Institute in Israel, appeared under the title *Proteins and Nucleic Acids: Structure and Function*, covering much the same themes as the original document (Perutz 1962b).

Perutz was keen also that the plan for the new laboratory should quickly cease to be confidential. Even before the university was officially informed of the proposal, he wrote to the Secretary of the MRC, 'I think our case should be publicly pleaded and I should like to suggest to you that Crick and I publish our document "A Case for a Laboratory of Molecular Biology" in the *Cambridge Review*'.[51] As could be expected, Himsworth chose a more prudent way to approach the university (see below). But in the following years, *The New Scientist* in particular, founded in 1956 to inform a wider public critically on developments in science and its place in society, periodically featured articles by Perutz and others on the laboratory and its history (see especially Perutz 1962a, 1980, 1987 and other contributions in the same special issue dedicated to the fortieth anniversary of the laboratory). From the late 1950s onwards the work of the Cambridge molecular biologists also started featuring in new television programmes like *Eye on Research* and later the *Horizon* series designed to make a wider public acquainted with scientists' work and their results. Television presented itself as a much more congenial medium than radio (or for that matter any written form of communication) to convey complicated structural relationships as studied in the Cambridge unit. As argued earlier, molecular models played a key role in all these performances which provided just the kind of publicity for the work of the unit Perutz was aiming for to gain public support and legitimise tax spending (see above, chapter 5).

University politics

After the Council's decision to support the plan for a laboratory of molecular biology at Cambridge, Himsworth resolved personally to discuss the

---

[49] 'The architecture of proteins', *Report of the Medical Research Council for the Year 1956–1957* (London: Her Majesty's Stationery Office, 1958).
[50] Interview with M. Perutz, Cambridge, 25 June 1992.
[51] M. Perutz to H. Himsworth, 15 October 1958, file Himsworth, LMB Archives.

project with the Vice-Chancellor of the University of Cambridge, Lord Adrian, Professor of Physiology and Master of Trinity College, the richest and biggest among the Cambridge colleges.[52] Elected for two years from among the Cambridge professors, the Vice-Chancellor is the principal academic and administrative officer in the university, while the Chancellor, usually a public figure, fulfils mainly representative functions. While undoubtedly familiar with the research of the MRC unit, as well as with Sanger's work, Adrian showed no enthusiasm for Himsworth's plan. In view of pressing needs of other departments, including especially biochemistry, the Vice-Chancellor expressed particular doubts about the possibility of finding a central site for the new laboratory. He also suggested that the unit might be better off not being part of the university, citing the example of the Strangeways Laboratory, which thrived as an independent research institute.[53]

Perutz reckoned that it was better to launch a scheme through the single Faculty Boards and the General Board, the central body responsible for all matters regarding teaching and research, rather than from the top through the Vice-Chancellor. If they could enlist the support of some professors, the others would probably turn around. The group also speculated that the new Vice-Chancellor who would take office in the autumn, Herbert Butterfield, Professor of Modern History at Cambridge and Master of Peterhouse (the college of which Kendrew was a fellow), would be easier to win for their plan.[54] Over the summer Himsworth was active sounding out various Cambridge professors for their views. He also engaged Bragg in this activity, asking him to approach some of his former colleagues in the Cavendish. In addition Bragg volunteered to approach

[52] All colleges are represented on the university governing bodies, with the biggest colleges able to draw on more support. College interests add an additional layer to the complexities of representative bodies and decision making in the University of Cambridge. For a brief 'unofficial guide' to the organisation and procedures of the University of Cambridge, although incorporating some minor later changes to the system, see Leedham-Green (1996, appendix 5).

[53] Lord Adrian to H. Himsworth, 10 June 1958, and H. Himsworth to Lord Adrian, 12 June 1958, file E243/109, vol. 1, MRC Archives. The particular reasons for Adrian's resistance to the plan of a new research laboratory were not spelled out, but probably rested on the more general view that teaching in the traditional disciplines rather than new-fangled research was the primary mission of the university. Similar reasons were given by the Professor of Biochemistry for his opposition to the laboratory project (see below).

[54] M. Perutz to H. Himsworth, 11 June 1958, file E243/109, vol. 1, MRC Archives. The General Board consisted of the Vice-Chancellor, four appointed members, and two representatives of each of the four groups of faculties. Unlike in other governing bodies of the university, college affiliations did not play a direct role in the composition and functioning of the General Board. Among the professors most involved in the fate of the MRC unit only Mott, but not Todd or Young or any of the professors of biology, was a member of the General Board in the years 1958–60. They could, however, certainly exercise some influence on their representatives in the central body.

H. M. Taylor, the Secretary General of the Faculties, 'a close friend and key man'.[55] He reported favourably. Himsworth also found his approaches 'less gloomy' than he had feared. Mott and Todd continued to look at the scheme sympathetically; the newly elected Regius Professor of Physic, Joseph Mitchell, who was included in the consultation, was 'almost belligerently enthusiastic'.[56] But biochemistry in particular remained hostile to the plan.[57]

Young laid down his objections to the planned research institute at Cambridge in a letter to the Secretary General of the Faculties. He was obviously not happy to lose a distinguished and productive researcher in his department, but also felt that a new research laboratory in a closely related field would represent a more general threat to the future development of biochemistry in the University. The argument he presented was one of principle and regarded the creation of research institutes in small university towns such as Cambridge. Taking a strong stance in a longstanding debate, Young pointed out that research institutes tended to suck up the best researchers from university departments and to compete for resources and assistant staff. They created unfair competition between full-time researchers and researchers who also had a full share of teaching and administrative duties. This necessarily led to resentment. In addition, research institutes, even if lively at the beginning, were at risk of losing their innovative capacities. Instead of driving apart research units and university structure, both the Research Council and the university should seek to integrate them better. Researchers who worked at Cambridge should contribute to the association of teaching and research which Young saw as the 'essence of University life'. Otherwise they should move to research institutes which were independent of universities, like the MRC Institute at Mill Hill.

Regarding the case at hand, Young was 'not at all impressed' by the argument that Perutz's and Sanger's groups needed to move under one roof to cooperate fruitfully. If Perutz's group could continue to work in the Cavendish, Young saw no reason for action, since he was happy to keep Sanger in Biochemistry. If this was not possible, Young suggested the creation of a sub-department of molecular biology, perhaps with the institution of courses leading to a Certificate of Postgraduate Studies on the model of the Masters course in Biophysics at King's College London. The possibility that Sanger could join the sub-department was not raised.[58]

---

[55] W. L. Bragg to H. Himsworth, 30 May 1958, file E243/109, vol. 1, MRC Archives. Among the physicists Bragg approached was George Thomson, Master of Corpus Christi College.

[56] H. Himsworth to W. L. Bragg, 3 July 1958, file E243/109, vol. 1, MRC Archives.

[57] H. Himsworth, 'Proposed Institute of Molecular Biology, Cambridge' (note of 8 August 1958), file E243/109, vol. 1, MRC Archives.

[58] F. G. Young to H. M. Taylor, 20 September 1958, file GB120 (1), University Archives, CUL.

It should be noted that Young expressed a position he had defended earlier. When Sanger's postdoctoral fellowship came to an end in 1951, Young strongly supported Sanger's application to the MRC for a personal grant. He considered it 'a national tragedy' if Sanger should be forced to leave Britain, but insisted that it was against his policy to keep on research staff who were not involved in teaching on a permanent basis and that this would also apply to Sanger.[59] It is also clear, however, that beyond these more general concerns Young felt that the new research institute would represent a particular threat to his own department, luring away the best people.[60]

With his letter to the Secretary General Young intended to move the relevant university bodies to consider the question of the planned laboratory of molecular biology before it would be confronted with a final decision by the MRC. The Council, Young insinuated, was 'a highly efficient unified administrative organisation', able to act quickly on its decision, while the university machinery of consideration, decision making and action moved much more slowly. The university could thus easily fall behind and be faced with a situation which was not in its best interest.[61]

Young's initiative fulfilled its aim. The future of Perutz's research group and the proposal for a new institute of molecular biology were put on the agenda of the first meeting of the General Board after the summer break. Following a common procedure, the Board decided to appoint a committee to consider 'the desirability of providing new facilities within the university for these research workers'. The three professors who were more directly concerned, the Professors of Physics, Organic Chemistry and Biochemistry (i.e. Mott, Todd and Young) were appointed to serve on the committee.[62] The day after the General Board meeting, Himsworth officially informed the Secretary General of the Council's decision to bring the groups working under Perutz and Sanger under one roof. He also expressed the hope of collaborating with the university to this end.[63]

The General Board committee met five times between October 1958 and February 1959. Mott was elected to the chair, but the Secretary General

[59] F. G. Young to H. Himsworth, 12 February 1951, and Himsworth, 'Interview with Sir Charles Harrington, 21st April 1952 and telephone conversation on 24th April 1952', file P19/104, MRC Archives.

[60] On Young's continuing problem regarding the establishment of the LMB see Randle (1990, 591–2). Also interview with Philip Randle, Oxford, 2 June 1993.

[61] F. G. Young to H. M. Taylor, 20 September 1958, file GB120 (1), University Archives, CUL. See also F. G. Young to H. Himsworth, 13 October 1958, file E243/109, vol. 1, MRC Archives.

[62] General Board Note 17,524 and General Board Minute 17,507 of 15 October 1958.

[63] H. Himsworth to H. M. Taylor, 16 October 1958, file E243/109, vol. 1, MRC Archives and file GB120 (1), University Archives, CUL.

of the Faculties, in his role as secretary of the committee, was its driving force. The committee interviewed Perutz, Kendrew, Crick and Sanger and invited Himsworth to one of their meetings. A dense correspondence accompanied the meetings.

The committee had some trouble in pinpointing the basic problem. Was it only a question of space or did it concern bringing the two groups under one roof? Was it a problem of research versus teaching priorities and needs? Mott had no doubt. The Cavendish group was most strongly pressing for space, but did not want to move out of the centre of Cambridge.[64] Himsworth pressed Perutz to make very clear precisely how important the common scheme was, and how important it was to be on a central site.[65] Still in their final report the committee members mused that if more space could be found in the Physics and Biochemistry Departments for Perutz's group and Sanger's group, respectively, this would represent a satisfactory solution.[66] But neither Mott nor Young saw himself able to promise more space.

There was no real discussion about the quality of the research workers. The desirability of keeping them in the university surfaced a few times, but was never really on the table. Rather, Perutz, Kendrew, Crick and Sanger seemed to command considerable negotiating power and were confident enough to pose firm conditions.[67] The fresh announcement of Sanger's (first) Nobel Prize certainly helped their cause. Some space concession in the short run and a firm promise for a permanent solution in four or five years were necessary if the unit was not to disintegrate. A site more than $3/4$ mile from the centre was deemed unacceptable.[68] Practical problems of transportation and habits of life and work, as well as scientific reasons, were at the basis of this request. Moving out of the centre meant giving up the proximity to the Cavendish workshops and stores as well as to other facilities of the university like the computers of the Mathematical Laboratory. It meant giving up the daily interactions with researchers from other departments and the discussions over lunch in the pub and made it more difficult to be involved in teaching at the university or in the colleges.

---

[64] Discussion notes in Taylor's handwriting, 21 October 1957, file GB120 (1), University Archives, CUL.

[65] H. Himsworth to M. Perutz, 16 October 1958, file E243/109, vol. 1, MRC Archives.

[66] 'Report to the General Board of their Committee on the future of the M.R.C. Unit of Molecular Biology and associated matters', General Board Paper 4992, University Archives, CUL.

[67] This also surprised the Secretary of the MRC. After talking to Perutz about the interview, he noted: 'Curiously enough he [Perutz] was not challenged on the scientific case for bringing the group together in one place. The discussion was concerned with ways and means of so doing'; note by Himsworth, 6 November 1958, file E243/109, vol. 1, MRC Archives.

[68] Handwritten notes by secretary of committee on meeting of 29 October 1958, file GB120 (1), University Archives, CUL.

Asked to detail for the first time their plans for the new laboratory, Perutz and Sanger projected a four-storey building of around 18,000 square feet. This was three times as much space as both groups together were currently occupying. The laboratory would house three divisions, protein chemistry, protein crystallography and molecular genetics. Perutz also announced that the virus structure group from Birkbeck, originally headed by Rosalind Franklin, was going to join the Cambridge protein crystallographers.[69] The move had been discussed with Franklin before her untimely death that year. This event and Bernal's imminent retirement as head of the department gave the plan new urgency.

If Young's letter to the Secretary General had initiated the new round of discussion at Cambridge, he was not successful in impressing his point of view on the committee. His repeated affirmation that he would not like to see Sanger moving out of the Biochemistry Department did not carry much weight. All committee members, however, agreed that the new laboratory had to be integrated as closely as possible into the university. Different schemes were discussed. It was generally agreed that the term 'institute' was better avoided. As Himsworth explained to Perutz, this word, 'largely because of developments in other countries, has got the connotation of meaning a rival show to the university; and I do not wish to encourage that view in Cambridge'. MRC Laboratory of Molecular Biology was suggested as the name for the new institution.[70]

Even if the committee was not officially entitled to discuss possible sites for a new laboratory, the discussion always turned again to this crucial point. Finding space in an existing building seemed as hopeless as inserting the provision of a new building in the university building plan which was committed for many years ahead. A possible way out of the impasse was offered by Himsworth's suggestion that the MRC could use money from a legacy to fund the construction of a new building. Usually the MRC did not pay for bricks and mortar, but money from benefactions offered some flexibility.[71] In Himsworth's meeting with the General Board committee the available funds were for the first time mentioned officially, but whether

---

[69] 'Preliminary plan for Laboratory of Molecular Biology' (dated 7 November 1958), file E243/109, vol. 1, MRC Archives; 'Committee of the General Board on the M.R.C. Biophysics Unit and associated matters', General Board Office, 12 November 1958, file GB120 (1), University Archives, CUL, and 'Report of the General Board of their Committee on the future of the M.R.C. Unit of Molecular Biology and associated matters', Appendix A, General Board Paper 4992, University Archives CUL.

[70] 'Committee of the General Board on the M.R.C. Biophysics Unit and associated matters', General Board Office, 12 November 1958, minute 18, file GB120 (1), University Archives, CUL, and H. Himsworth to M. Perutz, 16 October 1958, file E243/109, vol. 1, MRC Archives.

[71] The new benefaction came from the Cusrow Wadia Fund, set up by the wealthy entrepreneur Cusrow Wadia from Bombay. The MRC's final share in the establishment of

it was really a new revelation is difficult to reconstruct. Even the money, however, did not solve the site problem; a piece of land near the central laboratory area, either of university property or of property to be bought by the MRC, was still needed.

The final report summed up the discussions, indicating that the committee regarded it as more important to make a firm offer for permanent accommodation rather than meeting short-term requirements. Regarding the two alternative schemes of a joint MRC–university building or a building fully funded by the MRC, the committee expressed a clear preference for the former.

The General Board discussed the report in March 1959 and referred it to its Committee on the Order of Urgency of Needs.[72] A possible site for the proposed new laboratory building seemed in sight. The university hospital was due to move from the centre of Cambridge to new grounds on the outskirts of the city and discussions were in course regarding the redevelopment of the hospital site for university use. The Committee recommended that high priority be given to the construction of a MRC Laboratory of Molecular Biology as part of the first stage of the development of that site.[73]

Negotiations over site

Despite this favourable decision, the site question proved to be long and thorny. Again, only the course and the length of the negotiations can explain the particular position of the Laboratory of Molecular Biology in Cambridge.

The plan to build the new laboratory on the new hospital site proved over-hasty. The Hospital Authority preferred keeping the last remaining ground on their site for their own use until the negotiations for the

the new laboratory amounted to £432,000, of which £312,500 were spent on the building, £90,000 for initial equipment, and £30,000 towards the Department of Radiotherapeutics which shared the same building (see below). The Wellcome Trust offered a further £70,000 towards the equipment of the new laboratory; Council Paper MRC 61/464 and Council Meeting of 19 January 1962, minute 19, MRC Archives. Even if money for the construction of the new laboratory did not come from the Treasury, political support for the decision to create the new laboratory, the future running costs of which would weigh on the MRC budget, was needed. According to Himsworth's testimony, Lord Hailsham in his multiple functions as Department Leader and Leader of the House of Lords (1957–60 and 1960–3), Lord President of the Council (1957–9 and 1960–4), Lord Privy Seal (1959–60) and Minister for Science and Technology (1959–64) was instrumental in getting the laboratory on its way; see M. Perutz to H. Himsworth, 17 October 1969, file Himsworth, LMB Archives.

72 General Board Minute 17,781 of 4 March 1959, University Archives, CUL.
73 Committee on the Order of Urgency of Needs, minute 190 of 7 March 1959, and General Board Minute 17,794 of 11 March 1959, University Archives, CUL.

7.5 Joseph S. Mitchell, Regius Professor of Physic at Cambridge, in a contemporary sketch by nuclear physicist Otto Fritsch. From O. Fritsch, *What Little I Remember*, p. 207. Copyright 1979 Cambridge University Press.

new hospital had made further progress. Discouraged by this new hurdle, Perutz felt the whole scheme was in danger of foundering and the unit would disintegrate.[74]

At this critical juncture, unexpected and decisive support for the plan of a laboratory of molecular biology came from the Regius Professor of Physic (Fig. 7.5). Since the war years Mitchell had been involved with the development of radiotherapeutics at Cambridge and, like Perutz, had been a member of the Biophysics Committee set up by the Medical Research Council in 1947 and in operation till 1954. The free site on the hospital ground had been earmarked for the construction of a new Department of Radiotherapeutics. This was now to be built on the new hospital site where the new Postgraduate Medical School was going to be developed. The hurdles encountered by the plan for the new laboratory of molecular biology gave Mitchell a chance to act as mediator. Engaging in discussions with all parties concerned, he suggested that the laboratory should be built next to his own department on the new hospital site. The proposed site was 2 miles out of town, but would develop into a major centre of university

---

[74] N. Annan to R. E. Macpherson, 2 May 1959, file Himsworth, LMB Archives. A desperate Perutz had turned to Annan, Provost of King's College and friendly with the molecular biologists, asking him to lobby Macpherson, a member of the Financial Board, on their behalf.

activity. Mitchell stressed how strongly he would welcome the laboratory on the site.[75]

Mitchell had opportunistic reasons for intervening in favour of the molecular biologists. The laboratory would raise the research profile of the new Postgraduate Medical School and its status in relation to other research centres at Cambridge, thereby strengthening Mitchell's own position. Other members of the Board scorned such self-interested behaviour, but for the molecular biologists his support proved decisive.[76]

A few months before, Mitchell's proposition would not have seemed attractive at all to Perutz and his colleagues. But worn out by year-long negotiations they concluded that, all things considered, this seemed the best solution.[77] As had happened ten years earlier, when Bragg first applied to the MRC for funding of the protein crystallography in his department, the medical connection was the last one to be brought into play. But it was not the attraction of the Postgraduate Medical School which made Perutz change his mind. Once the group had accepted the need to move out of town, the Madingley site to the west of the city earmarked as a development site for nuclear physics gained attraction.[78] This site had been proposed repeatedly since the very beginning of the negotiations in 1953, but Perutz and his colleagues had never considered it a viable solution. The fact that they brought this option up again at this late stage shows that the Cambridge molecular biologists were still attracted by a connection with physics.

Accepting a laboratory on the outskirts after six long years of negotiations did not only mean a defeat for Perutz and his group. In these years, the plan for a laboratory of molecular biology had matured: several groups had joined the project and the planned laboratory had grown considerably in size.

A special agreement regulated the relation between the newly planned institution and the university. Long negotiations had preceded the drafting of the document. Three points which were of crucial importance for the development of the laboratory proved particularly thorny: the regulation of the size of the laboratory staff, the employment of technicians and the formal relation of the laboratory staff to the university. In line with a policy of limiting its own expansion, the university demanded some measure

[75] J. Mitchell to N. Mott, 9 May 1959, file GB120 (1), University Archives, CUL.

[76] On Mitchell's ambitious plan for a close integration of hospital wards and university laboratories, including the LMB, in the new postgraduate medical school at Cambridge see below (chapter 11).

[77] H. Himsworth to H. M. Taylor, 13 May 1959, file GB120 (1), University Archives, CUL.

[78] General Board Minute 17,864 of 13 May 1959, University Archives, CUL. The Cavendish Laboratory moved to the Madingley site in 1974. Currently, work is under way to turn the site into a major centre for Cambridge science.

of control over the number of staff. Following an earlier agreement regarding outside-funded research bodies, the university also expected that
technical staff should be appointed by the university and be paid according to its rates. This regulation was to rule out the possibility of first-class
technicians being lured away by higher salaries. On both these points, the
Secretary of the MRC gained the upper hand. The responsibility for management and scientific policy of the new MRC establishment remained
with the MRC which was, however, to consult with the university. Similarly, the technical staff were to be appointed by the MRC. The third point,
however, concerning the university status of the laboratory staff and the
possibility of supervising graduate work – a point which much concerned
Perutz and his colleagues – was 'left to sort itself out in practice'. Several
regulations which applied to this question existed, and much good will
was professed.[79]

The biologists' protest

Himsworth and Taylor had hardly congratulated each other on the satisfactory conclusion of a 'highly complicated and kaleidoscopic series of
negotiations', when a letter of protest signed by a group of seven biologists
brought new turmoil to the planning process.[80] Their protest turned out
to be half-hearted, but is none the less significant as a first intervention on
the part of biological scientists and their interest in molecular biology as
teaching subject.

The signatories, among them the professors of Zoology, Physiology and
Biochemistry and some of their associates, intended to make sure that
the General Board was 'fully aware of the extent of the misgivings which

[79] The last quote is from H. M. Taylor to H. Himsworth, 4 June 1959, file GB120 (1),
University Archives, CUL. On the heads of agreement see also 'Institute of Molecular
Biology, Cambridge. Telephone conversation with Dr. H. M Taylor, 20.11. [notes by H.
Himsworth, 20 November 1958]; H. M. Taylor to H. Himsworth, 21 November 1958;
H. M. Taylor to H. Himsworth, 4 February 1959; H. Himsworth to M. Perutz, 9 February
1959; M. Perutz to H. Himsworth, 11 February 1959; H. Himsworth to H. M. Taylor, 12
May 1959; H. M. Taylor to H. Himsworth, 21 May 1959; H. Himsworth to H. M. Taylor,
30 May 1959; 'Report of the General Board on research wholly or partly supported by
funds from outside bodies', 4 May 1955 [published in Cambridge University Reporter of
18 May 1955, pp. 1397–1415]; file E243/109, vol. 1, MRC Archives [all correspondence
between Taylor and Himsworth also in file GB120 (1), University Archives, CUL]; H.
Himsworth to H. M. Taylor [undated, but probably 5 June 1959] and H. M. Taylor to N.
Mott, 5 June 1959, file GB120 (1), University Archives, CUL; General Board Paper 5077;
General Board Minute 17,864 of 13 May 1959; General Board Minute 17,881 of 20 May
1959; General Board Note 17,969 of 3 June 1959; General Board Minute 17,920 of 3 June
1959, University Archives, CUL. On the status of staff members and connected problems
see also M. Perutz to N. Mott, 15 May 1961, file Himsworth, LMB Archives [contains
extracts of all relevant documents regarding the negotions on this point].

[80] H. Himsworth to H. M. Taylor [undated, but probably 5 June 1959], file GB120 (1),
University Archives, CUL.

are felt throughout biological departments at the prospect of the Medical Research Council's Unit for Molecular Biology being moved to a laboratory away from the central sites'. They appreciated the difficulties faced by university planners, but believed that the work pursued by the members of the unit was of 'such fundamental importance to biology' that a special effort had to be made, even at this late stage, to find a central location, even if it meant upsetting plans already made for other departments. The biologists explained that the work which fell under the term of biophysics would almost certainly be included in university teaching courses within a few years. Some plans in this direction were already under way. But it was obviously much more difficult to plan such courses if those who could teach them were moved out of town.[81]

Above all, Young's signature put the other Committee members, and in particular Mott, into an uncomfortable position. 'I would have associated myself with any move to keep them here if I had been asked', he defended himself.[82] But to Taylor he confided, 'I am not happy about the way our planning has gone – particularly my own part in it!' Personal interests and ambitions had prevailed, 'and molecular biology, probably the best thing in Cambridge, had to go'. The problem, as Mott saw it, was that it was 'no one person's business'.[83] Mitchell, whose own plans were now most tightly linked to the future laboratory of molecular biology, was 'extremely concerned' by the latest developments.[84]

At this point also Fulton Roberts of Pathology, himself a member of the General Board and representative of the biological faculties to which the signatories of the letter of protest belonged, dissociated himself from the Board's decision to move the laboratory of molecular biology away from the centre. He seized the opportunity to attack more generally the politics of the General Board and its tendency to take decisions 'on considerations of convenience rather than those of educational policy'.[85]

---

[81] C. F. A. Pantin, J. W. S. Pringle, J. A. Ramsay, F. G. Young, E. F. Gale, B. H. C. Matthews, A. F. Huxley to H. M. Taylor, 12 October 1959; copied as General Board Paper 5201, University Archives, CUL. In a separate letter the Professor of Pathology, H. R. Dean, associated himself with the document signed by Pantin and others; H. R. Dean to H. M. Taylor, 3 October 1959 [but obviously of a later date], file GB120 (1), University Archives, CUL. A 'last ditch battle' was also attempted by Richard Keynes from Physiology who approached his college, Peterhouse, proposing to make one of its properties available for the new laboratory; M. Perutz to R. D. Keynes, 20 October 1959, file Himsworth, LMB Archives. On the review of biology teaching at Cambridge and the introduction of a new 'half-subject' cell biology, including much of 'biophysics', see above (chapter 3).

[82] N. Mott to H. M. Taylor, undated note, file GB120 (1), University Archives, CUL.

[83] N. Mott to H. M. Taylor, undated note, and N. Mott to H. M. Taylor, 20 October 1959, file GB120 (1), University Archives, CUL.

[84] J. Mitchell to H. M. Taylor, 23 October 1959, file GB120 (1), University Archives, CUL.

[85] F. Roberts to H. M. Taylor, 15 October 1959, file GB120 (1), University Archives, CUL.

But the protesters had not made a constructive suggestion as to how a central site might be obtained. The General Board threw the ball back to the Faculty Boards of Biology and Medicine, enquiring if they wanted to review the priorities of their own needs and encouraging them to suggest alternatives. The Boards repeated their discontent that they had not been consulted before. They proposed a couple of sites, but these proved unviable, especially so since the size of the newly planned laboratory had grown again, now reaching 24,000 square feet. Electron microscopist Huxley was going to rejoin the laboratory and more space for storage and workshops was needed than originally planned. Pressed once more to pronounce on the matter, Perutz confirmed: 'We decided we really want to remain together and that we would not be interested in a central laboratory unless it can be made of sufficient size'.[86] Also Sanger reckoned: 'While we would all be very pleased if a site nearer the centre could be found, it seems to me that the possibilities have been thoroughly explored and that the present plans are very satisfactory and have the great advantage that they can be accomplished in reasonable time'.[87] The building plans for the new laboratory were pressed forward. But the General Board accepted the provision, that the move was not to be regarded as permanent, and that it 'would keep in mind the need for providing some space on a central site for molecular biology'.[88]

The question was brought up again one year later, when plans for the reorganisation of one of the central laboratory areas of the university were discussed. The Council of the Schools of Biological Sciences recommended that some modest space on the site should be made available for the MRC Unit for Molecular Biology to allow the staff to take an active part in undergraduate teaching. But the first signatory of the letter of protest, Professor Pantin of Zoology, now reckoned that all space in his new building was already assigned. Nor was any other department of those concerned ready to renounce any of the allocated space.[89] In the following academic year, the Council of the School of Biological Sciences repeated its recommendation, but to no avail.[90] A few months later Perutz

---

Putting his dissent into writing at this moment was, at least in part, surely also a move to protect himself. Before, he had expressed 'some fears', but had not dissociated himself from the General Board's decision; see handwritten notes in Taylor's hand on General Board Paper 5201, file GB120 (1), University Archives, CUL.

[86] M. Perutz to N. Mott, 22 October 1959, file Himsworth, LMB Archives.

[87] F. Sanger to H. M. Taylor, 2 November 1959, file GB120 (1), University Archives, CUL.

[88] General Board Note 18,191 and Minute 18,127 of 2 December 1959; see also General Board Note 18,166 and Minute 18,106 of 25 November 1959, University Archives, CUL.

[89] General Board Minute 18,774 of 25 January 1961; General Board Note 18,924 and Minute 18,840 of 15 February 1961, University Archives, CUL.

[90] Barker to W. J. Sartain, 5 December 1961, file GB10 (1), University Archives, CUL.

7.6 The new MRC Laboratory of Molecular Biology in the building shared with the Department of Radiotherapeutics on the new hospital site (centre). The flag is at half-mast for the death of President John F. Kennedy on 22 November 1963. Courtesy of Medical Research Council Laboratory of Molecular Biology, Cambridge.

and his colleagues and Sanger with his group vacated their old laboratories and offices and moved to their new building on Hills Road (Fig. 7.6). Perutz enquired if they could keep a *pied à terre* in the Cavendish, but Mott was not agreeable.[91] Perutz has stressed repeatedly that Mott, initially hostile to the MRC researchers, once arrived in Cambridge, became their main supporter (Perutz 1998). However, Mott had his own view in this regard. In his autobiography he remarked: 'J. G. Crowther's book on the Cavendish says that I was very sad when [the MRC unit] left, but this was not so.' And he continued: 'The space they left enabled me to build up electron microscopy and develop the work on dislocations' (Mott 1986, 110).

The hope that the laboratory could eventually be relocated in the centre of Cambridge continued to be expressed from time to time in some

[91] N. Mott to M. Perutz, 14 March 1962, file Setting up Lab, LMB Archives. Mott did, however, agree to the continued use of some central facilities by Perutz and his group (*ibid.*). Anxious to keep some links with the Cavendish, Perutz also offered a position of honorary adviser to William Cochran, Reader and crystallographer in the Cavendish. Cochran held that position until his move to Edinburgh in 1964.

quarters, but became increasingly unlikely, not least because of the dynamics with which the new laboratory developed. When, in 1966, the Vice-Chancellor explored the possibility of a closer association of the laboratory with the university, the governing board of the laboratory expressed its preference for a 'brother–sister-relationship' and found the prospect of coming under university administration 'totally unacceptable'.[92]

Molecular biology in Cambridge

This chapter has argued that molecular biology in Cambridge was established in response to a protracted institutional crisis of the MRC unit. With Bragg, the unit's main patron in the Cavendish and in university governing bodies, moving to London, the group had to find new interlocutors and a new institutional home. Only when all plans to secure a new institutional home in the university failed, either for lack of support or for the growing needs of the unit, and when the prospects of keeping the group together became increasingly bleak, did Perutz and his colleagues change strategy. Joining forces with Sanger's group from Biochemistry, they applied to the Medical Research Council for an independent new Laboratory of Molecular Biology. The term 'molecular biology' first appeared in this connection. By that time, Crick, Ingram and Brenner had embarked on an ambitious new project in molecular genetics aimed at relating the chemical structure of genes to the structure of proteins. However, it was not this project, its viability still untested, but Kendrew's success in the structure determination of myoglobin and the agreement with Sanger that provided the decisive impulses for the new negotiations. The longstanding and continuing support from the MRC gave force to their plan. Sanger's work on protein sequencing, the crystallographic studies of proteins and Crick and Brenner's new research venture were all subsumed under the term 'molecular biology'.

The laboratory took shape in the ensuing negotiations between scientists, science administrators and university governing bodies. In the course of the negotiations the laboratory grew considerably in size. Rather than

---

[92] Laboratory of Molecular Biology, 6 May 1966, MRC.58/1017, and H. Himsworth, 'Laboratory of Molecular Biology and the University of Cambridge. Interview with the Vice-Chancellor of Cambridge, 9 May 1966', file E243/109, vol. 2, MRC Archives. The suggestion followed a report on the development and long-term needs of the scientific departments at Cambridge which explicitly mentioned the need to relocate the LMB. In the view of the reporting committee of which Kendrew was a member there was 'no overriding argument based on academic affinities' in favour of retaining the laboratory on the hospital site. On the contrary, every effort should be made to bring the laboratory 'into closer physical association with related fundamental disciplines' ([Deer Report] 1966, 589). The report more generally identified a great expansion of and interest in the biological sciences and, among other things, suggested providing central accommodation for the Department of Genetics.

in the centre of Cambridge, a firm condition at the beginning of the negotiations, it ended up being located far away from the main laboratory sites and the teaching institutions. Formal links to the university
as well as to the other university departments were looser than originally wished. Only conflicting interests and the course of the negotiations,
in which intermediate choices guided subsequent steps, can explain this
result.

Writing their own history, Cambridge molecular biologists have often
pointed to the strong resistance and the little cooperation presented by
the university to their plans (e.g. Perutz 1962a, 209). If the MRC more
readily supported the plan of an independent research laboratory than the
academic bodies involved, this was because the university complied with
other rules. Its tradition as a teaching institution and the disciplinary set-
up of the university, where departments keenly defended their territories,
made it difficult to accommodate a research unit which showed no loyalty
to a particular discipline. But this resistance, as well as the support of the
MRC, was inscribed into the structure of the future laboratory. In the end,
it was the refusal of the Cambridge professors to hosting an expanding
research unit, not grown in their own departments, which led to the plan
for the new laboratory. Inside the university, the laboratory could not have
grown to the size it did, especially not in the time of postwar congestion
and retainment politics of Cambridge University.

The loose connection to the university was acceptable since the main
interest of the Cambridge molecular biologists lay in research and they had
no problem in staffing the laboratory with postdoctoral graduates, mainly
from America.[93] There was, none the less, a spectrum of reactions regarding the peripheral nature of the links to the university. For some, as for
Perutz, there remained some regret about the missed university career.[94]
Personal disappointment was turned into a more general criticism of the
way the university responded to innovative fields of research and of the
personal interests governing university politics. Crick was most adamant in
seeing a career with the MRC as his best choice. Thanks to support by the
MRC he had been able to move from physics to biophysics. Subsequent
employment by the MRC had allowed him to devote all his efforts to
research instead of starting, at an advanced age, at the bottom of the university career ladder. Sitting on committees and participating in university
politics was not what he aspired to. Nor was he keen to teach, wonderful

---

[93] M. Perutz to H. Himsworth, 24 October 1958, file Himsworth, LMB Archives.
[94] Interview with M. Perutz, Cambridge, 3 July 1992. Perutz, however, was quick in
correcting this view, affirming that with a university post he would probably not have
been able to dedicate time to build up the LMB.

expositor though he was.[95] The same was true for Sanger, who did not rate himself highly as a teacher. Others, like Kendrew and Brenner, using the special opportunities offered at Cambridge, got involved in college teaching and in this way also recruited young researchers to the laboratory. Still others eventually left the MRC establishment to take up university positions. Most of them mentioned their wish to teach as one important reason for their decision.[96]

If the problematic integration of the new research institute into the academic structures appeared as a weakness, it was soon turned into an asset. With molecular biologists gaining access to advisory functions on governmental level, the promotion of molecular biology entered the governmental agenda and the Cambridge laboratory became the model according to which teaching and research in the biological sciences were to be reformed (see below, chapter 10). In other respects too, the position of the LMB as an academic institution, but independently funded and administered, became crucial for the establishment of the laboratory as a centre of the new science (see Part III).

It may here be useful briefly to compare the developments leading to the establishment of molecular biology in Cambridge in the late 1950s with the attempt, mentioned earlier, to set up an Institute of Mathematico-Physico-Chemical Morphology in Cambridge in the early 1930s. The scheme submitted to the Rockefeller Foundation by Needham, Wrinch and Bernal, who was expected to receive a position in the new institute, did not find the necessary support and had to be abandoned. Historian Pnina Abir-Am has discussed this (failed) initiative of a 'British avant-garde' as an early form of molecular biology which she defines as a cognitive enterprise as well as a social manifestation centred around new forms of collaboration between physical, chemical and biological scientists (Abir-Am 1987, 27). Especially through Bernal, who was Perutz's teacher, there were personal links between this attempt at establishing an interdisciplinary institute for biological research at Cambridge in the 1930s and the plan for a laboratory of molecular biology in the late 1950s. However, the search for origins remains problematic and I am doubtful how useful it is in this particular case. Never in the course of the negotiations for the Cambridge Laboratory of Molecular Biology was reference made to an earlier attempt to establish the field at Cambridge. And the relations of the sciences as well as the technologies involved were different in both cases. Without entering into a

---

[95] F. Crick to H. Himsworth, 15 June 1953, file P3/125, vol. 1, MRC Archives and interview with F. Crick, Cambridge, 25 May 1993.

[96] Interviews with D. Blow, London, 29 September 1992; B. Hartley, Elsworth, 28 September 1992; and V. Ingram, Cambridge, MA, 20 July 1993.

discussion of the respective merits of the two proposals, an analysis of the conditions under which the two plans were discussed can, none the less, highlight the changed opportunities for scientific endeavours in postwar Britain. In the case of the plan for the new laboratory of molecular biology, money came not from a foreign private foundation, but from a British state agency which, since World War II, had seen its status and budget steadily increasing. The university posed similar resistances to accommodating an enterprise which ran against and openly opposed the disciplinary set-up of the university. However, in contrast to the situation in the 1930s, a link to the university was considered desirable, but was not, as we have seen, an absolute condition, either for the researchers involved or for the funding body. This points to a changing role of universities as privileged centres of research. The newly acquired independence from academic dealings offered scope for more rapid expansion and new alliances, opportunities which molecular biologists were quick to seize upon.

The LMB was the response to a local situation. Constraints of space, alliances and resistances which modelled the new laboratory were as contingent as the combination of people, skills and projects which came to define the new field and found their place in the new research institution. As Perutz put it well, regarding the staffing of the laboratory: 'You take together the people you know'.[97]

How does this local account relate to the standard story of the origins of molecular biology? To answer this question, it will be useful to come back once more to Watson and Crick's DNA model and follow its fate.

---

[97] Interview with M. Perutz, Cambridge, 25 June 1992.

8

...

The origins of molecular biology revisited

If one visits the Laboratory of Molecular Biology today, one will find a ball-and-spokes model of the DNA double helix in a perspex case on the landing of the staircase to the second floor. However, this is not the original model built from metal bits and clamps by Watson and Crick in 1953. Fragile and difficult to transport, the latter was none the less moved from the Cavendish to the model room of the new laboratory.[1] This room has now been transformed into a computer working station, since modelling is today mainly done on mainframe computers. Among the models which still populate the room are the remains of a DNA structure. This dilapidated model, however, is again not the original one, but a replica built by Claudio Villa, a researcher in the laboratory, for the BBC production of *Life Story* in 1987.[2] Where has the original gone?

Reconstructing the fate of the actual Watson and Crick model as well as the history of the widely reproduced photographs representing Watson, Crick and the double helix will shed further light on the role of the structure in the making of molecular biology at Cambridge. It will also offer a way to relate the local account to what I presented as the standard history of the field. In the second part of the chapter I will explore to what extent developments at Cambridge, even if representing a local solution, responded to more general trends.

Constructing a discovery

When investigating the history of the original Watson and Crick model, a first, initially surprising finding is that the exact fate of the structure

---

[1] Because of its size and fragility this particular model had not toured through lecture halls, exhibitions, World Fairs and television studios like other DNA models. These models were not only more robust and colourful, but were also made according to the more accurate coordinates later established by Wilkins.

[2] Since this chapter was written, a stand and a few clamps are the only surviving bits of the BBC replica. Apparently students and postdocs took away the model parts bit by bit as 'souvenirs' (R. Henderson, pers. comm.).

is difficult to trace. Indeed, even what the 'original' model was remains unclear. Is the tall model we all seem to know Watson and Crick's working model or is it a later construction? Crick recollects:

Of course we built the hopelessly incorrect model we showed to the King's people, but after that we were forbidden to build any more models. After Pauling's paper arrived we got permission (both from Bragg and Maurice) to try again. Jim ordered metal models of the bases (of which we had none before) but we became impatient and made make-shift paper ones, with which we discovered the AT, GC pairs.

After that we immediately started to build models, but using only one sugar-phospate, plus one extra atom, and using a constraint to allow for the base pairs. We built at least two fairly similar models, of which one is the published one.

I think I was wrong to suggest that we went straight from this to the tall, open-day one. We probably built a shorter model, with both chains, and perhaps a couple of base pairs, but my recollection is that for the open day we used everything we had to build as big a one as possible.[3]

As Crick's recollection makes clear, a series of models preceded the tall demonstration model, but it is the latter which is generally referred to as the 'original' one. Certainly no trace exists of the earlier ones.

Once transported from the Cavendish Laboratory to the LMB, the model apparently became neglected. Parts of it broke off and many pieces disappeared. What happened next nobody in the laboratory knows for certain. One story goes that the model was given by Crick to Amy Pollack, when the Department of Molecular Biology opened at Edinburgh University in the mid-1960s (Yoxen 1978, 352). However, the model donated to Edinburgh was another, simpler model.[4] More likely is the following course of events. When Herman Watson (the common last name with the well-known protagonist of the double helix is purely coincidental) moved to Bristol, a number of pieces (possibly all those which were left) were mixed up with other model building parts packed up for him.[5] By the 1970s, however, Watson and Crick's work had acquired symbolic meaning as the origin of the new science of molecular biology and the story of the model was much more widely known. Watson's best-selling book had appeared in 1968, and in 1974 *Nature* dedicated a special issue to the model's 21st birthday. Thus, when in 1976 Ann Newmark from the Science Museum travelled to Bristol to discuss with Watson the reconstruction of the 'forest of rods' model of myoglobin found in the museum store and, on that occasion, discovered

---

[3] F. Crick to author, 5 January 1998.
[4] F. Crick to author, 2 June 1997.
[5] M. Fuller, pers. comm. and H. Muirhead, letters to author, 8 and 12 May 1997.

the original metal plates (the only parts which, because purpose built, could be identified as belonging to the Watson and Crick model), she arranged for the pieces to be acquired by the Museum.[6] Farooq Hussain, a research student at King's College London, volunteered to build a facsimile of the demonstration model, incorporating the few original bases. This model, later found to be taller than the original, has since served as prototype for further replicas, one of which (made to correct size) was part of the *Living Molecules* exhibition in the Science Museum (Fig. 5.1 above). Hussain's facsimile, held in the Health Matters Gallery in the Museum and faithfully labelled 'The nearest there is to the original model', has meanwhile undergone conservation, sealing its status as museum object (Fig. 8.1).[7]

The career of the photographs of the Watson, Crick and the double helix tells a similar story. They come as a pair. One shows Crick pointing up to the model with a slide-rule. Watson follows Crick's movement with an innocent and amazed gaze. In the second (slightly less staged) version of the same motif, the slide-rule points downward and Watson and Crick laugh somewhat embarrassedly. The pictures have acquired iconographic status. They appear on book covers and posters, celebratory items and advertisements. This, however, is a rather recent development.

The two pictures are part of a set of eight exposures taken by Antony Barrington Brown, once an undergraduate in Chemistry, then a freelance photographer at Cambridge (Fig. 8.2).[8] A graduate student, paid by *Time* magazine to look out for interesting stories about Cambridge science, had pointed Barrington Brown to the model. The photographs were to accompany an account of Watson and Crick's work for the magazine, but neither the article nor any photograph was published at the time. The

[6] Two of the original base pairs actually were incorporated in an *ad hoc* model Watson had built at Bristol. He later found six additional pairs; see F. Hussain to A. Newmark, 5 October 1976, file 996/45, Science Museum, London. A later memo records that as many as thirty-six base plates were collected by Hussain from the Science Museum to build the model; see A. Newmark, handwritten memo, 24 January 1977, file 996/45, Science Museum, London. Whether these were all original pieces from Bristol remains unclear. One original base pair remained incorporated in the Bristol model. The myoglobin model Newmark planned to rebuild had been deposited and forgotten in the museum store. Next to Kendrew who had moved abroad, H. Watson, Kendrew's former collaborator, was most familiar with the model.

[7] I thank Michael Fuller (LMB), Hilary Muirhead (Bristol), and Ann Newmark and Nicola Perrin (both Science Museum) for help in reconstructing this course of events. See also file 996/45, Science Museum, London.

[8] The photograph of Watson and Crick drinking tea in their office at the Cavendish, also commonly reproduced (see e.g. Watson 1980, 130), comes from the same series. The other five shots were never reproduced.

8.1 The 'nearest there is' to the original DNA model, on display in the Health Matters Gallery in the Science Museum in London. *Source*: Science Museum/Science & Society Picture Library, slide no. SCM/BIO/C1000271.

8.2 Contact prints of the photographs taken by Antony Barrington Brown of Watson, Crick and their model on 21 May 1953. Courtesy of Antony Barrington Brown.

photographer may well have forgotten about his pictures. The model had not impressed him very much. It looked like a piece of engineering rather than chemistry to him and he found it not at all photogenic. He therefore staged Crick and Watson in front of it.[9]

Fifteen years on, one of the photographs (the one with the slide-rule pointing upwards) first appeared in print in Watson's account of the 'race' to the double helix. The book also included another, unacknowledged photograph of the helix, placed in front of a dark curtain (Watson 1980, 121).[10]

---

[9] A. Barrington Brown, pers. comm.

[10] This photograph, together with the two by Barrington Brown, are the only pictures of the original model that are in public circulation.

Despite the fact that this picture offered a more close-up view of the model, it was Barrington Brown's photograph of the two scientists with their bewitched gaze in front of the model which better captured the heroic discovery account. After this first public appearance, demand for the picture rose steadily. Together with its companion, it was distributed by a number of institutions and was reproduced innumerable times all over the world. Barrington Brown realised how much he was losing on royalties for his 'historic' photographs and he is now pursuing the breach of copyright.[11]

The story of the model and its photographs is emblematic for the history of the double helix at Cambridge and elsewhere. In 1953 and throughout the 1950s the double helix did not play the role it does today, either in the public imagination or in accounts of the history of the field. As Crick himself once observed in conversation, 'I don't know what [the MRC] thought of us looking back. Because you remember, it was not that the people – that the structure of the helix made a big difference. Everybody was quite pleased. But it was not regarded as sort of, you know . . . the way it is nowadays, as an overwhelming breakthrough.'[12] Several scientists working in Cambridge or even in the Cavendish at the time have later confirmed with regret that they were not aware of the 'ground-breaking' work happening 'next door'.[13] Michael Fuller, the young laboratory assistant, recalled:

It appeared to take a long time before the excitement sank in at the Cavendish. There was much more excitement at the Cavendish at that time over the

[11] A. Barrington Brown, pers. comm., 27 July 2000, and letters to author 12 December 1994 and 21 May 1995. Barrington Brown did receive some royalties through Camera Press, which served as his agency from early on. But according to the photographer, this was 'but a drop in the ocean'. Unfortunately, neither he nor Camera Press was able to produce precise figures about requests for the pictures, since the early records were either not available any more or not itemised.

[12] Interview with F. Crick, House to the Golden Helix, Cambridge, 25 May 1993. Surprisingly, also forty years later, when a plaque commemorating Watson and Crick's endeavour was unveiled at Cambridge, the LMB was only sparingly represented. The Duke of Edinburgh, in his capacity as Chancellor of the University, and Watson (but not Crick) attended the ceremony. Apparently this was the first plaque ever to commemorate a scientific achievement at Cambridge (Holmes 1993). Following this example, J. J. Thomson also received a plaque in 1997, hundred years after he produced the first demonstration of the electron.

[13] For instance interview with R. Diamond, Cambridge, 1 July 1992. Diamond, then a Ph.D. student at the Cavendish, met Watson and Crick regularly in the cafeteria and knew about the DNA model, but at the time did not appreciate its significance. From 1963 Diamond worked in the LMB, developing mathematical tools for the interpretation of crystallographic data. Note, however, that already in the autumn of 1953, students of Harold Whitehouse's course on genetics were introduced to the structure of DNA; interview with D. Hopwood, Norwich, 18 November 1995. Among those who came to see the DNA model soon after it was built were Todd with some younger colleagues from chemistry; Linus Pauling who was visiting Cambridge; Wilkins from King's College London; Brenner, Leslie Orgel and Jack Dintzis who came over from Oxford. Visitors at the Cavendish open day in mid-July could also see the model.

mathematics of the Slinky wire frame. You know the Slinky, the little wire frame that walks down the stairs? This came out at that time and one of the scientists brought one back from the States and I remember seeing all the scientific staff come from the tea room and watch this walk down the stairs and then intense discussions on the mathematics of how it was working. That seemed to excite them a lot more than the DNA model actually itself.[14]

And while the press eagerly seized on the origin of life experiments performed by Stanley Miller in Chicago in the same year, the double helix certainly did not create a media stir right away. The first article in the press to describe Watson and Crick's work appeared in the *News Chronicle*, a liberal daily, two weeks after the scientists' first publication in *Nature*. The brief report by the renowned science editor Richie Calder did not even mention Watson's and Crick's names (Calder 1953).[15]

As we have seen, for the MRC unit at Cambridge, where Watson and Crick had presented their model, the autumn following their celebrated collaboration was above all the beginning of a serious institutional crisis. Molecular biology did not yet exist. In the negotiations for a new Laboratory of Molecular Biology in 1957 and 1958, the DNA structure was mentioned as an important, but by no means the only or most outstanding achievement of the group. The work underlined the potential of X-ray diffraction, coupled to model building, as a method for structure determination (even if in this particular case the data were not collected at Cambridge).[16] The model of DNA also exemplified how structure

[14] Interview with M. Fuller by J. Finch and the author, Cambridge, 4 August 1998.

[15] Calder started his career as science writer in the interwar years. A dedicated socialist, he promoted the harnessing of science for social needs, on the national and, through his services for UNESCO, on the international level. He took a leading role in establishing the Association of British Science Writers in 1949. His article on the double helix followed a talk by Bragg at Guy's Hospital Medical School dealing with the DNA work. Calder was well acquainted with Bragg. Two weeks after Calder's article, on 30 May, the University of Cambridge undergraduate newspaper *Varsity* carried a somewhat more sensational report. Under the heading 'X-ray discovery' a brief front-page article presented Watson and Crick's work as 'the biological equivalent to crashing the sound barrier' (Anonymous 1953b). An early and informed report on the double helix appeared in the Penguin *Science News* (No. 29, August 1953, pp. 109–12). The article dealt with the newly proposed structure for DNA in the context of other coiled structures proposed for proteins and nucleic acids. On the presentation of the double helix in the media see Yoxen (1978, 231–75) and Turney (1998, 135–59). On the assimilation of Watson and Crick's work into the scientific literature see Winstanley (1976).

[16] This fact must have caused considerable embarrassment at the time. In the letter in which Perutz, according to his habit of reporting important progress in the unit, first described Watson and Crick's work to the Secretary of the MRC, he mentioned that the two researchers used 'published X-ray data of Astbury and of the MRC unit at King's College, London, plus a certain amount of unpublished X-ray data which they had seen or heard about at King's'. He commented that 'all these data were either poor, or referred to a different form of structure, and while they indicated certain general features

determination could lead to an understanding of the function of biological molecules (this was true at least regarding the replication of genetic material). Kendrew's first model of a globular protein, or for that matter any protein structure proposed later, did not suggest as promptly a molecular mechanism for its action. But the way Watson and Crick had achieved their model, even if not just as play-like as later suggested by Watson, was certainly not exemplary for the kind of industrious and meticulous research pursued by Perutz, Kendrew and Sanger. On the instrumental level, Crick and Brenner's work on the genetic control of protein biosynthesis represented a much bigger innovation.

A letter addressed to the Secretary of the MRC by Bragg at the end of the 1950s confirms the relative oblivion surrounding Watson and Crick's work on the double helix. Following a conversation with Himsworth, Bragg in this letter put on record his recollection of the first announcement by Crick and Watson of the DNA structure. Bragg summarised the basic features of the structure and the evidence on which it was based. He recalled 'the great occasion' when Todd and Pauling, then visiting Cambridge, were invited to see the model. 'Much beautiful experimental work' done by biochemists, Bragg noted, has confirmed the rightness of Watson and Crick's conclusions. 'But so that the contribution of these younger scientists should not be forgotten', he concluded, 'I feel we should acknowledge it and refer to it as the Crick–Watson model of D.N.A., because it was quite novel and unexpected when they first put it forward.'[17]

Himsworth was glad to have 'this record of an actual eye-witness of the situation, because, as time goes on, these things get forgotten and people may be deprived of the full justice of their claims'. Himsworth also

of the structure of DNA they did not give a guide to its detailed character'. He also mentioned that while Watson and Crick were building their structure, Franklin and Gosling at King's obtained a new and very detailed picture. When Watson and Crick sent the first draft of their paper to King's, Perutz continued, they had only heard of this photograph, which would probably have led Franklin and Gosling to build much the same model (M. Perutz to H. Himsworth, 6 April 1953, file FD1/426, MRC Archives). As we now know, Watson did have a glance at this picture before (Watson 1980, 98; Olby 1994, 396). Even if Perutz did not know this at the time, his mentioning these details suggests to me that there was the necessity to explain the fact that the decisive data came from *another* (friendly) laboratory. In Himsworth's copy of the letter, the paragraphs relating to this point were underlined. Regarding the importance of Watson and Crick's work, Perutz commented: 'I believe that this discovery will provide a great stimulus to the structural and chemical interpretation of genetics. I am also very pleased on personal grounds, because the discovery justifies the Council's continued support of Crick in spite of initial failures and upsets' (*ibid.*, this paragraph was marked as particularly important by Himsworth).

[17] W. L. Bragg to H. Himsworth, 10 November 1959, file S18/1, MRC Archives. Some years before, Bragg had written to Himsworth to excuse and correct his former harsh judgement on Crick; W. L. Bragg to H. Himsworth, 22 November 1956, P3/125, vol. 1, MRC Archives.

8.3 DNA model with (from left to right) Max Perutz, John Desmond Bernal and Francis Crick at the opening party of the MRC Laboratory of Molecular Biology in 1962. Courtesy of Medical Research Council Laboratory of Molecular Biology, Cambridge.

mentioned that 'it was likely that from time to time in the next few years we shall have need to make reference to this discovery and how it came about'.[18] This was the autumn that Severo Ochoa's and Arthur Kornberg's Nobel Prize for their biochemical work on the enzymatic synthesis of RNA and DNA was announced. By that time the plans for the new laboratory at Cambridge were also well advanced. Himsworth probably hinted at this with his remark. Sure enough, at the official opening of the laboratory by the Queen in May 1962, a model of the double helix was on display (Fig. 8.3).[19] In the autumn of that year, the Nobel Prize for Watson, Crick and Wilkins, as well as for Perutz and Kendrew, was announced.

The full background to this decision will not be disclosed before the year 2012. For sure Bragg, a veteran in Nobel politics with his Nobel

---

[18] H. Himsworth to W. L. Bragg, 16 November 1959, file S18/1, MRC Archives.

[19] The model displayed was one built by A. Barker, superintendent of the workshop of the Department of Engineering at Cambridge, for presentation at the Brussel's World Fair in 1958. This model incorporated the corrected coordinates following Wilkins' latest measurements. Built for display, it was more colourful, stable and easy to transport than the original model.

Prize from 1915, played a decisive role in the decision. In 1960, he nominated Perutz, Kendrew and Dorothy Hodgkin for the prize in physics and suggested that the chemistry prize should go to Watson, Crick and Wilkins for their work on the structure of DNA. Bragg also canvassed the support of other scientists for his nomination (Judson 1994, 566–7).[20] Two years later, Monod, invited by the Nobel Committee to make a nomination, again proposed Watson, Crick and Wilkins for the prize in chemistry, this time with success.[21]

As already mentioned, the combined prize and the public ratification of the eminence of the MRC Laboratory of Molecular Biology were perceived as an important step in the establishment of the new science. In the disputes on the origins of molecular biology fought out between participants in the late 1960s, the double helix acquired its central historical role. Watson's discovery account followed suit, heightening interest in the event. But in Cambridge, in 1962, the double helix still stood side-by-side with the models of haemoglobin and myoglobin as emblems of the new discipline and its successes (Fig. 8.4).

Local strategy or general trend?

So far I have stressed the local and contingent character of the construction of molecular biology at Cambridge. To what extent were developments at Cambridge really unique and how did they relate to developments elsewhere?

[20] On Bragg's role in convincing the Nobel committee to include Wilkins among the winners of the DNA prize see W. L. Bragg to A. Tiselius, 13 December 1962, RI MS WLB, item 32E/31. See also M. Wilkins to W. L. Bragg, 30 November 1962, RI MS WLB, item 32E/17, and Judson (1994, 195). Bragg's role in promoting the recognition of Watson and Crick's work on DNA and smoothing controversies surrounding it did not end here. In 1968 he wrote a generous foreword to Watson's controversial account of the events leading to the construction of the DNA model. Touching on the dilemma encountered by Watson and Crick of trespassing on another researcher's turf, he confirmed his 'deep satisfaction' that 'in the award of the Nobel Prize in 1962, due recognition was given to the long, patient investigation by Wilkins at King's College (London) as well as to the brilliant and rapid final solution by Crick and Watson at Cambridge' (Watson 1980, 1–2). Interestingly, Bragg's original proposal did not include the candidature of Wilkins and Hodgkin. Pauling, whose help Bragg had solicited, agreed to write in support of Watson and Crick, but held Perutz and Kendrew's candidature as 'premature', suggesting the nomination of D. Hodgkin and of the Dutch crystallographer J. M. Bijvoet instead; see W. L. Bragg to L. Pauling, 9 December 1959, with enclosed draft letter and forms of recommendation addressed to the Nobel Committee of Physics, and L. Pauling to W. L. Bragg, 15 December 1959, Pauling Papers, file 7.1, B corr, Oregon State University. On the 1962 Nobel Prizes see also below, p. 261, note 2.

[21] J. Monod to Swedish Royal Academy of Science, 29 January 1962, fonds Monod, Mon. Cor. 27, AIP; see also F. Crick to J. Monod, 31 December 1961, with annex 'The structure of DNA and the replication mechanism' which gives Crick's account of the DNA story, fonds Monod, Mon. Cor. 04, AIP.

8.4 John Kendrew (in academic robe) showing Queen Elizabeth II a small-scale wire model of myoglobin on the occasion of the official opening of the laboratory. On the table three-dimensional density map of myoglobin built from perspex sheets. Onlooking (from left to right): Mavis Blow, Gisela Perutz, Charlotte and Harold Himsworth and Lord Shawcross. Courtesy of Medical Research Council Laboratory of Molecular Biology, Cambridge.

In their early negotiations with the Medical Research Council, Perutz and his colleagues had not made explicit reference to other laboratories which could serve as a model to follow for the Cambridge one. However, soon after the plan for the Laboratory of Molecular Biology had received Council agreement, Perutz communicated to Himsworth that he was receiving 'many reports of biophysical laboratories being established at American Universities'. Perutz also mentioned to Himsworth a proposal, sponsored by Lwoff, Jacob and others at the Pasteur Institute, to establish a European Institute at Paris. With little modesty, perhaps justifiably so in view of the leading position the unit already occupied at least in the field of protein crystallography, he declared, 'I think our Laboratory should act as a focus and model for such laboratories'.[22] The fact that the creation of the new laboratory depended on local contingencies

[22] M. Perutz to H. Himsworth, 15 October 1958; see also M. Perutz to H. Himsworth, 24 October 1958, file Himsworth, LMB Archives.

certainly did not mean that its founders did not pursue more ambitious aims and that the Cambridge model could not be exported. The founding of new centres abroad, however, meant also that attractive offers reached Cambridge molecular biologists. To mention just some: before the new Cambridge laboratory was finally inaugurated, Brenner was offered the directorship of the Carnegie Institute of Genetics at Cold Spring Harbor, Kendrew the chair of Biophysics at Ann Arbor and at Los Angeles, and Crick the directorship of a new laboratory of molecular biology at the National Institutes of Health. Later, Monod tried to attract Brenner to Paris to head the new Institute of Molecular Biology to be built there. The threat of losing key scientists abroad was often and successfully played out by the Cambridge molecular biologists in applying for more support.[23]

This, however, was only one way in which developments elsewhere impinged on local developments. The newly defined Cambridge molecular biologists were in constant contact with researchers in other localities. People, technologies and results were moving in and out of the unit, linking investigations at Cambridge to other places, people and projects. These exchanges were an integral part of the construction of the local research cultures, as we saw for instance in the case of Brenner's import of phage genetics into the unit. The establishment of phages, mutants and codes as new research objects in Cambridge was connected to the establishment of new networks, which linked research at the unit to other places. These (old and new) relations were equally part of what made the Cambridge unit a distinctive place on the general map of research institutions. It has been argued that molecular biology – profiting from an increased mobility of people created especially by new science policies and funding schemes in the Cold War era – constituted itself in an international space (Abir-Am 1992b). My view is that the increase in international exchanges modified the relations between local settings, and thus the local settings themselves, but did not do away with them. It is only by looking locally that exchanges between different cultures and the construction of an international space can be studied. The establishment of the Laboratory of Molecular Biology, which soon became one of the centres of the new science and attracted a continuous stream of, in particular, American postdoctoral researchers, was part of the construction of this international space (see Part III).

---

[23] M. Perutz to H. Himsworth, 9 February 1961, file S18/1, and 'Possible accommodation developments at the Molecular Biology Laboratory', note for file by B. Lush (MRC), 10 June 1963, file P2/326, vol. 1, MRC Archives. In the 1960s, the 'brain-drain' of British scientists to the US became a hot political issue (see below, chapter 10). The expression was first used with respect to the emigration of scientists from Nazi Germany.

As indicated in Perutz's letter to Himsworth, in the late 1950s and then especially in the 1960s many new institutes and departments of biology, molecular biology or biophysics were created, in the United States as well as in Europe. A general account of these developments, which were part of a more general expansion of science teaching and research in these countries, goes beyond the aims of this study. But a brief look at parallel developments in Britain and abroad will none the less help to put the Cambridge story into context as well as to highlight its distinctive local features.

In Britain the Cambridge unit occupied a unique position in the development of the new science. For many years, molecular biology simply was the Laboratory of Molecular Biology. Other biophysics and protein crystallography groups which were set up on similar terms after the war, and which had personal, scientific and institutional links with the Cambridge unit, developed in different directions or were eventually integrated either into the LMB itself or into institutions modelled on the LMB. The example of the Unit for Biophysics at King's College London is especially instructive here. Despite its decisive role in the structure determination of DNA, it always remained 'biophysics'.

Comparison with developments in France where the Pasteur Institute acquired a similarly dominant position as the Cambridge laboratory in the establishment of molecular biology, nationally and internationally, provides further important insights into the early phase of institutionalisation of the new science and into the different local stories involved. In the late 1950s and 1960s American postdoctoral fellows who, unlike their European colleagues, had money to travel, would visit either the Laboratory of Molecular Biology in Cambridge or the Pasteur Institute in Paris, or both, for training in molecular biology. But despite the fact that the term 'molecular biology' appeared in Cambridge and Paris around the same time, namely in 1957/8, both institutions represented distinct research traditions. Interactions between the Cambridge and Paris researchers developed only in subsequent years. Similarly, the institutionalisation of the field in France followed a very different pattern and involved different political decisions than in Britain.[24]

Biophysics at King's College London

As discussed earlier, the Biophysics Research Unit at King's College London was set up by the MRC in the same year as the crystallography

---

[24] On the history of molecular biology in France see Gaudillière (1991; 1993; in press). On the move from biophysics to molecular biology in a number of American institutions starting in the late 1950s see Rasmussen (1997a). For developments at the Californian Institute of Technology and at Berkeley see Kay (1993a) and Creager (1996) respectively.

unit at Cambridge (see above, chapter 3). As with the Cambridge unit, the King's College unit was also part of the Physics Department. However, in London, Randall, the director of the unit, was also Professor of Physics and Head of the Department of Physics. Many physicists in the department also engaged on biophysical projects, receiving funds from a variety of sources including the University Grants Committee, the Agricultural Research Council, the Nuffield Foundation, the British Empire Cancer Campaign and the Rockefeller Foundation. Thus, under the direction of Randall, physicists at King's College more generally moved towards biophysics. As a consequence, the MRC unit was much more integrated into the Physics Department, both institutionally and with regard to shared research interests, technologies and skills. At King's, biophysics also formed an integral part of the physics curriculum, starting at undergraduate level.[25]

While the Cambridge unit started off as a pure protein crystallography unit, biophysicists at King's College London applied a wide range of physical technologies to the study of biological material. With Wilkins, Franklin and Raymond Gosling, the King's College group contributed decisively to the structure determination of DNA. Wilkins shared the Nobel Prize with Watson and Crick for this work. But despite this recognition and other distinctive contributions to molecular structure analysis, the unit was not generally viewed as a key centre of the new science of molecular biology[26] (which once more confirms that the double helix alone did not make molecular biology). In more recent discussions the re-evaluation of Franklin's contribution to the structure determination of DNA and a feminist reading of her difficult position at King's has, if anything, thrown a negative light on the role of this institution in the history of molecular biology.[27] In a recent publication, dedicated to the participation of

---

[25] This notwithstanding, Randall, like Bragg, had to defend his support of biophysics from the attacks of physicists who dismissed the unconventional activities flowering in the Physics Department at King's by referring to it as 'Randall's Circus' (Gosling 1995, 51; see also Wilkins 1987, 517–18). On the same lines, Bragg was seen as running down the prestigious Cambridge physics laboratory by supporting a 'bunch of clowns', chief among them Perutz, Crick and Martin Ryle, working on 'mysterious radio sources in the sky' (Dyson 1970 and above, p. 79).

[26] This was partly so because, at least in Britain, molecular biology was in many ways reserved to describe the work of the Cambridge laboratory. Note, however that in a 1968 report on molecular biology by the Council for Scientific Policy, the King's College unit figured as one of the very few centres where 'biology at the molecular level' was successfully pursued (see below, chapter 10, especially note 42).

[27] The general situation for women at Cambridge was hardly better than at King's. When, during the war, Perutz was able to hire an assistant, he lamented that all men were called to war service and he had to content himself with a woman assistant. Expressing a view which must have been common at the time, he explained to his correspondent at the Rockefeller Foundation: 'I should have preferred a graduate, and even better a man, because I find that the lack of initiative and originality of most woman research workers

the King's College group in the 'discovery' of DNA, the authors, who were all connected to Randall's enterprise at some stage, tried to correct this view (Chomet 1995).[28] However, merely stressing the contribution of the King's College researchers to what molecular biologists have celebrated as the key discovery of their field, this attempt was less successful than it could have been, had the authors taken a more critical stance towards the standard account of the history of molecular biology more generally. An equally important contribution of King's in building up the new field, for instance, consisted in biophysics teaching and in the preparation of a new generation of 'physically minded' molecular biologists (Wilkins 1987, 521).

Significantly, biophysics at King's developed in a very different way from biophysics at Cambridge. Despite the smooth integration of physics and biophysics at King's College London, in the early 1960s Randall took steps to transfer biophysics (including the MRC unit) from physics to biology. This he achieved by using his position as Physics Professor. First he established a Sub-Department of Biophysics within the Physics Department. Only when the sub-department became an independent department did Randall give up his chair in Physics and transform himself into the first Professor of Biophysics which he chose to be part of the Faculty of Biological Sciences. At that time the department was still housed within the space of Physics. It was not until 1964 that it moved to its own premises in Drury Lane. The MRC unit continued to exist as an integral part of the new department (Fig. 8.5 and 8.6).

Randall did not adopt the term molecular biology to name his new enterprise, even if this could have been a possible strategy to distinguish the more biological direction of his work from the biophysics further pursued in the Physics Department. That Randall did not follow this possibility

(there are some notable exceptions!) is rather a drawback'; M. Perutz to F. B. Hanson, 1 August 1944, folder 565, box 44, series 401, record group 1.1, RAC. The first woman scientist joined the unit in 1954, but until well into the 1970s there were very few women among the scientific staff as well as among the junior researchers. In contrast, from the beginning, there were several women scientists on the King's College unit staff list, among them Honor Fell from the Strangeways Laboratory at Cambridge who acted as external biological adviser and senior biologist Jean Hanson. According to Wilkins, the 'large proportion of women' working in the laboratory was probably due to the fact that women had less choice of jobs than men and therefore were more prone to join an enterprise which could well appear risky (Wilkins 1987, 517). Wilkins' implication that it was difficult to encourage good male scientists to join the laboratory, however, stands in contrast to the recollections of a onetime Ph.D. student in Randall's laboratory who recalled that 'people inside and outside the University of London were queuing up to get into his Department' (Gosling 1995, 51).

[28] In a parallel move, King's College has named one of its new establishments the Franklin-Wilkins Building.

8.5 John T. Randall, director of the MRC Unit of Biophysics at King's College London (*c.* 1964). Courtesy of King's College London.

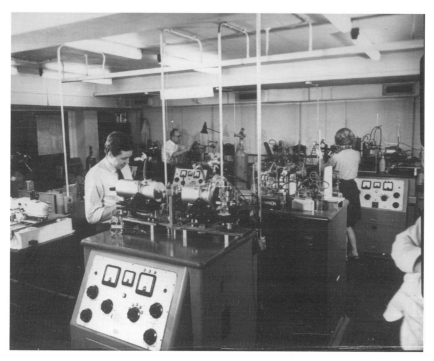

8.6 View of the new Biophysics Laboratory in Drury Lane at King's College London (*c.* 1964). Courtesy of King's College London.

indicates that molecular biology had not yet become the general umbrella term used to attract funding. In some ways it seemed reserved to designate the Cambridge laboratory. Randall had also more specific reasons to adhere to the old designation of his unit. He had assumed a leading role in the organisation of the newly founded British Biophysical Society which aimed at holding together the various factions of biophysics, comprising the 'muscle–nerve people', the 'molecular biophysicists' and the radiobiologists (see above, chapter 3). Changing the name of his own unit would have given the wrong signal.

There was, however, a more decisive difference between the Cavendish and King's College groups. Randall's biophysics enterprise was from the start better integrated into the university and always remained part of it, finally becoming established as an independent department. In contrast, the Laboratory of Molecular Biology at Cambridge was only loosely connected to the university. As I will argue below, this situation, perceived as a drawback by many, was turned into an asset and decisively contributed to the construction of the Cambridge laboratory as a centre for the new science.

Besides the King's College group there were two other biophysics groups in London with direct links to the Cambridge unit: Bernal's Biomolecular Research Laboratory at Birkbeck College and Bragg's crystallographic laboratory at the Royal Institution. Both these groups remained purely crystallographic groups, but were eventually integrated into larger institutions.

In 1938, after the failed attempt to create the Institute of Mathematico-Physico-Chemical Morphology which would have offered him an institutional home at Cambridge, Bernal became Professor of Physics at Birkbeck. After the war, Bernal, together with Charles H. Carlisle, was able to build up a group working on proteins, viruses and small biological molecules in his new Biomolecular Research Laboratory. Funds were provided by the Nuffield Foundation. When Franklin left Randall's laboratory and DNA to join Bernal's laboratory, she took over the virus group and together with Aaron Klug and their students produced important work on the structure of tobacco mosaic virus. With Bernal's age of retirement approaching, it was agreed early on that the Birkbeck crystallography group would join Perutz's group in the new laboratory of molecular biology. Franklin prematurely died while the negotiations for the new laboratory were still in course, but Klug, John Finch, Kenneth Holmes and Ruben Leberman moved to Cambridge in 1962. Anxious to secure the continuing existence of crystallography at Birkbeck itself, Bernal suggested that the Department of Physics be divided into two and that a new chair in Crystallography be created. Bernal became its first and indeed only occupant.

After his move from Cambridge to London in 1953, Bragg quickly built up a protein crystallography group at the Davy-Faraday Laboratory in the Royal Institution. In the first years, they closely collaborated with the Cambridge crystallographers. Indeed Kendrew spent several days a week at the Royal Institution, and much of the work for the structure determination of myoglobin was done in Bragg's laboratory (see chapter 4). But competition soon drove researchers in the laboratory to embark on independent projects, the central focus of the work becoming the structural determination of lysozyme. Before his retirement in 1966, Bragg orchestrated the move of most of his collaborators to Oxford where, with finances from the MRC and the Nuffield Foundation, they set up a new Laboratory of Molecular Biophysics in the Zoology Department.

Besides John William Pringle, called from Cambridge to Oxford as new Professor of Zoology to rescue a depressed department, the biochemist Hans Krebs, head of the MRC Unit for Research in Cell Metabolism, played a crucial role in the negotiations for establishing biophysics at Oxford. As early as 1956 Krebs, who was an active member of the Board of the Faculty of Biological Sciences Sub-Committee on the Development of Biology, tried to attract Randall's group to Oxford. Later he offered Perutz a floor in his newly built laboratory. He also initiated the contacts with Bragg regarding the move of Phillips and his group to Oxford.[29] In his inaugural lecture at Oxford, Pringle distinguished 'two biologies': an 'organismic' one, concerned with ecology, evolution, behaviour and genetics, and an 'atomistic' one which included physiology and biochemistry, but which had come of age with molecular biology and more precisely with Watson and Crick's publication of the structure of DNA (which he mistakenly dated as 1948). According to Pringle, the first brand of biology was already well represented at Oxford, while the second one needed to be reinforced (Pringle 1963).

This brief overview will suffice to indicate that, at least in Britain, the move from biophysics to molecular biology represented a local strategy.

---

[29] Royal Commission on Historical Manuscripts, *Report on the Correspondence and Papers of Hans Krebs (1900–1981)* (Bath, Contemporary Scientific Archives Centre, 1986) and letters M. Perutz to H. Himsworth, 8 November 1958, file E243/109, vol. 1, MRC Archives, and 29 November 1958, file Himsworth, LMB Archives. For Krebs' own account of these events see Krebs (1981, 207–13). Krebs' interest in building up biophysics at Oxford is particularly interesting in view of his later involvement in the controversy between biochemists and molecular biologists in Britain. See [Krebs Report] (1968) and below (chapter 10). On Hodgkin's role in bringing Phillips to Oxford and the relations of the two groups see Ferry (1998, 299–305).

The Cambridge molecular biologists represented a unique combination of skills including, above all, protein crystallography, protein sequencing and phage genetics. They pioneered institutional strategies followed only later in other centres. I will discuss the impact of the Cambridge laboratory on the establishment of molecular biology in Britain and internationally in the next part. However, it should be noted that, as the construction of molecular biology at Cambridge responded to local contingencies, so it did in other places. The first Department of Molecular Biology instituted in Edinburgh in 1965, for instance, while supplanting biophysics, built on a strong local tradition in genetics, turned molecular by the appointment of bacterial geneticists Martin Pollock from the National Institute of Medical Research in London and Bill Hayes, former Director of the MRC Microbial Genetics Research Unit at Hammersmith Hospital.[30] A wider overview of molecular biology in Britain would have to include these developments.

Action concertée and molecular biology in France

Negotiations for the Laboratory of Molecular Biology in Cambridge were still under way when the news spread of plans for the creation of a European Institute of Molecular Biology at Paris. What was this plan about? And were there any connections to the Cambridge initiative?

The frame of the negotiations in Paris was certainly different from the Cambridge case. On 19 November 1958, a subcommittee of the US Senate headed by Senator Hubert Humphrey visited Paris to discuss the creation of a European Institute of Molecular Biology with Lwoff, Jacob, François Gros and George Cohen from the Pasteur Institute (Monod could not attend). This plan was to be part of an ambitious 'Marshall Plan for Medicine' proposed by the American Senator to 'immunize the World from War'. According to Humphrey's vision, American funds, administered by the National Institutes of Health, would allow scientists from East and West to cooperate in the field of medical research and to work towards a common objective, the well-being of mankind. This would override Cold War divisions of distrust and aggression.[31] The plan came just a year after the Sputnik launch. Despite the internationalist rhetoric the choice of Paris for investment in this project was consonant with the more general

---

[30] On the institutional developments leading to the creation of the Department of Molecular Biology at the University of Edinburgh see Birse (1994, chapters 7 and 8).

[31] Letter J. A. Shannon (Department of Health, Education and Welfare) to J. Monod, 14 October 1958, fonds Monod, file EIMB, AIP, and H. Humphrey, 'Bold "Marshall Plan for medicine" proposed by Senator Humphrey to "immunize world from war"', attached to letter J. Cahn to J. Monod, 1 January 1959, fonds Monod, file EIMB, AIP.

political strategy of the United States to strengthen the competitiveness of the Western European countries (Krige 1999). For the proposed institute it meant that instead of being a national prestige object like the laboratory project in Britain, it was to be a 'European' project funded by American tax money.

The Senator's visit had been prepared by Julius Cahn, staff member of the same senatorial subcommittee. On this occasion, after general discussion of the organisation of French biomedical research and possibilities for promoting it, Lwoff had handed over to Cahn the plan for the proposed Institute. The memorandum was drafted overnight and was tailored to fit the American programme.[32] The proposed scheme, however, rested on longstanding discussions between Lwoff and his colleagues at the Pasteur Institute aimed at improving the status and funding of their research in bacterial genetics and biochemistry. They struggled to achieve better research conditions at the Institute, but also – and this marked a sharp difference from the position of the Cambridge molecular biologists – to achieve changes in the biological and medical curriculum at French universities, where genetics and biochemistry played little more than a marginal role. At the Colloquium organised in Caen in 1956, left-wing scientists, industrialists and politicians had met to discuss measures to be taken for the promotion of French research. Monod, as chair of the committee on fundamental research and university teaching and co-author of the final report, had played an outstanding role in this meeting which was crucial for the formation of new links between the participants. Thus, from the beginning, the Pasteurian researchers acted on a higher political level than their Cambridge counterparts.[33]

In the proposal presented to the Americans the definition of the new field of molecular biology was given in as general and fundamental terms as in the Cambridge case. On a first reading it could appear strikingly similar: the central aim of the new biology was to interpret the cell, the essential living unit common to all organisms, in terms of its macromolecular constituents. It comprised studies of structure, function and development.[34]

[32] A. Lwoff to J. Monod, 20 October 1958, and 'A European Institute for Molecular Biology' [undated, unsigned], fonds Monod, file EIMB, AIP.

[33] The 1956 colloquium is generally seen as an important milestone in the postwar development of a science policy in France (Jacq 1995). On the role of the scientists of the Pasteur Institute in these discussions see Gaudillière (1991, 71–6; 1996, 421). For an illuminating discussion on science policy not as an *explanans* but as an *explanandum*, that is on the formation of science policy as itself an object of a contested history in postwar France, see Jacq (1995).

[34] Here and in the following see 'A European Institute for Molecular Biology' [undated, unsigned], fonds Monod, file EIMB, AIP.

Both proposals stressed the important medical implications of the study of fundamental life processes. A more attentive reading, however, reveals that the definition of the new science as well as the actual proposal for the new institute were closely tailored to the experiences, projects and needs of the Pasteurian researchers.

According to Lwoff's memorandum prepared for the American subcommission, the new science of molecular biology was based on recent advances in genetics, cell physiology and biochemistry which brought these formerly separated approaches to merge. This view of a fusion of disciplines reflected the actual dynamics at the Pasteur Institute, where Jacob from Lwoff's laboratory and Monod had recently initiated a collaboration deriving from their earlier and quite separate work on bacterial genetics and on the regulation of enzyme activity to study the mechanism and regulation of gene activity.[35] Technologies and skills like X-ray crystallography or protein sequencing which the Cambridge molecular biologists regarded as central to their project were hardly mentioned in the French proposal. The terms of structure and function used by both teams thus acquired quite distinct meanings.

The institute to be funded by the Americans was to be a European institution. The rationale for such an institute offered by Lwoff is reminiscent of the 'brain drain' argument used by the Cambridge researchers. Lwoff affirmed that the new biology was most vigorously expanding in the United States, where it was already an accepted discipline. In Europe traditional administrative and educational structures did not favour the formation of interdisciplinary groups and posed serious problems to the development of molecular biology. Lack of space and of funds were other negative factors. European molecular biologists were therefore strongly attracted to the United States. But a 'sterilisation' of European biology, Lwoff reckoned, was not in the interest of America. The creation of a European Institute of Molecular Biology would be an effective measure to prevent such a situation. The success of CERN exemplified what European collaboration could achieve in the scientific field.

Despite the European rhetoric, local biases were strong. 'At the present time', Lwoff surmised, 'the largest European nucleus of molecular biologists appears to be in Paris, at the Pasteur Institute'. The best place to build such an institute, therefore, seemed to be Paris, possibly on an available site near the Pasteur Institute.[36] Lwoff estimated the costs for the

[35] For a description of this work see Jacob (1972).
[36] Indeed, one is left to puzzle why the Institute had to be European rather then French. The idea of a European laboratory, however, would become an important issue a few years later (see below, chapter 10).

construction of the institute to be $3,000,000; the total annual running cost $800,000.

The question remains why the French chose the term molecular biology to name their proposed institute. The Cambridge project had already been discussed and in principle accepted by the MRC. But this news was hardly public and there was little interaction between the Cambridge and Paris groups until some time later.[37] The investigators of the Pasteur Institute interacted much more closely with American colleagues, and especially with some members of the phage group, from where they had probably imported the term (Gaudillière 1996, 425). As one participant commented, it was no chance that the term was adopted by people in Cambridge and Paris about the same time. Even if not formally agreed, it was commonly assumed that molecular biology stood for the 'new biology' against the 'classical biology of universities'.[38] More than delineating a specific research programme, the term thus marked off a territory that could be occupied by different groups in different ways.

Lwoff's plan was well received by the American visitors. The scientists at the Pasteur Institute played out their connections in the United States, seeking comments and active support for the proposed institute. But the bill died before reaching Congress, and the correspondence with the responsible subcommittee ended abruptly.[39] The plan for an Institute of Molecular Biology, however, was soon resumed on a national level.

---

[37] In a revised plan for a laboratory of molecular biology to be built in Paris under the auspices of a new state institution for the promotion of science in France (see below), the Cambridge laboratory was listed as one of the new laboratories in the field. But this document also reveals how poorly informed Lwoff, Monod and the other Pasteurians were on the situation in Cambridge. Crick was described as a crystallographer who now also pursued investigations in enzymology; close connections to Hayes' group for bacterial genetics in London were mentioned, while the name of Sanger was only added in a corrected version; see 'Annex II. Récents développements de la biologie moléculaire à l'étranger', fonds Monod, file IBM/IP [1961], AIP. Key examples of the later interactions between the Paris and Cambridge groups for molecular biology were Brenner and Jacob's common experiments with Matthew Meselson in Caltech in the summer of 1960 proving the role of an instable RNA (later called messenger RNA) in protein synthesis and Monod's use of Perutz's observation of a conformation shift in haemoglobin to elaborate the concept of allosteric changes in proteins (Creager and Gaudillière 1996).

[38] Interview with E. Wollmann, Pasteur Institute, Paris, 26 June 1996. Significantly, where molecular biology was introduced as the 'new biology' into mainstream biological departments, as for instance in the newly founded Department of Biology at the Massachusetts Institute of Technology to which Ingram migrated from Cambridge, the term 'molecular' did not necessarily appear in the title.

[39] The last letter by the project director to Monod is dated 26 February 1959, fonds Monod, file EIMB, AIP. It announced that hearings on a related bill had started but that it would be a long time before money would really become available. No further communication followed.

The occasion came in 1960, when De Gaulle created the Délégation Générale à la Recherche Scientifique et Technique (DGRST), a state instrument to promote research in France on an unprecedented scale. Molecular biology was chosen as one of the areas of special intervention (*action concertée*). This choice was not obvious, but was made possible by the leading role of the biologists of the Pasteur Institute in the decision making process, and by the convergence of views between them and their political interlocutors on the matter. This convergence, as well as the very presence of the scientists in decisive positions, went back to links created since the mid-1950s, and especially since the Colloquium at Caen. Thus, the designation of molecular biology (together with, among other things, oceanography, electronics, cancer research and nutrition) as area of special intervention was not so much the choice of a discipline as the choice of a group of people who had acquired trust and authority in political as well as scientific circles (Gaudillière 1991, 76–81). The plan for a Centre of Molecular Biology in Paris was only one of a series of initiatives in a number of places agreed in a first five-year plan. Molecular biology was now much more succinctly described as the study of the structure, function and biosynthesis of nucleic acids and proteins. The project proposed by the Pasteurians for their new centre focused on the question of biosynthesis, using microorganisms as research material.[40]

The realisation of the *action concertée* project at the Pasteur Institute took a tumultuous course (Gaudillière 1991, 87–92). The first plan to build a centre with money from the DGRST, but on the land and with the support of the Pasteur Institute, was blocked by the Director of the Institute on the grounds that molecular biology did not contribute to medical innovation and that the Institute was not there to step in for the inadequacy of the university in funding fundamental research. Accepting the plan also meant giving up part of the autonomy which the Pasteur Institute enjoyed as a private institution. This situation led to a split among the prospective molecular biologists at the Institute. Monod, not wanting to put at risk the important state support, now suggested the creation of an Institute of Molecular Biology in the Science Faculty, turning it into a truly academic institution, even if preserving its interdisciplinary and flexible structure. Lwoff and Wollmann and their groups refused to leave the Pasteur Institute; the others joined the new enterprise. While the negotiations between the university, the CNRS and the DGRST were still under way, a deep financial crisis and internal quarrels in the Pasteur Institute resulted in

---

[40] 'Rapport préliminaire sur la situation présente et les actions concertées à envisager dans le domaine de la biologie moléculaire' [draft, 1961] and 'Project de centre de biologie moléculaire à l'Institut Pasteur', fonds Monod, file IBM/IP, AIP.

structural changes which were more favourable to the establishment of an institute for fundamental research on its premises. Since an independent institute was still the favoured solution, Monod and Jacob quitted the negotiations at the Science Faculty, where the plan continued to be pursued through the recruitment of other scientists and under slightly changed priorities. The Institute for Molecular Biology at the Pasteur Institute got final approval in 1965, just before the Nobel celebration for Lwoff, Jacob and Monod.

The institutional position of the molecular biologists at the Pasteur Institute thus ended up being quite similar to the Cambridge one: they were state-funded, full-time researchers outside the university. But the Institute of Molecular Biology in Paris was part of a national project for the reorganisation of biological research. The Pasteurians dominated the state-directed plan of intervention politically and conceptually (and in this sense defined a 'French tradition'), but on the national level biochemists were the ones who benefited most from the new state funds under the *action concertée* for molecular biology (Gaudillière 1993). This is markedly different from the British situation where, throughout the 1960s, biochemists felt that the funding of molecular biology happened to their detriment (Abir-Am 1992a and below).

What molecular biology was thus always depended on the people involved, their respective research programmes and the institutional strategies pursued. The history of molecular biology is the collection of all these local histories, including the different textbook traditions and the standard history with its heroes and intellectual breakthroughs which has been constructed out of and concurs with the many other stories. Yet events either in Paris or in Cambridge do not support the central role of Watson and Crick's model of DNA (or, for that matter, of any other single intellectual 'breakthrough') for the formation of the new field. The establishment of molecular biology rather preceded and indeed precipitated the authority disputes among self-elected molecular biologists which, in turn, led to the construction of the double helix as icon of the new science.

The disputes among molecular biologists hinged to an important extent on the growing prestige and the rising stakes of the science they were involved in. The newly founded Laboratory of Molecular Biology at Cambridge became a key site for the practices and the strategies pursued to further the new science. By the late 1970s, however, the same strategies led to a diversification and expansion of the field which challenged its central position. The following final part charts these changes.

PART III

· · ·

Benchwork and politics

The Queen had only just inaugurated the new MRC Laboratory
of Molecular Biology on the outskirts of Cambridge when the announcement of the Nobel award to three members and one ex-
member of the laboratory propelled it again into the limelight.
The laureates were Max Perutz, chairman of the board which ran
the new research institution; John Kendrew, deputy chairman and
head of the division Structural Studies; Francis Crick, head of the
division of Molecular Genetics; and James Watson, an ex-member
and regular visitor to the MRC unit, honoured for his work with
Crick on the structure of DNA. Fred Sanger, head of the third and
last division of the laboratory, had won his (first) Nobel Prize in
1958. 'The laboratory', Crick remembered chuckling, 'was run by
a committee formed exclusively of Nobel laureates.'[1] This came
close to Bragg's altogether unusual proposal some years earlier
that the then MRC Unit for Molecular Biology 'as a whole' should
be awarded the Nobel Prize.[2]

Other Nobel Prizes were to follow, making the laboratory the
recipient of more awards than most other institutions on the
globe.[3] The series of accolades, regarded as the most prestigious
in science, is deeply imprinted into the history and public image of

[1] Interview with F. Crick, Cambridge, 25 May 1993.
[2] L. Bragg to L. Pauling, 9 December 1959, with enclosed draft letter and forms of
recommendation addressed to the Nobel Committee for Physics, Pauling Papers,
file 7.1, B corr., Oregon State University Library. Bragg's actual nomination,
which, following suggestions he had received, included the names of Hodgkin
and Wilkins, was not successful, but the suggestion to recognise jointly the
nucleic acid and protein structure work which had come from the unit survived.
From the available documents it transpires that the Swedish biochemist Arne
Tiselius might have had a decisive role in synchronising the chemistry and
medicine awards; see A. Tiselius to W. L. Bragg, 18 December 1962, RI MS
WLB 32E/35; also interview with F. Crick, La Jolla, CA, 14 October 1998. On
Bragg's role in securing a prize for Wilkins see above, p. 245, note 20.
[3] Prizes were awarded to Fred Sanger in 1980, to Aaron Klug in 1982, to César
Milstein and Georges Köhler in 1984 and, in 1997, to John Walker.

the laboratory, leaving sociologists, science policy makers and the media to wonder what exactly it was that produced such a creative environment and leading them to ask how these conditions could be reproduced.[4] But this line of inquiry forgets that Nobel Prize winning achievements are selected retrospectively, not on the basis of some intrinsic standards of distinction, but in the process of complex negotiations involving scientific peers and the Nobel committee. The prizes did much more than sanction scientific excellence; they created it.[5]

For similar reasons, the series of Nobel Prizes, especially when taken at 'face value', is not the best guide to the history of the laboratory. Such a history tends to celebrate individual scientific achievements rather than to present a contextualised account of scientific practices, and presupposes rather than explains the prestige of the field. This is not negating that the Nobel Prizes had an important impact on the history of the institution and the careers of the recipients. But these effects, too, call for historical inquiry.

In the following I will investigate the place of the new laboratory in the establishment of molecular biology in the 1960s and its role in the subsequent transformations of the field. To this end I will focus on the institution of research practices, on the place of the laboratory in the international traffic of research students and the training of a new generation of scientists, on political discussions in which the laboratory was presented as the model according to which teaching and research in the biological sciences in Britain were to be reformed, and on resistances to these developments. In the 1970s, the expansion of the field propelled by the same laboratory, together with changes in funding policies and commercial interests, marked a new era for molecular biology which also affected the place of the Cambridge laboratory. As will become clear, this periodisation is rather different from the Nobel saga, which suggests an unbroken success story.

Chapter 9 starts with a brief discussion of the organisation and 'culture' of the laboratory. It moves on to analyse the role of the laboratory in fostering interdisciplinary collaborations, often seen as the hallmark of molecular biology, and the part played by the young postdoctoral researchers visiting the laboratory in

---

[4] See for instance Silcock (1982) and the celebration of the laboratory as a 'nobel tradition' in the issue of the *MRC News* (No. 76, Winter 1997) dedicated to its fiftieth anniversary.

[5] On the history of the Nobel Prizes see Crawford (1984; 1992).

the establishment and export of new research tools and practices. Brenner's project on the nematode worm or his attempt to fashion a new model organism for the study of development, extending from the 1960s to the recent genome sequencing project, further illustrates the local creation and the export of a research tool as well as the participation of the laboratory in some of the transformations the field of molecular biology underwent. Chapter 10 investigates the ways in which the promotion of molecular biology entered the governmental agenda in Britain, the place of the Cambridge laboratory in these negotiations and the debates which ensued. The chapter also includes a discussion of the role of members of the LMB in the creation of the European Molecular Biology Laboratory (EMBL), an important step in the establishment of molecular biology in Europe. The final chapter starts by looking at a major review of the laboratory in the mid-1970s which coincided with changes in governmental policies requesting higher accountability for government research grants. These changes concurred with a growing competition in the field, new expectations of medical pay-offs and a push towards commercialisation. The scandal which developed around the apparent failure to patent the monoclonal antibody technology first developed in the laboratory serves to highlight the changes in research culture. The chapter analyses the responses of members of the laboratory as well as of the laboratory as a whole to these changes which, it was said, marked the end of a postwar bonanza for the sciences.

9

...

Laboratory cultures

Of the new laboratory a *Sunday Times* journalist wrote: 'It would be difficult to find a less distinguished looking building. It lies five storeys high in anonymous Sixties brick on the outskirts of Cambridge, looking rather like the branch office of an insurance company' (Silcock 1982). If this was the impression from outside – at least at some years' distance, inside the modular structure offered rational and flexible accommodation for the various needs of the different research groups moving into the new building. The internal design, devised by London-based architects in collaboration with the scientists, has remained the same in the various extensions the laboratory has undergone in the following decades.

This chapter will analyse 'life' in the institution, the role of the laboratory in fostering new research practices and its place in the import and export of researchers, tools and materials. Work on the structure determination of proteins and Brenner's worm project, including the later genome sequencing project of the nematode *C. elegans*, will provide the main examples. Again, institutional and experimental strategies will prove to be closely linked.

Federative constitution

The first divisions which became operative in the new laboratory were the workshops and the stores. Well before scientists started occupying offices and laboratory benches, engineers, technicians and laboratory assistants were busy building and fitting equipment, ordering apparatus, and stocking chemicals and other consumables. Among the acquisitions in the new building were cold rooms, still unusual at the time, and big fermenters for growing microorganisms in large batches. The laboratory also received a host of new state of the art equipment, purchased on a special Wellcome Trust grant. The list was headed by a Siemens electron microscope, a Spinco Analytical Ultracentrifuge, an amino acid analyser and a Packard

Scintillation Counter.[1] The X-ray equipment, much of it purpose built in the in-house workshop, was moved from the old laboratory. It took months of tinkering in particular to get the X-ray tubes working again.

The organisation of all this was in the hands of Michael Fuller, the resourceful technician in the laboratory, who had surprised the leading scientists with a minute plan of how to organise the 'infrastucture' of the new laboratory. In the event he was promoted to laboratory steward, a position he was to hold until his retirement in 1997. Fuller's main responsibility for many years remained the stores, which he saw as the 'central clearing house' for all sorts of demands relating to the purchase and sharing of research materials and equipment. Among the jobs he united in his hands was the liaison with local engineers and manufacturers who took over the serial production of equipment developed at the LMB for which demand existed outside the laboratory (Fig. 9.1).[2]

The kitchens, supplying glassware, media and sterile equipment to all researchers in the laboratory, were designed by Brenner on the 'supermarket model', with quick and easy access for everybody. This did away with a lot of form filling and accelerated work. According to Brenner, it was one of the innovations which turned the laboratory into a recognisably 'modern' institution, distinguishing it from most other laboratories in Britain at the time. The design was widely copied, and kitchens at the LMB and elsewhere are still laid out according to the same basic plan (Fig. 9.2).[3]

The other central institution of the new laboratory, though coming into operation somewhat later, was the canteen, placed on the roof of the building. Unlike the three tearooms in the Cavendish which catered separately

[1] 'Equipment required for the Molecular Biology Research Unit 1962/63 (omitting items under £1,500)', appended to letter of H. Himsworth to F. Green (MRC), 3 January 1962, Wellcome Trust Grant File 2458. The enthusiastic response of the Wellcome Trust to Himsworth's call for help in financing the costly equipment followed after just two weeks. '[H]appy to be thus even indirectly associated with the activities of such an important Research Unit', the Wellcome Trust agreed to a grant of £70,000 in addition to the £30,000 paid by the MRC towards new equipment (the building costs amounted to £320,000). As the Secretary of the MRC noted in his letter, it was 'particularly in regard to big apparatus and the multiplicity of moderate-priced apparatus' that the new laboratory was going to be expensive; F. Green to H. Himsworth, 22 January 1962, and H. Himsworth to F. Green, 3 January 1962, Wellcome Trust Grant File 2458. I thank Isobel Hunter and the Grants Department of the Wellcome Trust for making this file available to me.
[2] Interview with M. Fuller, conducted by author together with J. Finch, Cambridge, 4 August 1998.
[3] Interview with S. Brenner, Cambridge, 17 July 1992, and with M. Fuller, Cambridge, 4 August 1998. The hierarchical and formal manners of other scientific establishments in Britain, for instance the National Institute of Medical Research, were a strong reason for Brenner to turn down otherwise attractive job offers at these institutions; interview with S. Brenner, Cambridge, 30 June 1993.

9.1 Laboratory steward Michael Fuller running the stores.
Courtesy of Michael Fuller and Medical Research Council
Laboratory of Molecular Biology, Cambridge.

for scientists and administrative and technical staff, the new cafeteria was
to be open to all. With the exception of the mechanical engineers, who
kept to their habit of having their cup of tea 'by machine' and never made
use of the central facility, all staff did meet for coffee or lunch, although
secretaries and technical staff tended to sit at their own table, as did other
groups in the laboratory. In Fuller's and many other people's eyes the common canteen, for many years run by the chairman's wife, Gisela Perutz,
was very important for the general 'morale' of the laboratory. Acoustic tiles
were fitted on the ceiling, after Crick complained that it was too noisy to
follow a conversation and that it was the wrong place to economise, given
the importance of scientific interaction over lunch. When the laboratory
grew too big for people to know everyone, Crick suggested that mug shots
were put up in the canteen, so that one knew whom one was talking to.
For Crick lunch in the canteen replaced the tradition of the Eagle lunches,
now phased out because of the distance of the new laboratory from the
city centre (Figs. 9.3 and 9.4).

9.2 Kitchen at the LMB built on the supermarket model. Photograph by Lesley McKane (2000). Courtesy of Medical Research Council Laboratory of Molecular Biology, Cambridge.

The canteen was cultivated and propagated as a distinctive part of life in the laboratory. For younger members and visitors to the laboratory, participation in the open discussions with the more senior scientists over lunch or tea fulfilled something of an 'initiatory' function, including its intimidatory aspects. An earlier research student in the laboratory, later himself a Nobel Prize winner, remembered it thus:

Everybody went to tea, according to the English custom, mid-morning and mid-afternoon. These 'gods' of molecular biology were there, sitting with everyone at tea. They encouraged everybody to participate freely in discussion and they treated everybody equally. You could be the youngest graduate student or a technician, it didn't matter. They took your ideas seriously...You had to develop yourself in two ways. On the one hand, you had to train yourself not to say anything that was superfluous or stupid. The 'gods' made it very clear if they felt that you were frivolous or you were thinking in a less than coherent way. One could become very discouraged very quickly if that happened. Some people succumbed. On the other hand, if you were able to engage in the discussions, take advice, and approach the gods only when you had really something to say, it was fantastic. They made you feel part of the enterprise. (Altman in Hargittai 1999, 15)

9.3 Francis Crick (centre) with James Watson (left) on one of his return visits to Cambridge in the LMB canteen. Courtesy of Medical Research Council Laboratory of Molecular Biology, Cambridge.

The same scientist later tried to run his own laboratory 'exactly the way the lab in Cambridge was run' (Altman in Hargittai 1999, 15). Another American fellow, remembering the 'unbelievable...number of trips' made to the canteen each day, reckoned that afternoon tea proved difficult to introduce in American laboratories but that the 'spirit and style of free

9.4 View of the new canteen soon built to accommodate the growing number of staff and visitors in the laboratory (1970). Courtesy of Medical Research Council Laboratory of Molecular Biology, Cambridge.

interactions' was 'frequently observable around the labs of former MRC fellows'.[4]

The common facilities and the 'myths' or identificatory tales surrounding them fulfilled an important role in bringing together a laboratory consisting of separate groups and divisions moved under one roof.[5] Since the very first discussions concerning the organisation of the laboratory, it was envisaged that it should comprise a series of more or less independent sections and be run 'by a kind of Soviet of directors'. This was to prevent the possible clash of interests and personalities some feared might occur.[6] Later proposals saw the 'body of workers' as 'more of a confederation than

---

[4] T. A. Steitz to S. Brenner, 29 June 1983. On the role of American fellows in exporting the LMB 'culture' see below.

[5] On the identificatory role of 'mythologies' in scientific communities see Latour and Woolgar (1986, 54ff). Informality and a distinctive discussion culture were also part of the self-portrayal of other twentieth-century scientific groups and institutions, for instance Bohr's circle in Copenhagen, the Rockefeller Institute in New York, Lwoff's laboratory in the 'attic' at the Pasteur Institute and the Phage School (see respectively Aaserud 1990, 6–15; Dubos 1976; Jacob 1971; Cairns, Stent and Watson 1966 and Kay 1993a).

[6] H. Himsworth, 'Proposed Institute of Molecular Biology, Cambridge', note of 26 August 1957, file E243/109, vol. 1, MRC Archives.

a single entity'.[7] Details of the 'constitution' of the future laboratory were discussed among the senior researchers at a private meeting in Kendrew's house in Tennis Court Road well before the move to the new laboratory. The document drafted for approval by the MRC proposed the creation of three divisions, each of which remained independent with respect to its scientific programme and its internal policy. The divisions would also prepare separate budgets, although a common budget would be submitted to the MRC. This and other matters concerning the laboratory as a whole would be discussed by a board consisting of the heads of divisions and a (permanent) chairman. With Sanger as most senior researcher not showing interest in the chairmanship, the role fell to a grateful Perutz, while the other three senior researchers, Sanger, Kendrew and Crick, were designated as heads of the divisions of protein chemistry, protein crystallography (soon changed to structurel studies to include virus research and electron microscopy) and molecular genetics respectively (Fig. 9.5).[8] This organisation, and especially the lack of a 'director', while often justified as 'democratic', was in fact designed to strengthen the role of the senior researchers, guaranteeing each one of them as much independence as possible. The distribution of the three divisions, including their special equipment, on three separate floors in the new building only confirmed that these were separate territories.[9] Indeed, each division head followed his own style of leadership and each floor developed its own 'research culture', a term employed by the scientists themselves. It comprised the way research was conducted as much as what was researched.[10] Collaborations across divisions were nevertheless much encouraged. How were these achieved? And what did they look like?

---

[7] H. Himsworth to M. Perutz, 16 June 1959, file Himsworth, LMB Archives.

[8] 'Laboratory of Molecular Biology, Cambridge. Proposed constitution', Council Paper MRC.61/464, MRC Archives.

[9] That this was not a trivial distribution is confirmed by the fact that for instance in the Institute of Molecular Biology at the Pasteur Institute the divisions were vertical rather than horizontal. A banal incident indicates how jealously divisions at the LMB were guarded. When moving from the hut to the new laboratory, Allen Edmundson, an American postdoc working on the amino acid sequence of myoglobin, while hired by the structural studies division, was expected to work alongside the other protein chemists. Yet that division was not prepared to give up any of its space. Edmundson, together with his bulky amino acid analyser, finally joined the molecular geneticists on the middle floor, but the solution caused some resentment.

[10] Crick and Brenner (who soon joined Crick as division head), for instance, described their leading style as not *impeding* younger people in doing their work. This meant actively removing obstacles to make it possible for things to happen. The design of the kitchen (see above) was carried out with the same intention. In their eyes, the work organisation in other divisions was more traditionally structured; interviews with S. Brenner, Cambridge, 17 July 1992, and with F. Crick, Cambridge, 25 May 1993.

9.5 Governing board of the LMB in 1967, including the chairman, heads of divisions and, by that time, staff members being elected Fellows of the Royal Society. From left: John Kendrew, Francis Crick (standing); Hugh Huxley, Max Perutz, Fred Sanger and Sydney Brenner (sitting). Photograph by John Hedgecoe. Courtesy of Medical Research Council Laboratory of Molecular Biology, Cambridge.

The first step, everyone agreed, was to encourage people to talk to each other. 'In cramped laboratories', Perutz reflected, 'people speak much more to each other. The problem [in the new laboratory] was to get people to talk to each other. That is why I insisted from the beginning on the common cafeteria, workshops, instruments. To save money, but also to get people to speak to each other.'[11] Another attempt to get people from different divisions to engage with each other was the institution of a week-long seminar at the beginning of the academic year when the laboratory would 'teach itself'. All researchers were expected to attend and seminar

[11] Interview with M. Perutz, Cambridge, 25 June 1992.

papers were to be pitched so that everyone (represented in the person of Max Perutz) could understand them. Crick, who intitiated this tradition, explained: 'The whole point of that was to get communication across the divisions. That additional mechanism was invented to assist that. So the fact we had to invent it shows that the spontaneous mixing was not quite enough.'[12] The 'Crick weeks' carry on to the present day, though increased competition and commercial interests have affected the openness with which information is exchanged.

Did talking lead to collaboration at the bench level? The elucidation of the structure and reaction mechanism of chymotrypsin, an enzyme of the pancreas, by David Blow from the structural studies division and Brian Hartley from the protein chemistry division counts as a primary example of the interdivisional and interdisciplinary collaborations that, with time, developed in the new laboratory. This cross-feeding was seen to lead to spectacular results and to serve as powerful attractor of young researchers to the laboratory. An analysis of this case, then, will allow us to study the kind of collaborations that developed and to determine the role of the laboratory in fostering interactions between researchers of different research cultures. Interdisciplinarity, which is often seen as a defining feature of research in molecular biology and as the key to rapid advances, is here analysed on the level of research practices and the material cultures that go with them.[13]

Molecular mechanism

Hartley and Blow started working on the sequence and X-ray analysis of chymotrypsin quite independently and while still working in different institutions.[14] Hartley, a Cambridge graduate with a Ph.D. from the University of Leeds, began work on the sequence analysis of chymotrypsin as a postdoctoral fellow in the Biochemistry Department in Cambridge in 1953. Chymotrypsin had long been commercially available in pure form. That made it attractive to protein chemists who had used it as model to

---

[12] Interview with F. Crick, Cambridge, 25 May 1993.

[13] Following Rheinberger's introduction of the term, much recent work has focused on 'experimental systems' in the life sciences (Rheinberger 1997). While very useful especially to address epistemological questions regarding experimental practice and the status of objects and representations, experimental systems as discussed by Rheinberger function somewhat in isolation from social and institutional factors (de Chadarevian 1994). I have therefore preferred to use the looser but more embracing concept of research culture. Experimental systems can be seen as integral parts of research cultures.

[14] The following reconstruction profited greatly from interviews with Brian Hartley and David Blow in September 1992 as well as from discussions with Richard Henderson, who joined Blow's group as a research student in 1966.

study the chemical properties and the reaction mechanism of enzymes. Hartley himself, before embarking on the sequencing work, had performed extensive kinetical studies of the enzyme with the aim of elucidating the catalytic mechanism.[15] But it was the time when Sanger, working on the same corridor, had just published the complete sequence of one of the two insulin chains, a pioneering achievement which opened a completely new area of inquiry. Knowledge of the sequence promised new insights into the reaction mechanism of hormones and enzymes and Sanger's work had provided the tools to pursue this aim.[16] Hartley became convinced that in order to 'really know' how his enzyme worked he had to embark on a sequence analysis (Hartley 1962, 85). A Czech group was already working on the same project, but in face of the enormity of the task – chymotrypsin was three times the size of insulin, the only protein sequenced so far in a ten-year effort – a parallel attempt seemed warranted. The local expertise counted to Hartley's advantage. To make full use of it, Hartley associated himself with Sanger, while keeping his bench in the enzyme laboratory. After a two-year stint in Hans Neurath's protein chemistry laboratory in Seattle where he refined his techniques – Neurath himself was working on the sequence analysis of trypsin, another pancreatic enzyme – he was offered an MRC appointment in Sanger's group and moved with him to the LMB.

Hartley published the complete sequence of chymotrypsin in 1964 (Hartley 1964). To the group in Prague it only remained to confirm the data, suggesting a correction in a single position (Meloun *et al.* 1966). By that time Hartley's interest in sequence data had switched from enzyme function to comparative evolutionary studies, an area he helped create. He gathered evidence indicating that enzymes like chymotrypsin, trypsin and elastase which showed a common catalytic mechanism, but differed in their substrate specificity, nevertheless presented a very similar sequence. From here he proceeded to develop genetic models to account for the evolutionary history of enzyme families like the one he had studied and to produce ancestor trees.

[15] Hartley first came in touch with chymotrypsin while studying the effects of nerve gases on locusts during his doctoral research at Leeds (1949–52). Chymotrypsin was among the enzymes inactivated by the nerve gases. The study of the physiological effects of nerve gases later turned into an interest in the reaction mechanism of the enzyme.

[16] On Sanger's move from endgroup determination to sequencing and the impact of new fractionation techniques see de Chadarevian (1996a). By the mid-1950s Pehr Edman's method of stepwise degradation of the polypeptide chain and the development of an automatic amino acid analyser by Stein and Moore at the Rockefeller Institute in New York, as well as new separation methods, provided powerful new tools for the sequence analysis of larger molecules.

Blow took his first X-ray picture of chymotrypsin in 1955 or 1956 while working in Perutz's group towards his doctorate.[17] The crystals Blow analysed were given to him by Hartley, who had found the egg-shaped crystals growing on the bottom of one of his tubes which he had left in the refrigerator, uncorked.[18]

Both Hartley and Blow agree that this event marked the beginning of their collaboration. However, they also confirm that at that time biochemists and X-ray crystallographers working on proteins were rather sceptical of each other's work. Biochemists thought that protein crystallographers were wasting their time. They were convinced that proteins were a lot more complicated than crystallographers could imagine and that the analysis of the diffraction spots would never yield an adequate representation. X-ray crystallographers in their turn held that knowledge of the amino acid sequence of proteins was meaningless without information on their three-dimensional structure. However, it was generally expected that complete knowledge of the sequence would yield the three-dimensional structure of the protein. Sanger's publication of the complete sequence of insulin had brought this prospect nearer and represented a severe threat to the future of protein X-ray crystallography. As we have seen, Kendrew, working on the X-ray analysis of myoglobin next to Perutz and Blow, reacted swiftly to this new situation by actively seeking the collaboration of a sequencer. When refined techniques allowed crystallographers to determine the position of single amino acids, the picture turned around again and protein sequencing was under threat of becoming obsolete. In fact the opposite happened and crystallographers became the best clients of sequence data.

At the time, Blow suspected that the crystals Hartley gave him were crystals not of chymotrypsin but of ammonium sulphate, which shows once more how little crystallographers trusted biochemists in the matter of crystal growing. He was surprised to find out that they were indeed gamma-chymotrypsin. But he did not follow up the X-ray analysis of the enzyme then, nor did Hartley follow up what Blow was doing in the following years.

Blow started work on the structure of chymotrypsin only in 1960. Like Hartley, he too spent two years in the United States before being offered a position in the unit, which hired new researchers in view of the move to the

---

[17] The first X-ray picture of chymotrypsin was taken by Perutz in 1937 when searching for a topic for his thesis (Perutz 1940, appendix). But chymotrypsin showed a diffraction pattern which was difficult to interpret and Perutz settled on haemoglobin instead.

[18] Interview with B. Hartley, Elsworth, 28 September 1992. Good fortune by neglect is a common trope in discovery accounts.

new laboratory. A decisive reason for taking up chymotrypsin was again that it was commercially available (peroxidase, the haem protein suggested by Perutz, had to be prepared starting from tons of fresh horse radish, a procedure which deterred Blow). That the sequence was being worked on was a further advantage. By that time Kendrew's work on myoglobin had shown that, while X-ray analysis could indeed aim at a resolution of single amino acids in a polypeptide chain, sequencing data were crucial for a complete interpretation of the structure (Kendrew *et al.* 1961).

Blow recalled that once he seriously started working on chymotrypsin he was in fairly regular contact with Hartley. According to Hartley they started to talk together *properly* only once they both moved to the new laboratory. Then they met daily in the canteen. Despite the daily exchanges, however, both continued to pursue their own research agendas and struggled with quite different experimental problems. While the basic lines of attack for protein sequencing and protein crystallography were established by the pioneering work of Sanger and Kendrew respectively, the task in both cases was still immense. The methods required constant adjustment and refinement and there was vast scope for the development of new techniques to cope with the complexities of the structures and the mass of data to process. Still in 1967 Bragg compared the decision to initiate the structure analysis of a protein to the decision to build a new type of aeroplane – so massive being the investment in manpower, time and money, and so uncertain the success (Bragg 1967).[19] Sequencing equally proceeded at a very slow pace and Sanger's feat remained unparalleled for a long time. Only in the early 1960s were longer sequences published, but every sequence involved several man years' work (Hartley 1970). In a preliminary report on his sequence analysis, Hartley stressed the difficulties and tediousness of the task as well as the embarrassment of having to present an unfinished picture which he compared to a jigsaw puzzle in which large holes remained to be filled in (Hartley 1962). And while he struggled with enzymatic digestion products, columns and paper chromatograms, Blow's crystallographic analysis hinged on the development of new computational methods to calculate and refine the structure. There was little in the way of practical experience to exchange.

Only when it came to model building did Blow make use of, and indeed depend on the sequence data which Hartley had published a few years

---

[19] In the early 1960s Bragg estimated the costs for the structure analysis of a protein to be 'fifty man-years of first-rate researchers' time and a quarter of a million pounds'. This estimation, based on the very first protein analysis, soon turned out to be excessive (Bragg 1967).

9.6 David Blow (sitting) with part of the chymotrypsin group, including some of the 'computors'. In front, a few of the cardboard boxes containing the 10,000 IBM cards with the diffraction data. On the left, part of a Joyce Loebl densitometer. Courtesy of Brian Matthews and of Medical Research Council Laboratory of Molecular Biology, Cambridge.

earlier. He had to rely on them especially around the active centre of the enzyme where an inhibitor, put there for technical reasons, blurred the electron density map. Blow built the first atomic model of chymotrypsin following the peptide chain through the map and transposing the structure on a rigid framework using 'Kendrew-type' skeletal model parts (Matthews *et al.* 1967) (Fig. 9.6). Two years later, Blow together with Jens Birktoft, a Danish postdoctoral research fellow, built a second model, based on a refined map. By that time, 'Fred's folly', an ingenious new device consisting of a half-silvered mirror which projected an image of the electron density map on the model while it was being built, greatly facilitated the task.[20] The more accurate model, however, prompted a doubt in regard to a particular amino acid in the sequence. The amino acid in position 102 formed a

[20] On 'Kendrew models' and their commercialisation see above (chapter 4). The new optical device was developed by Frederic Richards from Yale while on leave in Oxford (Richards 1968). The invention followed Richards' publication of a first atomic model of ribonuclease in 1967. Ribonuclease was the third protein structure (after myoglobin and lysozyme) to be 'solved'.

hydrogen bond with the histidine which was known to have an important function in the reaction mechanism of the enzyme. This amino acid was asparagine in Hartley's sequence, but aspartic acid in the reported trypsin sequence.

In their recollections of the events leading to the correction of the sequence and the proposal of a reaction mechanism for chymotrypsin, Hartley and Blow offer slightly different versions. According to Hartley, once he saw the model, it struck him 'at once' that the asparagine residue under the histidine residue could not be right and that he could invent a 'marvelous mechanism' if the residue had been aspartic acid. After all, he had been thinking about a possible mechanism for about fifteen years. He also knew best that it was perfectly possible that he had got the sequence wrong. So he went back to his lab and in only forty-eight hours did the sequence again, confirming that there was indeed an error.[21] In Blow's recent reconstruction of the 'tortuous story' leading to the explanation of the reaction mechanism of chymotrypsin, it was Birktoft who carefully checked the electron density map and the different sequences of homologous proteins and first pointed to the possible error in position 102. Only at that point did Blow confront Hartley with their findings, prompting him immediately to re-do the sequence (Blow 1997 and interview with D. Blow, London, 29 September 1992) (Fig. 9.7).

A paper on the postulated mechanism for publication in *Nature* was rushed out (Blow *et al.* 1969). Even before it was submitted, a 'thrilled' Perutz dispatched a letter to the Secretary of the MRC announcing the success.[22] Chymotrypsin was only the fourth protein of which the atomic structure was known, and only the second structure, following Kendrew's pioneering achievement, coming from the Cambridge crystallographers. The postulated mechanism gave the work particular importance.

The work had an interesting sequel. After the common work on the mechanism, Blow allowed Hartley to 'play' with his model. In his only attempt at model building ever, Hartley tried to built the different side chains of elastase and other proteins he had shown belonged to the same family into the active centre of the chymotrypsin model. He found that this could be done without any significant distortion. This confirmed his prediction that enzymes with a similar sequence had a similar three-dimensional structure, even if they differed in their substrate specificity. Hartley thus integrated the model into his own research, turning it into a tool that could represent the structure of all the enzymes he had studied. Shortly

[21] Interview with B. Hartley, Elsworth, 28 September 1992.
[22] M. Perutz to H. Himsworth, 4 April 1967; Himsworth file, LMB.

9.7 Jens Birktoft with the atomic model of chymotrypsin.Courtesy of Jens Birktoft
and of Medical Research Council Laboratory of Molecular Biology, Cambridge.

afterwards, David Shotton, a doctoral student in Hartley's group, embarked on a complete structure analysis of elastase to confirm Hartley's
observations and to unravel the subtle structural differences responsible
for the substrate specificity of the enzyme. Originally, he set out to do only
the sequence, but on Hartley's suggestion later embarked on the X-ray
analysis as well. This was highly unusual, and the work could only be
achieved through the energetic help of people from Blow's group, who
lent their support on all stages of the analysis.

When it came to model building, Shotton, together with Herman
Watson from Kendrew's group, who had collaborated in the elastase work,
had moved to the new University of Bristol where the Department of
Biochemistry was creating a Molecular Enzymology Laboratory.[23] The two

[23] Watson had joined the Cambridge laboratory in 1959 as Kendrew's collaborator. For his
role in the fate of the DNA model see above, pp. 237–8. Hilary Muirhead, Perutz's
collaborator, followed Watson and Shotton to the Enzymology Laboratory at Bristol. The
creation of the new university at Bristol was part of a more general expansion of British
universities in that period, concerning especially the science departments. Against the
backdrop of continuing animosities between biochemists and molecular biologists (see
below, chapter 10), the Bristol department made a strong move to integrate the new
structural approaches into biochemistry. Philip Randle, who had moved from the

researchers used the same model building technique employed for chymotrypsin. A batch of new 'Kendrew-type' skeletal models was urgently ordered from Cambridge Repetition Engineers, the small precision engineering firm which, following Kendrew's original specifications, had been producing the models since the late 1950s.[24] To compare the structure of elastase with that of chymotrypsin the bulky model was transported to the LMB. The results of the analysis were published in a common paper by Shotton and Watson (Shotton and Watson 1970).

Boundary objects

What does the story tell us about interdivisional collaboration in the new laboratory? Working under one roof certainly made it easier for Blow and Hartley to communicate. This helped and accelerated research. Borrowing a vocabulary developed by Galison to describe the interactions between different professional subcultures in high-energy physics in postwar America, the canteen can be seen as the 'trading zone' where scientists, acculturated in different traditions, developed a common language and trust was built (Galison 1997, 781–844). The model was one of the 'boundary objects' around which interaction developed (Star and Griesemer 1988). Walking just two floors down and looking at Blow's model, Hartley could convince himself that something 'looked wrong' in the active centre of the molecule (this is possibly what remained inscribed in his memory and explains his version of the events). He could also discuss with Blow the possible reaction mechanism of the enzyme. This required a familiarisation with the model which would have been difficult to achieve at a distance. While sequence data can easily be circulated in print, models are much

Department of Biochemistry in Cambridge to the new chair at Bristol, commented that he did at Bristol what was impossible at Cambridge; interview with P. Randle, Oxford, 2 June 1993. Gutfreund who, while a Research Fellow in the Department of Colloid Science at Cambridge from 1947 to 1957, had been closely associated with the MRC group in the Cavendish, was appointed director of the Molecular Enzymology Group. While his own expertise lay in the field of solution physical chemistry as applied to proteins, he became instrumental in establishing protein crystallography at Bristol. On the formation and work of the group see H. Gutfreund, 'The history of the Molecular Enzymology Laboratory at the University of Bristol' [typescript, November 1999]. Several years earlier, Krebs had attempted to attract the Cambridge protein crystallographers to the Biochemistry Department in Oxford. Later Phillips and some of his colleagues from the Royal Institution moved to Oxford. But in contrast to the set-up in Bristol, the new Laboratory of Molecular Biophysics at Oxford ended up being part of the Zoology Department (see above, p. 253).

[24] See correspondence file at Cambridge Repetition Engineers, Cambridge, UK. See also above (chapter 5). According to Shotton and Watson (1970), the model parts used for the elastase structure were designed by Watson 'in conjunction with Dr J. C. Kendrew', but John Rayner, who followed his father as director of the firm, could not confirm that the original design of the model parts had indeed changed.

less mobile (Latour 1990; de Chadarevian, forthcoming). These direct interactions between protein crystallographers and protein biochemists were possibly unique at the time. The LMB facilitated and institutionalised these instead of other possible exchanges.

It is none the less significant that collaboration did not necessarily mean joining hands at the laboratory bench and that the different cultures remained in place. Indeed, the fruitfulness of the interactions between Hartley and Blow very much rested on the fact that they both remained committed to their own experimental approaches and, to a large extent, to their own projects. The chymotrypsin model, though partly based on Hartley's sequencing data, was built by the crystallographers. Only Hartley could repeat the sequence and he did this 'away' from the crystallographers in his own laboratory.[25] Blow mentions as a main lesson from the chymotrypsin story to distrust other people's results, even if he obviously depended on them (Blow 1997, 408). When a reunion was held at the LMB to celebrate the thirtieth anniversary of the chymotrypsin structure, Hartley and some researchers of his group were among the eighty invitees, but in the main it was a structural studies affair.[26] Blow and Hartley continued to interact closely on a number of issues, but apart from a brief note in 1973 and an article twenty years later there were no common publications.[27]

Shotton's 'double bill' might seem to contradict these observations, but apart of relying on the crucial help of crystallographers, the very fact that it remained unique at the time can be seen as a confirmation of the rule. More interestingly perhaps, Shotton's story highlights the role of the young doctoral or, more often, postdoctoral researchers as well as the role of technological developments in, at least partially, bridging the different research cultures and in the economy of the new laboratory more generally.

Ever since the early days of the MRC group in the Cavendish, the development of new technologies had been a priority and had given the group a lead over other laboratories (this was particularly true for protein crystallography). Many instruments were first introduced at Cambridge

---

[25] Letter from D. Blow to author, 2 February 2001.

[26] On the role of scientific reunions and anniversaries in building group identity see Abir-Am (1992c and 1998) and Abir-Am and Elliott (1999).

[27] The note concerned preliminary X-ray data of tyrosyl-transfer RNA synthetase, to which Hartley's group had contributed by providing the crystals (Reid *et al.* 1973). The transfer RNA synthetases, the enzymes responsible for specifically binding amino acids to their respective transfer RNAs, occupied Blow and Hartley and other members of their groups for several years. Blow and Hartley eventually left the LMB. Both took up positions at Imperial College London, Hartley as Professor of Biochemistry, Blow as Professor of Biophysics. Hartley's appointment was vehemently opposed by his predecessor, Ernest Chain, Nobel Prize winner for his contribution to the work on penicillin during World War II, who resisted the takeover of his brand of physiological biochemistry by molecular biologists (Olby 1991).

before they became commercially available.[28] The LMB had inherited the strong workshop tradition. All groups, if not to the same extent, did make use of the fantastic in-house facilities. Similarly, the Cavendish group had already attracted numerous postdoctoral researchers. But with the new laboratory and its enlarged facilities, numbers increased sharply. Several fellowship programmes, the number shooting up with the rising tensions of the Cold War and increased allocation for research, allowed in particular young American researchers to spend one to several years in Europe. For those interested in molecular biology, the LMB together with the Pasteur Institute were the most attractive places (some would spend time in both institutions during their stay in Europe).[29] But researchers also came from Canada, Australia and the European continent (including some East European countries). A further, though comparatively small contingent came from British universities (Fig. 9.8). For the LMB, which was not part of the normal university set-up, the influx of well-trained post-doctoral researchers represented a vital source of innovation as well as an important workforce. In the laboratory the 'postdocs' represented a distinct social group. They tended to socialise across divisions and thus played a crucial role in bringing the laboratory closer together. Moving on after some years, they were also instrumental in exporting the research culture of the LMB to new locations.

New technologies

Shotton performed the X-ray analysis of elastase, including the making of heavy atom derivates, the measurement of tens of thousands of intensities and the extensive calculations in the record time of only eight weeks. Not only for a newcomer in the field would this have been impossible to achieve without improved technology.

The most important technological development concerned a computer-linked densitometer, devised by Uli Arndt in conjunction with crystallographers and computer scientists in the laboratory (Arndt, Crowther and

---

[28] Examples include the rotating anode tube built by the group's own technician and improved by Hugh Huxley and Ken Holmes in the laboratory, before being licensed and commercially produced by Elliott Bros., and a semi-automatic densitometer, first used by Kendrew for measuring diffraction spots and later commercially produced by Joyce-Loebl. On the contributions of laboratory members in the continuing search of more powerful X-ray sources see Huxley and Holmes (1997).

[29] On the introduction of postdoctoral fellowships at the NIH in the 1950s see Strickland (1989), though unfortunately the book completely ignores international contexts. For the attraction of American postdocs to Paris see Cohen (1986). For memoirs of American fellows of their time at the LMB, especially in Brenner's group, see Steitz (1986) and the volume of letters dedicated to Brenner on the occasion of the MRC Fellows Symposium in 1983 (Brenner, personal collection).

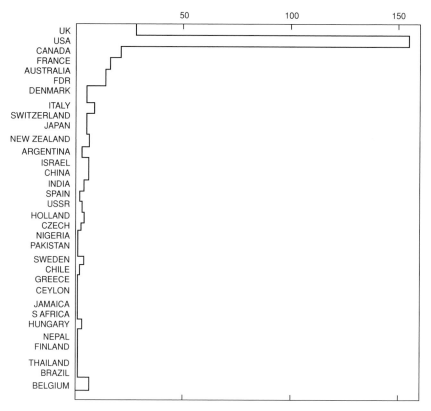

9.8 Nationality of visitors to the LMB, 1970–5. *Source*: Committee to Review Molecular Biology, 'The Laboratory of Molecular Biology: some facts and figures' (December 1975), file A147/14, vol. 1, MRC Archives.

Mallett 1968). Arndt, himself a trained crystallographer, was hired from the Royal Institution to the LMB with the exclusive brief to develop apparatus which facilitated crystallographic analysis. The new densitometer was only one in a long series of instruments Arndt developed in close collaboration with the technicians in the in-house workshop. It dramatically accelerated data collection.[30] Elastase became the test case for the new computerised data collecting equipment.

[30] In the late 1950s, while still at the Royal Institution, Arndt had devised an automatic diffractometer. It was commercially produced by Hilger and Watts and quickly became part of the standard equipment for protein crystallography, though it was soon superseded by a computer-linked diffractometer, developed by Arndt in collaboration with technicians at the Atomic Research Establishment at Harwell. The diffractometers replaced photographic techniques of data collecting by directly measuring the diffracted X-ray intensities. However, the operation of the instrument could be tedious, and for analysis with many spots photographic techniques remained the preferred option. Here the computer-linked densitometer found its place. Arndt later contributed to the development of new X-ray cameras, new detector systems and more powerful X-ray tubes. On Arndt's contributions to the development of crystallographic apparatuses see his

In protein crystallography it became a rule that work which was worth a Nobel Prize, as for instance the full structure determination of a protein, five years later was at the level of a Ph.D. Automation and acceleration of more and more steps in the analysis made protein structure determination more available to researchers outside the small circle of professional crystallographers. But despite this trend, for the reasons already pointed out, Shotton's case remained rather an exception.[31]

In and out of the laboratory

That the first combined sequencing and X-ray analysis of a protein was performed by a young researcher (with the help of more skilled colleagues from different divisions) is no accident. Young researchers came to learn new techniques and they were freer than senior staff to move between divisions. Indeed, it was seen as one of the attractions of the Laboratory of Molecular Biology in Cambridge that people who came to do their doctoral or postdoctoral research in the laboratory could learn a combination

'The development of X-ray crystallographic techniques in molecular biology' [unpublished typescript, 1987], and interviews with U. Arndt, 15 October 1998 and 1 October 1999.

[31] The extent to which the different divisions remained committed to their own research agendas is exemplified by the work performed in the laboratory on soluble (later transfer) RNAs, the molecules implied in the transfer of information from nucleic acids to proteins. Throughout the 1960s and 1970s the laboratory produced a constant flow of papers on the subject. The work of the laboratory contributed to making transfer RNAs one of the most intensively studied molecules in molecular biology. Yet despite the sharing of a general understanding of the function of transfer RNAs in protein synthesis and often the use of the same research material and despite the occasional collaborative paper, transfer RNA meant different things to people working in different divisions. Crick and Brenner's group was interested in the mutant forms of the molecules to study their function. For Sanger and his collaborators pure fractions of transfer RNAs were the first material on which nucleic acid sequencing methods could be developed. Finally, Aaron Klug's work on the structure of transfer RNAs in the mid-1970s was part of his more general effort to study protein–nucleic acid interactions and to develop techniques which allowed the visualisation of large molecular assemblies arising from these interactions (Klug 1983). When trying to account for his 'conversion' from protein to nucleic acid sequencing, Sanger did not fail to acknowledge the 'atmosphere' in the Laboratory of Molecular Biology, the importance of the canteen where ideas were exchanged and the direct influence of people like Francis Crick and John Smith, the 'nucleic acid expert' in the laboratory. But he also noted that, in his case, it never came to common publication with people from other divisions (Sanger 1981, 142; 1988, 14; interview with F. Sanger, Cambridge, 19 March 1992). While working with Sanger on transfer RNA, Kjeld Marcker, a visiting researcher, identified a modified form of the soluble RNA which bound methionine. This led to a collaborative effort with Brian Clark, a postdoctoral student in the molecular genetics division, to establish the function of the molecule in protein synthesis. Experiments showed that the modified RNA worked specifically as chain initiator in bacteria. This was heralded as an important result. But significantly, Sanger described Marcker's feat as 'the most successful *diversion*' [my italics] from the sequencing work (Sanger 1988, 25). The first transfer RNA was sequenced by Robert Holley, while Sanger's group soon shifted its focus to ribosomal and bacteriophage RNA.

of techniques, elsewhere taught in separate academic departments. This did not only regard different techniques for structure determination. According to Brenner, for instance, in the 1960s the laboratory was the only place where students and postdoctoral researchers could take up structural biochemical and genetic techniques. Senior researchers, in contrast, were encouraged to continue to do 'what they were good at'. Some of them perceived this as a restriction and it became a reason for them to leave.[32] In the longer term it posed the problem of innovation in the laboratory.

Shotton's move to Bristol points to another important function of the young researchers in the history of the laboratory. Necessarily moving on after a few years, they played a crucial role in the export of research techniques from the laboratory in Cambridge to other laboratories and in building up mixed groups of the kind they had encountered and profited from during their apprentice time (even if in this particular case Shotton followed Watson, who was a senior researcher). Instruments which were not commercially available the LMB workshop would build for them. Watson and Shotton not only exported the parts and techniques of model building, but the whole gamut of techniques devised at Cambridge for protein X-ray analysis. The aim was to create an equally effective protein crystallographic unit at Bristol. Adapting to the research interests of other groups in the department, the Molecular Enzymology Laboratory focused work on the structure analysis of the enzymes responsible for glucose breakdown in the cell.[33]

The creation of a new research tool and the role of (post)doctoral students in the export of a local research culture as discussed for Shotton's case are exemplarily illustrated by the history of Brenner's worm project. Technical innovation here concerned not the design of a new apparatus, but the crafting of an organism into a laboratory tool for the study of development. The project was drawn up in connection with the first extension of the laboratory and from its very beginning became part of the renegotiations of what molecular biology was. These renegotiations and transformations of the field also affected the role of the laboratory itself.

---

[32] Interview with B. Hartley, 28 September 1992.

[33] Although the enzymology group was closely integrated in the Biochemistry Department in Bristol, protein sequencers were not locally available and protein crystallographers had to look for collaboration elsewhere, which represented a drawback in relation to the situation at the LMB. After the work on elastase and the much less speedy progress on the refinement of the structure, Shotton himself moved on to Oxford, where he took up structure determination by three-dimensional electron microscopy. Watson and Muirhead carried on X-ray crystallography at Bristol.

9.9 Nobel celebration in the laboratory (autumn 1962). Max Perutz (left) and Sydney Brenner (right). Courtesy of Medical Research Council Laboratory of Molecular Biology, Cambridge.

Expansion

Not even a year after the different research groups had established themselves in the new, lavishly funded four-storey laboratory, B. Lush, the Principal Medical Officer of the MRC, came to discuss plans for future expansion. He indicated that the MRC wished to build the laboratory up to what the principal researchers considered its 'final optimum size' until their retirement. This meant planning ahead for about fifteen years. The surprising move was doubtless prompted by the triple Nobel award, recently presented to members of the laboratory (Fig. 9.9).

Not all board members of the laboratory favoured expansion. The fear was voiced that a much expanded laboratory would make communication and interdivisional collaboration more difficult. Kendrew argued that dispersion of young researchers to other centres, rather than expansion of the existing laboratory, would be 'in the national interest'. But in the end these thoughts were brushed aside and an extension of about 20,000 square feet that would double the size of the existing laboratory was discussed.[34] This plan proved too ambitious, and in the formal proposal presented to the

[34] 'Minutes of Meeting of Molecular Biology Board. 26.4.63' [by M. Perutz], file on first extension, LMB Archives.

MRC a few months later the projected extension shrank to little more than half that size.[35] Plans for the new space included moderate expansion of some current lines of research, additional storage room, space for the installation of a lab-internal computer and some 'uncommitted space'. The most ambitious programme for the new space, however, was put forward not by one of the laureates, but by Brenner, Crick's collaborator since 1957.

The MRC inquiry had stimulated intensive talks between Brenner and Crick about their future research. In a memorandum to Perutz, Brenner summarised the outcome of their conversations. The two researchers agreed that nearly all the 'classical' problems of molecular biology either had been solved or would be solved in the next decade. This meant that 'most of molecular biology had become inevitable' and that it was time 'to move on to other problems of biology which are new, mysterious and exciting' (Wood 1988, ix). The most promising fields for a new attack, Brenner reckoned, were development and the nervous system. He appreciated that this was not an original thought; many other molecular biologists were thinking along similar lines. However, the great difficulty with these fields was to define clearly 'the nature of the problem' and to find 'the right experimental approach'.[36] Brenner held that molecular biology had succeeded in its analysis of genetic mechanisms because complicated phenomena could be reduced to simple units and because simple model systems had been devised. Building on this experience, he proposed to start attacking the problem of development by studying the process of cell division both in bacteria and in the cells of higher organisms. With respect to work with higher organisms, Brenner saw 'a great need to "microbiologize" the material' so that cells could be handled as conveniently as bacteria or viruses. As a longer-term possibility he also suggested 'taming' a small multicellular organism so as to study development directly.[37]

Only a few months later, when the actual proposal was submitted to the MRC, Brenner's 'fluid' ideas regarding this last point had solidified. The plan to create a new model organism for the study of development had moved centre stage. The proposal read:

We should like to attack the problem of cellular development . . . choosing the simplest possible differentiated organism and subjecting it to the analytical

[35] M. Perutz, F. H. C. Crick, J. C. Kendrew and F. Sanger, 'The Laboratory of Molecular Biology. Proposal for extension', October 1963; reprinted in Wood (1988, xii). Delays in the realisation of the plan and later considerations on the part of the MRC and the University increased the size of the extension even above that in the original proposal.

[36] S. Brenner to M. Perutz, 5 June 1963; reprinted in Wood (1988, x).

[37] S. Brenner to M. Perutz, 5 June 1963; reprinted in Wood (1988, xi).

methods of microbial genetics. Thus we want a multicellular organism which has a short life cycle, can be easily cultivated, and is small enough to be handled in large numbers, like a micro-organism. It should have relatively few cells, so that exhaustive studies of lineage and patterns can be made, and should be amenable to genetic analysis.

We think we have a good candidate in the form of a small nematode worm, *Caenorhabditis briggsiae*.[38]

At first, the plan met with resistance from the MRC officers. They felt that this was 'pure' rather than 'molecular' biology and would lead the laboratory away from its present emphasis 'on the physics, chemistry and genetics of simple biological mechanisms'.[39] Brenner, however, insisted that his aim was to 'molecularize' the approach to development and differentiation.[40] Project and extension were approved without much further delay, though the extension took five years to be completed. Brenner started preliminary experiments which led him to settle on *Caenorhabditis elegans*, a free-living species 1 mm long, as the model organism.[41] With the new laboratory space becoming available, the project also gained momentum.

A new laboratory tool

Caenorhabditis elegans was chosen out of around sixty nematode species, some of which were collected in Brenner's backyard or around the laboratory, and all of which were tested for their aptitude as laboratory creatures. The nematode finally selected for use in Brenner's laboratory was the Bristol strain of *C. elegans*, originally sent from Berkeley by Ellsworth C.

---

[38] M. Perutz, F. H. C. Crick, J. C. Kendrew and F. Sanger, 'The Laboratory of Molecular Biology. Proposal for extension', October 1963, Appendix I; reprinted in Wood (1988, xii).

[39] 'Interview with Sir Harold Himsworth and B. Lush on 6 June 1963' [notes by M. Perutz], file on first extension, LMB Archives and B. Lush, 'Possible accommodation developments at the Molecular Biology Laboratory', note for file, 10 June 1963, file P2/326, vol. 1, MRC Archives.

[40] M. Perutz to B. S. Lush, 14 June 1963, file on first extension, LMB Archives. Perutz and the other board members of the laboratory considered the extension a key means of inducing Brenner to stay. The new project also gave Brenner more independence from Crick, who was formally his senior.

[41] *C. elegans* was a model organism in so far as the processes of development in the small nematode worm were viewed as a simple representation of processes which, in principle, were the same in all organisms. I speak of *C. elegans* as a laboratory tool to indicate that the work of cultivation and representation that went into *C. elegans* turned it not only into a new scientific object, but also into a means to study development. For a similar discussion in relation to other laboratory organisms see Clarke and Fujimura (1992, 22) and Kohler (1994, esp. pp. 53–90). For a historical and philosphical examination of the use of *C. elegans* as model organism see Ankeny (1997); for a post-modernist reading of work on the worm see Doyle (1994).

Dougherty who had been investigating this particular worm since the 1940s. The possibility to build on this experience was considered an important advantage. Previously reported problems of cultivation could be overcome by feeding on *E. coli*.[42] Other properties which seemed attractive at the time were its small size (100,000 worms could live in a petri dish), its brief life-cycle (just three days) and, above all, its reproductive biology. The worm occurred in two sexes: self-fertilising hermaphrodites and males which reproduced by cross-fertilisation with hermaphrodites. This offered great advantages for genetic analysis. The smallness of the worm also meant that it 'interfaced' well with the electron microscope. Other 'unforeseen advantages', like the transparency of the body and the extremely small genome, were exploited only later (Hodgkin 1989) (Fig. 9.10).[43]

That 'taming' *C. elegans* was not easy is dramatically illustrated by the fact that when Jacob wanted to introduce it as a laboratory organism at the Pasteur Institute in Paris, his attempt failed.[44] Cultivating *C. elegans* in the laboratory, however, was only the first step in turning it into a tool for research on development. Brenner's plan was to produce mutants to study by deficiency the steps in the development of the worm. But in order to study the effect of mutations it was necessary to gain detailed knowledge of the genetic make-up and the normal development and anatomy of the organism. Brenner and his colleagues embarked on what grew into an ever-expanding effort to map *C. elegans*.

Brenner dedicated several years to describing the basic genetic features of the nematode (Brenner 1974b; Sulston and Brenner 1974). He estimated the size of its genome (100 million base pairs) and the number of essential

---

[42] The irony here was that *C. elegans* fed on the organism which was the standard experimental organism of bacterial genetics.

[43] In the 1960s several molecular biologists tried to develop new model systems or adapt old ones for research on development and neurobiology. Benzer, for instance, worked on *Drosophila*, Jacob on the mouse (see below), Stent on the leech, Streisinger on the zebra fish, W. Dove on slime moulds. A comparative investigation of these different systems and their success as model organisms would represent an interesting contribution to the discussion of the relations between 'jobs' and organisms; see Clarke and Fujimura (1992), Lederman and Burian (1993), and other papers in the special section 'The right organism for the job' in the same issue of the *Journal of the History of Biology*.

[44] Interview with S. Brenner, Cambridge, 30 June 1993. The explanation offered by Jacob and his team for abandoning the study of *C. elegans* referred to 'technical' difficulties regarding physiological and embryological experimentation with the organism (cf. *Annuaire du Collège de France. Résumé des Cours et Travaux* 1968–9, 195–6; 1969–70, 207–9). It should also be noted, however, that at that time Brenner was not yet keen for other groups to work on 'his' organism (see below). Jacob turned to the mouse instead. This organism better matched the medical research tradition in the Pasteur Institute. Jacob's plan to build an Institute of the Mouse for research on developmental biology, however, did not receive the necessary funding. See Gaudillière (1991, 535–40), Jacob (1998) and Morange (2000).

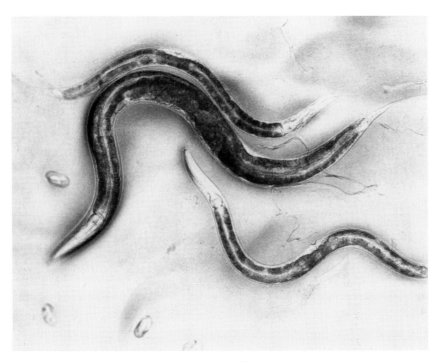

9.10 The nematode *Caenorhabditis elegans* in different developmental stages (adult, larvae and eggs) on a petri dish, enlarged about 150 times. Courtesy of John Sulston.

genes (about 2000). He isolated around 100 mutants. Embarking on a long series of classical crossing experiments, he mapped them onto six linkage groups which corresponded to the worm's six chromosomes. Recruits to his project tackled two other major tasks.

John Sulston, a chemist by training who had worked on the origin-of-life problem with Leslie Orgel at La Jolla before joining Brenner's group, embraced the ambitious project of tracing the cell lineage of each individual cell of the developing embryo.[45] Armed with a microscope, he sat down to observe the cells developing in the transparent body. The complete cell lineage of *C. elegans* was published in 1983 (Sulston *et al.* 1983). The work permitted unprecedented precision in experimental manipulation. A single cell, the 'fate' of which was known exactly, could be ablated with a laser beam, and the effects of the manipulation studied in detail.

A few years later, John White, an electrical engineer who had worked on computers before joining the worm project, together with Nicol Thompson,

[45] Orgel, who had spent several years at Cambridge, had been a frequent participant at the Eagle lunches. He was friendly with Crick and a close associate of the LMB. In the mid-1970s, Crick would join Orgel at the Salk Institute in La Jolla. Together they published a popular book on the theory of 'directed panspermia' which postulated an extraterrestrial origin for life on earth (Crick and Orgel 1981).

a technician specialised in electron microscopy, and Eileen Southgate, a long-term technician in the laboratory who did most of the painstaking work, presented the complete 'wiring diagram' of the worm's nervous system. Analysing 20,000 electron micrographs and matching the different series, they traced the approximately 8000 connections of the 302 nerve cells of the worm. Again, this was an unprecedented feat which was feasible only because of the small size of the organism. 'The mind of a worm' was published in a single 340-page article in the *Philosophical Transactions of the Royal Society* (White *et al*. 1986).[46] The material was displayed in a way that 'facilitated quick access', diagrammatically representing each single nerve cell with all its connections next to a series of electron micrographs on which the evidence was based. What was in fact a mosaic of several nervous systems was presented as a '"canonical" nervous system' (White *et al*. 1986, 4).[47]

The mapping efforts of the worm workers did not end here. No sooner had Sulston completed the cell lineage than he started work on a complete physical map of the worm's genome, again pioneering such a venture. The physical mapping effort grew into a British–American collaboration to sequence the genome. Initially the project met with harsh resistance from within the worm community, who feared that too much money would be detracted from other worm projects. Resistance, according to Sulston, disappeared when the community realised that the sequencing project, pursued with the most advanced technologies, far from reducing its funds, made the worm 'famous' and attracted more money and more researchers into the field.[48] The work ranked as a pilot of the Human Genome Project, the most ambitious mapping effort ever. It became the flagship project of the newly founded Sanger Centre in Cambridge, one of the largest genome centres in the world, of which Sulston was appointed director (Sulston *et al*. 1992; Aldhous 1993; Cook-Deegan 1994, especially pp. 48–55 and 333–5). The complete sequence of *C. elegans*, the first of a multicellular organism, became available in 1998.

---

[46] 'The mind of a worm' was the running title of the article. Originally White and Brenner intended to computerise the work using a graphics program they devised. Although pioneering in its kind, it did not prove sophisticated enough, and it was decided to take turns at labelling and tracing the endless series of micrographs. In the end, Southgate did it all, dedicating twenty years to this work, including some following up projects involving mutants. See interview with R. Fishpool and E. Southgate by author in collaboration with J. Finch, Cambridge, 13 January 1999.

[47] The paper also introduced a uniform system of nomenclature for naming the neurons of *C. elegans*. An appendix listed the equivalences between the new system and the various nomenclatures previously in use (White *et al*. 1986).

[48] Interview with J. Sulston, Hinxton, 20 January 1999.

Each of the mapping efforts just described not only pioneered new technologies, but also created a new description of *C. elegans*. The series of representations produced by the various technological approaches did not map onto each other (they could not be combined to a map of maps); rather they created *C. elegans* as a laboratory tool open to a series of interventions.

Exporting the worm

Brenner's project on the worm was not necessarily planned to become 'big'. At first Brenner hoped that looking at one ganglion might be enough to understand some of the principles involved in the development and function of the nervous system. At that point Brenner was also ambivalent about other people working on 'his' project. In the 1970s, however, he began actively recruiting new people. The change in strategy was in large part a response to an increasingly competitive climate in the field of molecular biology. For the first time there was a slump in the number of postdoctoral fellows applying to the LMB. This was largely the result of cuts in American federal funding for research. The cuts not only made young researchers more apprehensive about leaving their country and the local job market, but also effectively limited the availability of fellowships which, since the 1950s, had allowed an increasing number of American researchers to come to Europe for one or more years. This represented a serious threat to the LMB which, as argued earlier, very much depended on the influx of well-trained postdoctoral researchers, both as a source of innovation and as an important mechanism to keep up and enlarge international networks. Brenner, touring the States, used the worm to attract people to Cambridge. But not all the people Brenner recruited came from outside. Some of them had come to the laboratory to work on other projects and were then attracted to work on the worm. When moving back to their home countries, they set up their own worm groups. This was particularly true for the American fellows, who represented the largest group of postdoctoral researchers and, on their return to the United States, had better chances than their European colleagues of building up new and independent research groups. By this mechanism the community of worm workers grew.[49]

With the new recruits, the work on the worm changed. To accommodate the growing number of researchers, the work was parcelled off.

[49] Interviews with S. Brenner, Cambridge, 30 June 1993, and with J. Hodgkin, Cambridge, 28 June 1994.

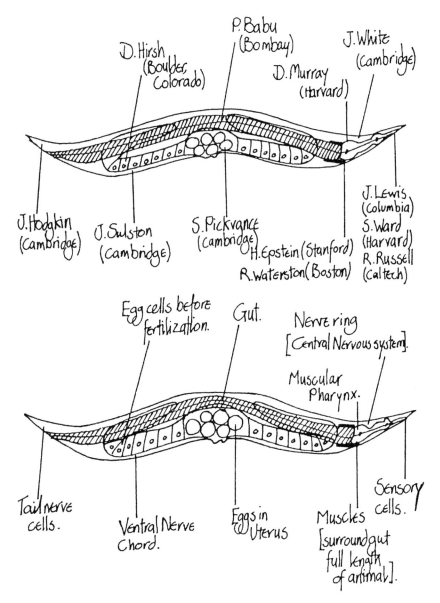

9.11 The subdivision of *C. elegans*. From *Radical Science Journal* 4 (1976), 20. Reprinted with permission.

One researcher who joined Brenner's group as a doctoral student at the beginning of the 1970s recalled: '[The worm] was subdivided. For example, the gut was given to one worker to work on, the head to another, the tail to another, the muscles to another, and by the time I arrived there were a few portions left. I was given the eggs to work on' (Pickvance 1976, 18) (Fig. 9.11).

The subdivision according to organ systems, in many respect a relic of an older anatomical tradition, was soon superseded by a subdivision according to different aspects of the worm's development and behaviour, with the cell lineage and the wiring diagram, later the sequencing project, providing more general tools for work on the worm. The principle, however, remained the same. Subdivision not only made research projects manageable, it also became a strategy to avoid competition among the growing number of worm workers, as well as allowing adaptation to different local ecologies. At the same time the *C. elegans* researchers represented a tightly knit community. As one of them commented: 'There is better co-operation than in any field of biological research I am aware of. And there is a simple reason. The field is quite small, and it started with one person, Sydney Brenner. Many people who now head labs were friends 15 years ago in England. It is a community' (Roberts 1990, 1311).[50] In addition to the common period of training in Brenner's lab, the worm community was held together by the services of the Caenorhabditis Genetics Center at the University of Missouri; it maintained and dispensed mutant strains and, twice a year, issued the *Worm Breeder's Gazette* (Fig. 9.12). *C. elegans* meetings were held biannually.[51]

Similar 'exchange networks' (Kohler 1994, 133–70) existed among drosophilists and the phage group.[52] What was new in the worm project, however, was the combined effort to arrive at a 'total description' (Hodgkin 1991, 951) of *C. elegans*. This plan was not there from the beginning, but 'gradually it became clear that it was both feasible and desirable' to achieve this aim (Hodgkin 1989, 2). Experimental and political reasons concurred in strengthening this project which became part of a more general shift of practices in the field.

---

[50] The interaction of the worm workers is also attributed to the worm itself which, in many instances, resisted subdivision, or at least encouraged collaboration; see Hodgkin (1991) and interview with J. Hodgkin, Cambridge, 28 June 1994.

[51] Both the *Gazette* and the stock collection of *C. elegans* were started in an informal way in the 1970s before a NIH grant allowed them to be developed on a bigger scale. A uniform genetic nomenclature, modelled on the one introduced by Brenner in the early 1970s, was agreed upon at the first *C. elegans* meeting at Woods Hole in 1977; see Horvitz (1977). Cells, nerves and their connections were also uniformly named.

[52] Significantly, Robert Edgar, who started the worm newsletter, was part of the phage group before joining the worm project. On the 'phage influence' on work on the worm see Hodgkin (1989, 2). For a long time the worm researchers, like other organism groups, formed a fairly separate community. This means that there was more interaction among people working on separate aspects of the same organism than, say, among researchers working on the development of the nervous system in *C. elegans* and the mouse. This throws important light on the question of model organisms. More recently, however, the growing importance of comparative sequence data is drawing the different communities closer together.

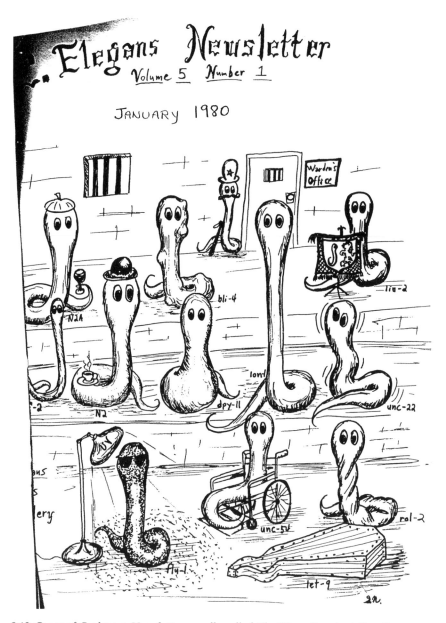

9.12 Cover of *C. elegans Newsletter*, usually called *The Worm Breeders' Gazette*, featuring various mutants of the worm. Design by Greg Nelson. Courtesy of Jonathan Hodgkin. Reprinted with permission.

Renegotiating molecular biology

In the early 1960s molecular biologists were starting to think 'big'. This is clear from the plans for a European Molecular Biology Laboratory (EMBL) to be modelled on CERN, discussed in a series of meetings around that time but brought to fruition only a decade and a half later (see below, chapter 10). Among the research schemes considered for the laboratory was 'Project K: The complete solution of *E. coli*', proposed by Crick and based on conversations with Brenner (Crick 1973).[53] The aim was to arrive at a 'full' description of the bacterial cell. The project as envisioned by Crick involved a huge workload and the pooling of results from many laboratories. A central laboratory could assist and coordinate this work by developing advanced technologies, and by producing and circulating mutants, chemical components of cells and results, and thus reducing waste and avoiding overlaps.

Crick's Project K was never embraced as such, but much of its vision was realised in the worm project, especially in the genome mapping and sequencing project. It marked a shift in the way molecular biology was practised and knowledge produced.[54] A brief example will illustrate the point. The *Nature* article announcing the complete genome sequence of *C. elegans* in December 1998 was authored by 'The *C. elegans* Sequencing Consortium'. A footnote referred to a website for the list of authors (*C. elegans* Sequencing Consortium 1998). Here around 400 authors, divided into two groups according to their affiliation to either the Sanger Centre or the Genome Sequencing Center at the University of Washington, St Louis, were listed in alphabetical order. Group leaders were not recognisable.

When setting up his project on the worm, Brenner was ambiguous as to whether he intended to leave molecular biology or to expand its domain. Arguably, for many years, work on the worm followed more the lines of 'classical' than molecular biology.[55] But for Brenner the choice of simple organisms and the reduction of complex phenomena to basic pinciples defined molecular biology as much as its specific achievements. When embarking on his new project, Brenner aimed to follow the same strategy. He expected that there was a genetic programme, the study of which would reveal the 'logic' (or universal structure) of development, very much as the genetic code represented the general key to an understanding of

[53] Crick credited Brenner with having invented the project title (Crick 1973, 67, footnote). The letter K referred to the K12 strain of *E. coli*, the strain most commonly used in the laboratory.

[54] For discussion on a new 'paradigm shift' in molecular biology see Fujimura and Fortun (1996).

[55] This was the term by which molecular biologists used to refer – mostly in disparaging ways – to non-molecular approaches to biology. Molecular biologists' own use of these approaches necessarily shifted the meaning of this demarcation (de Chadarevian 2000). As I have argued elsewhere, molecular approaches often relied in crucial, if unacknowledged, ways on natural history types of collections and non-molecular functional knowledge (de Chadarevian 1998a).

gene function. In this respect, the worm project clearly built on Brenner's earlier work in molecular genetics.[56]

Work on the worm, and especially the study of the cell lineage of *C. elegans*, which showed an invariant but completely 'illogical' pattern of development, made Brenner give up the idea of a genetic programme. He concluded that there was 'hardly a shorter way of giving a rule for what goes on than just describing what there is' (Lewin 1984, 1328). The 'grammar' of development which Brenner now invoked lay in the principles of molecular assembly and interactions. Brenner still believed that, ultimately, the organism had to be explicable in terms of its genetic information, but he now held that the representation of 'genetic space' onto 'organismic space' would not be a direct and explicit one, or, as he also put it: 'It is not a neat, sequential process, like the linking together of amino acids in a protein. It is everything going on at the same time' (Lewin 1984, 1327–29). What was needed was 'a way of getting to the biochemistry of gene products' (Brenner 1984, 172). New techniques, in particular recombinant DNA technologies and DNA sequencing methods, offered powerful tools for this kind of analysis and had in fact already been applied in a detailed genetic and biochemical study of muscle proteins in *C. elegans*, initiated in Brenner's group.[57]

The use of these technologies bound work on *C. elegans* back to molecular biology. Yet in the worm these technologies were put to new uses, ranging from the study of developmental processes to the sequencing of full genomes and thus considerably expanding molecular biologists' field of action. Developmental biologists, who had taken the place of 'classical' embryologists, also increasingly adopted molecular biological approaches to the study of development.[58]

Today around 200 laboratories around the world work on *C. elegans*. Most of those heading a group have spent some time in Brenner's laboratory. But the very success of the Cambridge laboratory in crafting

[56] On the notion of a genetic 'programme', its import from electronic computing and debates about its usefulness in studies of development see de Chadarevian (1998b). Continuities between Brenner's work in molecular genetics and the worm project also existed on the institutional level. Like other molecular biologists who moved to study development or the nervous system, Brenner too did not relocate to one of the biological departments where development was traditionally studied, but continued to work in his old institution. Funding for the work came forward on the basis of his earlier achievements (and that of the LMB more generally) in the field of molecular biology.

[57] Ironically, these same tools would later also lead to the description of a group of genes which control certain aspects of development in *C. elegans* as in other organisms. Some researchers see in these genes and their function the 'logic of development' which Brenner originally set out to find. On this point see Morange (1995; 1996).

[58] This indicates that the 'molecularisation' of developmental biology represented a *rapprochement* rather than a takeover on the side of molecular biologists. For a similar view see Morange (1998b).

9.13 The Sanger Centre on the Wellcome Trust Genome Campus at Hinxton. Courtesy of Sanger Centre.

and exporting a new research tool also affected its position as the hub of worm research. From the only centre it became one place among many on the map of the worm world. By far the largest number of worm workers are active in the United States (followed by Japan and Europe). The central facilities serving the worm community also operate in the United States. The Cambridge group involved in the sequencing of the worm genome, on which many of the worm's future prospects as laboratory tool hinged, moved out of the LMB into a newly equipped and highly automated sequencing centre on the outskirts of Cambridge, funded by the Wellcome Trust (Fig. 9.13). Here data are produced on an industrial scale.[59] Only a few researchers, working on several aspects of the worm, including sex determination and development, remained in the LMB. These are also now poised to move. Brenner himself has abandoned the LMB as well as the worm and, among many other projects, launched the puffer fish as a new model organism.

The relation of the relocated researchers to their 'mother institution' in Cambridge remains ambiguous, reflecting a more general and continuing

[59] On work practices in the new sequencing centres see Fortun (1998; 1999).

9.14 John Sulston (right) with Alan Coulson, his close colleague on the mapping and sequencing project of *C. elegans*. Both researchers moved from the LMB to the Sanger Centre. Courtesy of Wellcome Library, London.

debate on the role and function of the new sequencing centres: are they industrially operating service centres supplying research tools or do they represent the biological research centres of the future?[60] For some the move was definitive, others have meanwhile rejoined the LMB, while Sulston, the first director of the centre, kept a double affiliation (Fig. 9.14). The sequencing data are freely available on the net, redrawing again (and theoretically lifting) the boundaries of a (virtual) community of worm workers. The site of the Sanger Centre has become home of two other important institutions of genomics research: the European Bioinformatics Institute (EBI), an outstation of the European Molecular Biology Laboratory (EMBL), which manages its DNA data library, and the MRC Human Genome Resource Centre, which collects and distributes genomic

---

[60] Changes in work practices and in the understanding of what counts as biological knowledge spill over from dedicated sequencing centres to biological centres of research. At the Max-Planck Institut für Molekulare Genetik in Berlin, for instance, data production and their electronic storage and elaboration are the main imperative. Hans Lehrach, the director of the institute, also heads the human genome sequencing centre in Berlin. A visit of the laboratory was included in the programme of the conference on *Postgenomics?*, held at the Max-Planck Institute in Berlin in July 1998.

information to the research community. The Wellcome Trust is further pressing for a huge expansion of the site to allow spin-off companies to settle in direct proximity to the scientific centres, but so far the district council has rejected the proposal, which involves cutting into the green belt.

Following the worm and considering the shifting role of the LMB in the expanding biotech world we have moved far ahead. In this chapter I have investigated the role of the LMB in creating and establishing particular research traditions. We have looked at the organisation of the laboratory, the interactions between divisions, the strong workshop tradition, including in-house engineers hired with the particular brief to develop new technologies, and the role of the numerous postdoctoral researchers in the life of the institution. The laboratory, we have seen, provided a unique mix of research traditions. While the distinctions between the different divisions remained in place, interaction was encouraged. Postdoctoral researchers profited most from these opportunities and played an important role in exporting local research traditions. Much of the impact of the Cambridge laboratory lay in a string of new technologies which attracted researchers to Cambridge and found its way into other laboratories around the world. Besides numerous innovations in crystallographic instrumentation, only some of which I have mentioned in this chapter, and the nematode worm, in particular Sanger's nucleic acid sequencing techniques, which provided the basic tools for today's large-scale sequencing projects, would warrant more detailed investigation. The monoclonal antibody technology, which also originated at the LMB, became the subject of heated debates regarding the patenting and industrial exploitation of scientific innovations. I will deal with this case and the reflections on the Cambridge laboratory in the final chapter. To appreciate that discussion, however, we need to engage with another issue first. Cambridge molecular biologists used not only scientific networks but also the science policy arena to establish their science. The next chapter will investigate how, in the mid-1960s, molecular biology became an item on the governmental agenda in Britain.

10

...

On the governmental agenda

Cambridge molecular biologists were not only successful in attracting a large number of research fellows and exporting their research culture. They were also instrumental in building a political argument that, in the 1960s, put the promotion of molecular biology on the governmental agenda. This happened at the national as well as at the international level with the creation of the European Molecular Biology Organisation (EMBO) and the plan for a European laboratory, which saw Cambridge molecular biologists pivotally involved. Links between scientists and between scientists and government established during World War II were instrumental for these moves. This chapter, therefore, carries on themes introduced in Part I. The political support enlisted by molecular biologists in the 1960s became crucial for the later developments of the field.

However, this chapter is not only about the making of molecular biology, but also about the making of science policy in Britain and its uses by scientists for their disciplinary projects. Science policy, we are reminded, is not a fixed entity but an issue which organises and stimulates reflection around the development of science and its place in society (Jacq 1995). It played a very specific role in the 1940s to 1960s when the place of science and technology in reconstruction and the 'modernisation' of Britain was hotly debated (Vig 1968). Changes in scientific practices, like the push for more and costlier machines, the development of interdisciplinary and international collaborations (equally part of the politics of détente) and growing contacts with industry, also required and forged new policies. Interpreting the mood of the time, the Labour Party fought and won the 1964 election putting the scientific and technological development of Britain at the centre of its political platform. In this it drew on and revived earlier discussions among radical and left-wing scientists regarding the social relations of science and the link between planning, science and socialism (Horner 1993).[1] A soaring science budget and an increasing pressure on

[1] Though these ideas owed most to Bernal, he himself was excluded from much of the discussions regarding the 1964 platform for which Blackett's more revisionist line, which

government to legitimise spending produced new questions. The continuous reorganisation of the administrative and policy-making apparatus of both military and civilian science in the British government in these years, evident for instance in the frequent change of name, terms and membership of the consulting government body on science policy – from Advisory Council on Scientific Policy which covered the whole of civil science and was responsible to the government; through Council for Scientific Policy, responsible only for academic, not industrial research, in the new Department of Science and Education; to Advisory Board for the Research Councils – was not just a trivial expediency of successive governments, but indicates the ongoing debates on the very subjects of science and science policy.[2]

Molecular biologists, I will show, used science policy as a tool for their disciplinary projects, thereby shaping policies as well as the direction their science was taking.[3] From the beginning, Cambridge molecular biologists linked their demands for a new laboratory to current preoccupations with the scientific 'brain drain' to America. While the new laboratory was quickly gaining an international reputation, its integration into the university, and especially the relations to the biological departments, remained problematic. Biochemists in particular complained that the laboratory sucked up their best researchers without contributing to teaching in the field. Molecular biologists turned this apparent weakness into a strength, calling for policy interventions which would make space for interdisciplinary research in universities and proposing the LMB as the model according to which research and teaching in the biological sciences were to be reformed. The creation of a European Molecular Biology Laboratory (EMBL) on the model of CERN, the European Laboratory for nuclear physics, was a further step in the establishment of the discipline based on intergovernmental agreement. After nuclear physicists (and often emulating their example), molecular biologists were particularly successful

---

opposed nationalisation on a big scale, became more influential (Horner 1993). For a critical evaluation of Labour's rhetoric of the 'White Heat' of the scientific revolution as well as a critique of the standard historiography of British R&D and related policies in the 1960s see Edgerton (1996c).

[2] On the reorganisations of the administrative apparatus concerning military R&D in this period see Edgerton (1991; 1992; 1996c). Throughout the period under examination and despite ongoing discussions, especially by the Wilson governments, regarding the shift of resources from military to civilian research, the budget for military R&D by far exceeded the budget for civilian research. Another issue of debate was the relation of (fundamental) science and technology. Under Wilson the advisory apparatus for technology was separated from the one for science and a new Ministry of Technology, responsible for industrial development, instituted (Edgerton 1996c).

[3] This was not unique to developments of the discipline in Britain. On parallels to the French history see below.

10.1 John Kendrew in a contemporary drawing by W. Lawrence Bragg. Courtesy of the Medical Research Council Laboratory of Molecular Biology, Cambridge.

in using the science policy arena for their disciplinary ends. In this they also built on the legacies of postwar biophysics.

In the 1960s Kendrew was the key figure in the debates on molecular biology in Britain (Fig. 10.1). Especially after his Nobel Prize, he became increasingly involved in science policy and administration. Here I will focus, first, on his involvement with the Council for Scientific Policy (CSP) set up by the new Labour government and, second, on his leading role in the negotiations for a central European laboratory (EMBL). Kendrew was a member of the CSP from its inception in 1964 to its demise in 1972; he served as deputy chairman from 1969 to 1972 as well as member and later chairman of its Standing Committee on International Scientific Relations

(1970–2).[4] It has been argued that the CSP, and its predecessor the Advisory Council on Scientific Policy, were rather impotent bodies (Wilkie 1991, 51–2). Kendrew, however, successfully used his membership of the Council to put molecular biology on the political map. He convinced his colleagues at the CSP that something was 'wrong' with molecular biology and that it needed support, on the national as well as the European level. He himself engaged in separate campaigns on both fronts, inside and outside the CSP. As a member of the CSP he was also involved in the preparation of the *Dainton Report* on the organisation and support of scientific research. This report, together with the *Rothschild Report* which recommended the application of the customer/contractor principle, was published in a Green Paper in 1971. It laid down a new framework for the organisation of government (basic and applied civil) R&D which has dominated British science policy ever since.[5]

Kendrew's role as a 'political broker' for molecular biology and, more generally, his move into science policy were not always appreciated. Colleagues saw him as 'deserting' the laboratory bench for politics[6] and, more than once, he was reprimanded for his involvement in policy debates regarding the very government body (the MRC) which funded his research. Kendrew consistently reacted to the accusations of conflict of interests by defending his right to speak up on policy issues. As a complementary strategy, in his mind he neatly separated his 'scientific' and his 'political' activities, his work 'at the bench' and in committees. To such compartmentalisation he was well accustomed from his secret military work.[7] Here I will attempt to bring science and politics back together again

---

[4] On the demise of scientific advisory councils in the early 1970s and their subsequent reinstatement, although not dealing specifically with Britain, see Brickman and Rip (1979).

[5] See *A Framework for Government Research and Development* (1971) and the following White Paper, *Framework for Government Research and Development* (1972), based on the recommendations of the two reports and the ensuing discussions. On the Rothschild and Dainton reports see Gummett (1980, esp. 195–206) and Wilkie (1991, 73–95).

[6] An exception is represented by Watson, himself an active 'committee man' and influential 'lobbyist' in Washington, who, in his memorial address, defended Kendrew's move away from the laboratory bench, suggesting that he rightly saw that there were 'more important things to do – to convince governments to do the right things'; quoted from notes taken during the Memorial Meeting in Honour of Sir John Kendrew (1917–1997) in Cambridge, 5 November 1997.

[7] This mental separation initially led Kendrew to withhold permission for me to consult his papers relating to his committee work on the grounds that this material was 'not relevant' for my project on the making of molecular biology. He later revised his position. Judging from the dedication he put into committee work, it is difficult to believe that he was not convinced it had an impact on the development of the field. On the compartmentalisation of scientific knowledge and the conditions under which it is produced as a powerful psychic mechanism (though with important social consequences) among nuclear physicists in Cold War America see Forman (1996). Even before entering

by studying how the political work governed the very direction molecular biology was taking and vice versa, how (local) developments of molecular biology impinged on science policy issues. Molecular biology was not only constructed at the bench, but also expanded into a political space. Rather than being imposed on science, science policy became a tool in the hands of molecular biologists to make the 'scientific' 'political'.[8]

Brain drain

In their negotiations for the first laboratory of molecular biology to be built at Cambridge, Perutz and his colleagues combined scientific with political arguments. 'If we fail to go ahead', Perutz urged the Secretary of the Medical Research Council, 'it will be very difficult to keep our staff together in the face of tempting offers from abroad.' However, if the plan went ahead, Perutz confidently asserted, the new laboratory would act as 'focus and model' for other developments in the field, in Britain and abroad.[9] These arguments spoke to rising concerns, fuelled by scientists themselves, about the 'loss' of Britain's brightest scientific minds to America, a phenomenon soon dubbed the 'brain drain' and used to nourish debates on Britain's (economic) decline.[10] As one observer remarked,

the movement of mature scientists...has existed throughout the period of the rapid development of science as a university subject and it has not in the main aroused any uneasiness. Countries were even proud when their citizens were called to other countries to fill posts of honour and influence...This was before the time when scientific research, technology and economic life became so intertwined. (Reports and documents 1963, 342)

The post-Sputnik expansion of the US research budget raised the alarm about a further loss of British scientists (Reports and documents 1963, 352). The threat was felt throughout Europe, but the impact was thought to be particularly strong in Britain where the common language made it easier for scientists to contemplate a move. In the early 1960s the Royal Society appointed a committee to examine the matter. The figures published in

the science policy arena, Kendrew, in various functions, advised the government in defence matters (see below). Despite his ability to compartmentalise, the need to keep his insights into defence matters strictly confidential did reflect on his social relations in the laboratory and, in the eyes of some of his colleagues, added to his secludedness.

[8] For a similar concept of the 'political' with respect to science policy see Gottweis (1998a).

[9] M. Perutz to H. Himsworth, 15 October 1958, file Himsworth, LMB Archives. See also F. H. C. Crick and M. F. Perutz, 'The case for a laboratory of molecular biology', Council Paper MRC.58/307, MRC Archives.

[10] For a critical historical perspective on the thesis of Britain's 'decline' and its dependence on investments in science and technology see Edgerton (1996b).

the report seemed to confirm the worrying phenomenon. The issue was debated in Parliament and caused widespread discussion in the press (*Emigration* 1963; Reports and documents 1963).[11] Reacting to the Royal Society report, Lord Hailsham, first Minister of Science in the Conservative government, in a speech to the House of Lords, accused the Americans, suggesting that they 'parasitically live on the brains of other nations' and 'buy British brains' in order to supply the needs their failing high school system could not cover.[12] Outraged scientists attacked this response as misleading and unhelpful and called for more concrete measures that would help create more attractive research conditions in Britain itself (Reports and documents 1963, 373–80). With an article in *The Times*, which gave ample space to the debate, Lawrence Bragg intervened in the discussion, for the first time drawing attention to the emblematic case of molecular biology at Cambridge. According to Bragg, the 'success story' of this venture – four Nobel Prizes had recently been awarded to the laboratory – illustrated several points 'bearing on the question of the emigration of our scientists, which is arousing so much interest'. In the first place, it was clear that 'on more than one occasion the continuance of the research was only made possible by generous financial support from America'. Thus, when talking about the 'one-way flow of scientists across the Atlantic' one also had to acknowledge the 'equally one-way flow of money in the opposite direction which has done so much to enable British scientists to do research in Great Britain'. Second, Bragg argued, the story showed that the existing machinery of universities and research councils could 'keep our best men in this country', if it was used 'imaginatively and quickly' to create jobs for promising scientists and support new lines of research, even if final success was not guaranteed.[13]

Independently of the Royal Society Committee set up to work on the question, Kendrew, himself a recently elected Fellow of the Royal Society and deputy chairman of the new Laboratory of Molecular Biology, started his own investigations regarding 'the export of British scientists' to America. His concerns were tightly connected to the future of his own research field. To Zuckerman, friend and colleague in the Ministry of

---

[11] This was but the first report on the issue. A report issued a few years later by the government Committee on Manpower Resources for Science and Technology depicted an even bleaker picture of the brain drain's damaging effects on Britain's economy (Committee on Manpower Resources 1967).

[12] Lord Hailsham, 'The emigration of scientists from the United Kingdom', *Parliamentary Debates (Hansard). House of Lords. Official Report*, vol. 247, no. 46, Wednesday, 27 February 1963 (London: Her Majesty's Stationery Office), columns 92–5 and 181–2; reprinted in 'Reports and documents' (1963, 363–4). On Hailsham see Vig (1968, 67–73).

[13] W. L. Bragg, 'The emigration of scientists', *The Times*, 5 March 1963; reprinted in 'Reports and documents' (1963, 372–3).

Defence (see below), deputy chairman of the Advisory Council on Scientific Policy and chairman of its Committee on Scientific Manpower, he confided:

In my own field there are almost no openings for good young PhDs in England, owing to the reactionary policy of all British universities towards Biology, to shortage of money and to the rigidity of the traditional departmental framework; in America there is a tremendous boom and no self-respecting university can now face the world without its Molecular Biology Department – and the (literally) dozens of such departments now springing up will absorb all the above mentioned British PhDs unless something drastic is done, and soon. It seems very sad that a new and important field largely started over here should now be flourishing on so large a scale in America.[14]

Kendrew collected figures and comments through English colleagues in America who distributed questionnaires among compatriots at selected American universities. His informants showed themselves concerned about the 'defection' of British scientists. Their results indicated that over 50 per cent of the British scientists in America did not plan to go back to Britain.[15]

Besides talking to Zuckerman and passing on his material to Gordon Sutherland, chairman of the Royal Society Committee whose report on the matter was about to be published, Kendrew used his own networks and acquaintances as well as his new prominence as freshly nominated Nobel laureate to circulate his results. 'I took the opportunity yesterday', he wrote to one of his informants, 'to spend an hour talking to the newly elected Labour M.P. about these problems and the other day I had a go at a high official of the Treasury.'[16] On his friend's return from America, Kendrew sent the *Evening Standard* reporter to interview him. However, his plan to appear in a BBC programme on the issue prepared by Gordon Rattray Taylor was thwarted by his paymaster, the Secretary of the MRC, who disapproved. Kendrew complied, but did not accept the principle. He insisted:

I simply cannot see why a Council employee should not speak publicly on Government policy towards universities or the financing and organization of research generally... And it seems to me that letters to newspapers, articles and

[14] J. Kendrew to S. Zuckerman, 10 August 1962, Kendrew Papers, MS Eng. c.2607, R.18, Bodl. Lib.

[15] W. Gratzer (Harvard University) to J. Kendrew, 19 November 1962, Kendrew Papers, MS. Eng. c.2607, R. 18, Bodl. Lib.

[16] J. Kendrew to W. Gratzer, 26 November 1962, Kendrew Papers, MS Eng. c.2607, R. 18, Bodl. Lib. The issues taken up and the political channels used by left-wing scientists like Kendrew in the 1960s certainly marked a striking difference from those chosen for political action by socialist scientists in the 1930s, despite many real and professed links to this earlier movement (see Werskey 1988).

television programmes all lie within the same definitions. You may argue that they are not the most effective ways of getting something done, and that these matters are better arranged, in the matter of the Civil Service, behind closed doors of Ministries or the Athenaeum; and this may often be true, though if one feels strongly about some problems (as I do about the emigration of scientists and the support accorded to research in this country, to the point of feeling that the present situation is little short of disastrous) one may conclude that all methods should be tried, including the mobilization of public opinion by way of the Press and B.B.C.[17]

This was but the first time Kendrew had to defend his involvement with science policy matters in front of the MRC.

A political role

The occasion to pursue these matters in a more official capacity came with Kendrew's appointment to the CSP in 1964. The body was set up by the new Labour government, but its constitution and terms of reference closely followed the recommendations of the *Trend Report* on the organisation of civil science commissioned by the preceding government ([Trend Report] 1963). The purpose of the new council was to advise the Secretary of State for Education and Science on matters of science policy. It was composed of fourteen members drawn from the universities and industry, most of whom had some previous experience of government advisory service. The first chairman was Harrie Massey, Professor of Physics at University College London and President of the Council of the European Space Organisation. During World War II he had been among the British physicists joining the Manhattan Project. In 1970 he was succeeded by Frederick Dainton, Professor of Physical Chemistry and Vice-Chancellor of the University of Nottingham and a member of the Council since its beginning. Also among the first members were Patrick Blackett, Lord Rothschild and Michael Swann, with whom Kendrew was well acquainted. With Blackett, who had been centrally involved in formulating the Labour platform, Kendrew had worked in Operational Research during the war; Rothschild and Swann, the two other representatives of the biological sciences, he had met at the Hardy Club meetings in Cambridge (see pp. 91–2). Meanwhile Rothschild had left a brilliant research career to join Shell as their world coordinator of research. Like Kendrew (see below) he had advised the government on military matters. Swann had moved to Edinburgh, successfully building

---

[17] J. Kendrew to H. Himsworth, 20 March 1963, Kendrew Papers, MS Eng. c.2607, R. 18, Bodl. Lib.

10.2 Solly Zuckerman, Chief Scientific Adviser in the Ministry of Defence (second from right), on the Polaris missiles purchasing mission in America (January 1963). John Kendrew was also involved in advising the government on the issue. From S. Zuckerman, *Monkeys, Men and Missiles*. Copyright 1988 Lord Zuckerman Estate.

up the Zoology Department there. He had just been elected Dean of the Faculty of Science.[18]

Kendrew himself was no novice in Whitehall. From the last months of the war until his decision to return to academic life, he had spent several months as scientific adviser at the Air Ministry (see above, p. 120). After a long spell, in 1961 he had returned to Whitehall as part-time Scientific Adviser to the Ministry of Defence, sharing his time between the laboratory and his government office. The invitation to this post had come from Zuckerman, then Chief Scientific Adviser to the Minister, who had learned to appreciate Kendrew's skills in their common involvement in operational research during the war (Fig. 10.2).[19] Kendrew had

[18] On the biographical and social profile of scientists who gained positions of influence as scientific advisers see Gummett (1980, 93–108).

[19] Though several of Kendrew's colleagues suggested that he never severed his ties with the military and continued to serve as adviser on defence matters throughout the 1950s, there is no evidence in his papers to confirm this activity. Evidence may, of course, have been destroyed. For a glimpse of Kendrew's activities in defence matters in the early 1960s see Fenton (1996). The newspaper article reports on the activities of JIGSAW, the Joint Inter-Service Group for the Study of All-Out War, at one time chaired by Kendrew. A picture caption identifies him as 'planner'. Among other things, the committee studied

resigned the appointment after only two years, but continued to serve as an independent member on numerous panels and committees such as the Defence Research Committee.[20] Kendrew himself suggested that it might have been this latter experience that gained him the invitation to serve on the CSP.[21] More than the Whitehall experience itself, personal networks will have counted. With the reshuffle of the organisational machinery for science and scientific advice by the new government, Zuckerman, Kendrew's chief at the Ministry of Defence, had become Chief Scientific Adviser to the Cabinet. As the first person ever formally to fill this key position, he wielded considerable influence in the Whitehall apparatus of scientific advice and may well have suggested Kendrew's name.[22]

But Kendrew also had other credentials. Though not a proper member, he was in contact with the group of scientists around the London businessman Marcus Brumwell (also known as the Gaitskell or VIP group) who had supported the Labour campaign and had played a significant role in formulating its scientific platform.[23] Kendrew had commented on the draft policy statement and had participated in the subsequent debates on the scientific platform. In response to a circular letter by Brumwell to group members (sent also to Kendrew though 'not really a member') reviewing the situation briefly before the election and expressing pleasure on 'the striking conversion to a scientific attitude', Kendrew explained his position thus:

I am now no longer a part-time employee of the Ministry of Defence . . . so I feel a little more free than I did when we talked before, and I have been glad to have one or two recent opportunities of discussing future Labour Party policy about

the potential effect of atomic bombing on British cities. The brief according to Kendrew was 'to make the military realise that these weapons could be a deterrent and nothing more' (*ibid.*).

[20] The official reason for his resignation was that he wished to devote more time to his research. To Zuckerman's repeated attempts to win Kendrew's full-time support in the Ministry, Kendrew rebutted: 'I find I still do not want to come right in . . . The alternative is to get right out, and I honestly believe this is the right course for me'; J. Kendrew to S. Zuckerman, 11 September 1963, Kendrew Papers, MS Eng. c.2555, J.99, Bodl. Lib. Though hardly the reason for his resignation, Kendrew was also involved in a long-drawn-out dispute regarding the emoluments for his government service. Because he received a salary as a full-time employee of a government body (the MRC), the Treasury judged that he was not entitled to further payments. Kendrew felt that this put him in an inequitable position in respect to university colleagues involved in similar work and tried to argue his case, but without success. A colleague in the Ministry judged this 'a deplorable piece of jiggery-pokery' (Alton 1989, 294).

[21] Interview with J. Kendrew, Cambridge, 6 May 1993.

[22] On Zuckerman's career as high-level scientific adviser, which continued unabated from the war through consecutive Labour and Conservative governments to the 1970s see Gummett (1980, esp. 102–4) and Zuckerman (1988).

[23] Brumwell's efforts to bring scientists and Labour Party members in closer contact with the aim to incorporate science more prominently in Labour Party policy reached back to the mid-1950s when Hugh Gaitskell was Labour leader (Horner 1993).

science. The other evening, for example, we had a most interesting talk about the future organisation of the U.A.C. – a group of Cambridge dons convened by Noel Annan together with Dick Crossman, Tim [*sic*] Dalyell and Bowden. I certainly look forward to having the Election behind us, so that hopefully we can get on with putting some of these ideas into practice.[24]

This certainly does indicate his interest in the Labour position on science and his willingness to collaborate on concrete issues.

At the time of his appointment to the CSP, Kendrew was elected first to the Council of EMBO and later as its Secretary General. Kendrew made these different functions work together by using the international plans to put molecular biology on the national agenda and by using his membership in the CSP to press for Britain's participation in the European project.

Biology census

Kendrew's first initiative at the CSP concerned the status of the biological sciences more generally, but it was to give him important ammunition for his later campaigns for molecular biology.

Together with his two biological colleagues on the Council, Rothschild and Swann, he embarked on two pilot studies surveying research in biology in the Cambridge and Edinburgh areas. This and other such surveys were to provide the background information necessary to make informed decision on where money in the biological sciences should be spent.

The preoccupation with biology was not new. The status, reorganisation and funding of the biological sciences had been a subject of discussion among biologists and science administrators for several years. A government report on scientific manpower which suggested that there was an overproduction of biologists had provoked biologists to point to the potentials of new developments in the biological sciences for medicine, agriculture and the environment, and to lament the neglect of biological research (much of which did not fall under the responsibility of any research council) with respect to physics. A report by the Royal Society *ad hoc* Biological Research Committee, commissioned by the Advisory Council on Scientific Policy (the predessor of the CSP) and published as part of its 1961–2 Annual Report, called for a thorough reorganisation of

---

[24] M. Brumwell to J. Kendrew, 6 August 1964 [circular letter with handwritten note to Kendrew], and J. Kendrew to M. Brumwell, 10 August 1964; Kendrew Papers, MS Eng. c.2607, R.19, Bodl. Lib. Shortly before, while still employed in the Ministry, Kendrew, against his conviction, had felt he needed to conform to the 'anonymous tradition of the Civil Service' and to decline that his name appear on the Cambridge Labour candidate election address; J. Kendrew to R. Davies, 6 February 1964, Kendrew Papers, MS Eng. c.2607, R. 19, Bodl. Lib.

the biological departments away from the traditional divisions and along multidisciplinary lines. More specifically, the report recommended making space for new subjects like biophysics, biochemistry, genetics and animal behaviour and strengthening chemical, physical and mathematical approaches throughout all the disciplines. The new universities were seen to offer the best opportunity to spearhead such reorganisation, as demonstrated by the new Schools of Biology at the universities of Sussex and Leicester. The report also suggested the creation of a new research council responsible for the biological sciences or the expansion of the Biology Subcommittee of the Department of Scientific Research and Development.[25]

If the focus on the biological sciences was not new, the approach chosen by Kendrew, Swann and Rothschild was. The one important difference between the CSP and its predecessor, the Advisory Council on Scientific Policy, and the place where the new Council could make its mark, Swann passionately argued, was its task to recommend how the budget for civil science was to be distributed. This was a particularly sensitive job at a time when the steady 13 per cent annual increase of the science budget, to which scientists had become accustomed, was expected to drop, spending on science having come to represent over 3 per cent of the total budget. To be able to make informed choices where money should go, Swann concluded, a 'synoptic view of the whole of science' was necessary. 'Curiously enough', he added, 'I don't believe that any such survey has ever been made.'[26]

Forms entitled 'Biology Census' and marked 'important, future funding will depend on it' were sent to university and industrial biological research establishments, requiring from every scientific staff member detailed information on current and future research projects, including necessary apparatus and a timetable for the research. Much discussion among the organisers of the survey went into the determination of the categories and the organisation of the statistics. To facilitate future analysis the use of Hollerith cards was suggested.[27]

---

[25] *Annual Report of the Advisory Council on Scientific Policy 1961–1962* (London: Her Majesty's Stationery Office, 1963) [Cmnd. 1920], pp. 6–8 and Appendix B, pp. 17–30. Kendrew and also Krebs, who later strongly opposed Kendrew's campaign to establish molecular biology in the universities (see below), were members of the Royal Society committee reporting on biology.

[26] M. M. Swann, 'The pattern of research in Britain: the need for a synoptic view' [February 1965], Kendrew Papers, MS Eng. c.2538, J.29, Bodl. Lib.

[27] To deal with the problem of research projects which were relevant to different areas, the procedure adopted by the Ministry of Aviation for the 'Pink Book' was used. It meant that the head of department was asked to balance the case. On Kendrew's interest in technologies for indexing and sorting information, which he carried on from operational research to his work in protein crystallography, see above (chapter 4). Already in 1945, Kendrew had envisaged the use of information technologies, especially punched card systems, as the basis for an effective use and the organisation of scientific resources.

Molecular biology appeared as a separate category, though counting just forty-five researchers, concentrated at the LMB, against, for instance, well over 400 biochemists.[28] The initiative, later stepped up to a full-scale survey, gave a foretaste of a new approach to science policy, which the same Council was to embrace more fully. As in other fields too, criteria of effectiveness and accountability, based on continuing assessment, would gain increasing importance. For Kendrew the exercise became a stepping stone for later studies and campaigns regarding the poor state of molecular biology in Britain.

Biology and Cold War politics

A Unesco document on the need for international scientific cooperation in biology presented the first occasion to put molecular biology on the Council's agenda. Founded in 1946, Unesco (the United Nations Educational, Scientific and Cultural Organisation) promoted scientific cooperation as part of its mission to bridge political divisions and cultivate peace in an increasingly polarised postwar world. Precisely because science was presented as apolitical and universal, it could fulfil its political role. The Unesco document discussed at the CSP stressed the necessity of a major funding effort for biochemistry and biophysics in view of the benefits such research would bring to humanity by tackling problems such as cancer, the effects of radiation, and nutrition (in that order). Among the projects listed in the document was the EMBO project for a European laboratory of molecular biology.[29] The Department of Education and Science, after taking advice from the MRC, responded to the report, taking the decided view that 'no good scientific case existed at present for

---

[28] 'The biological review. The present state and suggestion for future work', CSP(BR)(65)1, Kendrew Papers, MS Eng. c.2538, J.32, Bodl. Lib.

[29] 'Problems of international scientific co-operation. International co-operation in biology. The problem of choice. Summary report. Meeting 19 May 1965' (CMS-CI/65/35), Kendrew Papers, MS Eng. c.2537, J.23, Bodl. Lib. Other projects listed included an International Life Science Institute (eventually merged with the EMBO plan), a European Radiobiology Centre and De Gaulle's proposal for an International Cancer Research Institute to which advanced countries were to contribute through a 1 per cent cut in military budgets. In 1965 an international biology programme entitled 'The biological basis of productivity and human welfare' was also launched by the International Council of Scientific Unions. The sheer number of international biological programmes promoted in the 1950s and 1960s indicates the importance attributed to biological research at the height of Cold War politics. For another ambitious if unsuccessful East–West cooperation plan in the biomedical field launched by an American senator to promote world peace see above, pp. 254–7. For more background to the EMBO project see below.

the proposal for an international research institute in the basic biological sciences'.[30]

Kendrew prepared a point-by-point rebuttal to the Ministry's response, which he presented, together with a revised proposal for a European Molecular Biology Laboratory, to an *ad hoc* Subcommittee on Molecular Biology in the CSP. The revised proposal owed much to Kendrew's own input and, according to general agreement, argued the case much more cogently than in the Unesco document.[31] The proposal was not yet ratified by the EMBO Council, but time pressed. Just in time before an international meeting of science ministers, the subcommittee presented a summary report on the matter, urging the Ministry to take action.

The inititiative did not have the intended effect. In what Kendrew dubbed a 'depressing document', the Ministry reiterated its position that there was no case for an international laboratory. The Secretary also found his views confirmed by delegates from other countries.[32] But Kendrew's efforts were not altogether wasted. The Department of Education and Science did appreciate that the problem of 'high-level training' in molecular biology in Europe and the trend to take up opportunities in the United States instead – the most important rationale for the EMBO plan – was 'a real one', even though it thought that this situation could best be redressed by building up positions for molecular biologists in universities. Kendrew would soon take up this point. Meanwhile he had impressed on Massey and other colleagues in the CSP that something needed to be done for molecular biology.

'Modern biology'

In a first CSP Report on Science Policy molecular biology, and especially the LMB, were cited as examples to draw attention to the problem of utilising the scientific resources in the country. The report stated:

The MRC Unit for Molecular Biology has made outstanding contributions to this field and is an acknowledged example in this country of a centre with a

---

[30] 'Proposals for international scientific co-operation in the life sciences', CSP (65), 24, para 13 (ii), quoted in J. F. Kendrew, 'Notes on the European Molecular Biology Organization proposal' (4 May 1965), Kendrew Papers, MS Eng. c.2537, J.23, Bodl. Lib. On the consultations leading to this decision see also M. F. Perutz, 'To members of EMBO Council', 22 April 1965, Kendrew Papers, MS Eng. c.2419, F.23, Bodl. Lib.

[31] P. Blackett to J. Kendrew, 17 May 1965, Kendrew Papers, MS Eng. c.2537, J.23, Bodl. Lib.

[32] Council for Scientific Policy. Interim Committee's discussion on international cooperation in biology. Note by Sir Frank Turnbull on the discussion on EMBO (CSP (65) 30, and J. Kendrew to J. Wyman, 27 May 1965, Kendrew Papers, MS Eng. c.2537, J.23, Bodl. Lib.

world-wide reputation where researchers may be found from any country in the world. But it is apparent that the development of teaching in this subject in the UK is less active, than, for example, in the United States, and more outstanding workers have left the Unit to teach abroad than in this country.[33]

The report is also interesting for the candid acknowledgement that the issues of science policies were new and therefore not yet well defined:

It is because we are so close – in a historical sense – to the origins of the problems of science policy that they are so ill-defined. In 1939–40 Government spent £4 million on civil scientific research and development. In the current year (1965–66) the expenditure is over £220 million (or 3.9 per cent of total Government Civil Expenditure) and it is about double what it was six years ago ... While it has always been important to satisfy public opinion that science is worthwhile in each of its aspects – cultural, social, and economic – the scale of the resources devoted to science now makes this absolutely necessary if science is to continue to have the resources it needs. Ironically this problem may be the more pressing because the nation seems at last genuinely aware that our economic future rests upon advanced technology which itself depends on science for its fundamental concepts.[34]

The objective of science policy as described in the report was to arrive 'at a synthesis of views on how scientific expenditure grows and the criteria by which that growth can be measured and justified'. This knowledge could then be translated into measures which would guarantee a coordinated growth of science in place of 'the present pattern of uncoordinated development'.[35] The same principle we have already seen applied in the biology census. It marked a shift from the postwar preoccupation with 'growth' of scientific manpower and funding to issues of assessment and the efficiency of the system. Molecular biology, a new field which pushed for resources, was an ideal test ground for new policy interventions and a field in which the Council could make an impact.

On Kendrew's initiative and under his chairmanship, a working party was set up to investigate further 'the problem' of molecular biology. Swann strongly supported Kendrew's initiative, both for 'the importance to biology' and because he saw it as an 'interesting test case on what can be done to stimulate particular branches of science'. Further support came from Wilkins of King's College London, who himself was keen to conduct

---

[33] Council for Scientific Policy, *Report on Science Policy* (London: Her Majesty's Stationery Office, 1966) (Cmnd. 3007). The reference to the MRC unit instead of the LMB (the successor institution since 1962) is confusing and must be a mistake.

[34] *Ibid.*, p. 1.

[35] *Ibid.*, pp. 15–16.

a survey of employment possibilities in molecular biology, largely because of the problem he encountered in placing his own students. Hans Kornberg, Professor of Biochemistry at Leicester, was elected to join the group to represent 'the point of view of someone battling to get the subject established [in the university]'. Partly at the suggestion of Himsworth, the Secretary of the MRC who, on consultation, found the membership 'unduly weighted towards one end', another four members, all directors of MRC units and Fellows of the Royal Society, joined the group.[36] The official brief of the working party was to inquire into 'the present position of and future plans for, teaching, recruitment and research in molecular biology in the United Kingdom'. It met fourteen times over a period of two years and issued a report, known as the 'Kendrew Report'.[37]

The working party heard evidence from all major funding bodies in biomedicine (though these were at times reluctant to disclose information) and undertook a literary survey of fifteen journals considered to contain all relevant research reports. The basis for all this work, however, and, as it turned out, the most politically loaded and controversial aspect, was the very definition of the term molecular biology and hence the interpretation of the terms of reference of the working group. This became apparent even before the working party started work.

'I suspect that it would be necessary to give some attention to other disciplines allied to molecular biology – in fact I think it will be quite difficult not to get involved in a full-scale study of biochemistry and genetics as well', Kendrew confided to Massey when suggesting setting up the inquiry. 'Nevertheless, I think it would be advisable to try to restrict the enquiry to molecular biology as far as possible, bearing in mind that a good deal of molecular biology is done in departments with different names (genetics, virology, biochemistry etc.).' Massey and Kendrew agreed that 'one of the really valuable things' the group could do, was 'the preparation for general

---

[36] M. Swann to J. Kendrew, 17 December 1965; J. Kendrew to M. Swann, 26 January 1966 and 15 February 1966; J. Kendrew to H. Massey, 9 February 1966; H. Himsworth to H. Massey, 17 February 1966; Kendrew Papers, MS Eng. c.2539, J.35, Bodl. Lib. The other members were: J. L. Gowans, Henry Dale Research Professor of the Royal Society and Director of the MRC Cellular Immunology Research Unit; W. Hayes, Director of the MRC Microbial Genetics Research Unit at Hammersmith Hospital, London; A. Neuberger, Professor of Chemical Pathology at St Mary's Hospital Medical School, London, and Director of the MRC Research Group in Enzymology; and M. G. P. Stoker, Director of the MRC Experimental Virus Research Unit at the University of Glasgow. Other names proposed or considered by Kendrew, but finally excluded, were M. Pollock, K. Burton, J. Randall, P. Walker, J. Pringle, and J. R. S. Fincham.

[37] See [Kendrew Report] (1968). Others who have dealt with the report are Yoxen (1978), Olby (1983; 1991) and Abir-Am (1992a), though mainly in relation to the authority dispute between molecular biologists and biochemists.

acceptance [of] a rather rigorous definition of the subject which would at least make it difficult for those not really contributing to the subject to ride on this particular chariot'.[38] In an earlier chapter I have already referred to names as strategic and political tools (chapter 7). Here we witness a further turn to the 'politics of names'. In Kendrew's working group, the name not only became an issue of science policy, but the political forum was also chosen to confer authority on a certain definition of the term. It is interesting to note that at exactly the same time Kendrew was involved in a dispute with American phage geneticist Gunther Stent, aimed at establishing the 'origins' and demarcating the field of molecular biology (see above, chapter 6).

In the event the group failed to agree on a definition, as conceded in the final report:

Our terms of reference direct our attention to *molecular* biology. At an early stage of our deliberations we came to the conclusion that this was a restriction which would be nearly impossible to define, and which, even if it could be defined, would demarcate a field of enquiry which does not correspond to any real subdivision of biology. The fact is that the fashionable term of 'molecular biology' is unfortunate, on several grounds. ([Kendrew Report] 1968, 2)

Instead the group adopted the phrase 'biology at the molecular level' (as distinct from 'biology at the level of the individual organism' or from 'biology of population') to describe the field of its inquiry ([Kendrew Report] 1968, 2). However, it would be wrong to interpret this choice as an opening to other biological fields concerned with the study of biological phenomena on the molecular level like biochemistry or cell biology. Rather, the move was the compromise chosen to enlarge the terms of reference of the inquiry. In the final analysis this move gave the working party the opportunity to pass judgement on the work of biochemists which, while representing the bulk of research work of 'biology at the molecular level', was considered of mediocre quality. This, in turn, raised the value of research in molecular biology pursued in a few key institutions (though not always under that name). Kendrew admitted to much of that when, responding to Bernal's strong objection, on historical grounds, against the use of the term 'biology at the molecular level', he wrote: 'Of course I agree but politics intervened and we used the term...to give ourselves an excuse to talk about biochemistry as well, feeling that actually this

---

[38] J. Kendrew to H. Massey, 9 February 1966, and H. Massey to H. Himsworth, 22 February 1966, Kendrew Papers, MS Eng. c.2539, J.35, Bodl. Lib. See also H. Himsworth to H. Massey, 17 February 1966, file D747/1, MRC Archives.

was in a much worse condition in this country than molecular biology properly.'[39]

The retention of the term 'molecular biology' in the title of the report as well as in some parts of the text (e.g. when the four model institutions were presented as 'Molecular Biology Laboratories' ([Kendrew Report] 1968, 19)) and the angry response of biochemists (see below) confirm the strategic choice of the term and its political implications. However, before turning to the biochemists' reactions to the report and to its reception more generally, it is useful to have a closer look at the findings and recommendations of the working group.

The working party decided on a literary survey instead of collecting information through a questionnaire as in the case of the Biology Census, exactly because of the problem of establishing a 'satisfactory definition' of molecular biology. '[W]e did not believe it possible', the report stated, 'to incorporate such a definition in any questionnaire to be sent to individual laboratories in a way precise enough for the answers to yield the statistical information we wanted' ([Kendrew Report] 1968, 3). The analysis of papers published in the period 1964–6 revealed that 1500 papers or about 10 per cent of the total British output of published papers in biology lay within the field of 'biology at the molecular level' as defined by the group. An objective standard against which this proportion should be assessed was not available, but the working party agreed that: 'This proportion is so small, and in our judgement the potentialities of the molecular approach to biology are so great, that we have no hesitation in stating our firm conviction that it should be substantially larger' ([Kendrew Report] 1968, 4). What made the situation worse in the eyes of the working party was that much of the work which fell under the category of 'biology at the molecular level' was 'at the best dull and at the worst trivial' (a phrase eagerly taken up by media reports) ([Kendrew Report] 1968, 4). This, the working party thought, was in marked contrast to the situation in the USA where 'as well as a good share of work of the highest class there is a great deal of the sound and competent, perhaps uninspired but at least

---

[39] J. Kendrew to A. Rimel [Bernal's assistant], 19 February 1969, Bernal Papers, J.109, CUL. In defending the term molecular biology Bernal honoured the ground-breaking work of Astbury, 'the founder of protein analysis', who first introduced the term for his own crystallographic studies. His decisive contribution concerned the introduction of 'three-dimensional thinking' in molecular structure, and this, according to Bernal, represented the decisive innovation in biology. The new term proposed by Kendrew 'obscured' this fact; J. D. Bernal, 'Comments on the White Paper: Molecular Biology', February 1969, Bernal Papers, J.109, CUL. Both documents are also in the Kendrew Papers, MS Eng. c.2539, J.41, Bodl. Lib.

relevant, work which is essential to sustained development' ([Kendrew Report] 1968, 4).

As important as the size was the distribution of the work. Here the report found that most of the work was scattered 'in research groups so small that we have serious doubts as to their efficiency in carrying out research in a field which is essentially multidisciplinary' ([Kendrew Report] 1968, 5). Of the estimated 1800 researchers in the molecular field, only around fifty to a hundred were considered of the 'first-class'. These were concentrated in a few key institutions where large groups of researchers fruitfully interacted and interdisciplinary work was possible. An appendix listed four unnamed centres where research at a high standard was pursued.

The reason for the depressed state of 'biology at the molecular level' in Britain, the report surmised, was not an overall shortage of posts in biological departments, but rather 'that so many posts and so much money go to the support of activities that, whether at the molecular level or at any other level, are by any reckoning quite out of date' ([Kendrew Report] 1968, 6). After touching on the problem of the 'drain' of able biologists to the USA – 'one might seriously ask whether any other factor except sentiment prevents a wholesale emigration of the better workers' – the report went on to list a number of wide-sweeping reforms regarding both teaching and research, in biology and in the university system more generally, which could reverse the situation. These included the revision of the elementary syllabus in biology to introduce 'modern molecular concepts' early on, and the introduction of postgraduate course work and of reorientation courses for senior researchers as well as the relaxation of traditional departmental boundaries to allow the formation of large interdisciplinary research groups. These recommendations were not very different from those made by the Royal Society Biology Research Committee a few years earlier. What was different, however, was that molecular biologists put themselves at the centre of reforms, which botanists, zoologists and biologists more generally had themselves called for. Instead of from within the 'classical' biological disciplines and comprising various levels of analysis and integration (including for instance such fields as animal behaviour, ecology and marine and freshwater biology), modernisation was to happen under the banner of 'biology at the molecular level'. This identification of molecular biology with biology *tout court* was new.

The working party expected that universities would be 'rather slow' in accepting the proposed changes. They therefore recommended the creation of a number of 'special centres of research and advanced teaching' (or 'focal centres') in 'fundamental biology' in which the proposed changes could be put into practice more quickly (the term 'centres of excellence' was studiously avoided in the report but was widely used in the discussion papers as well as in press reports later on). These centres were to be closely linked to existing universities, but teaching and research would not fall under any of the rigid disciplinary divisions dominating conventional university arrangements.

With these recommendations regarding the policy and administration of research the working group had strayed far beyond its original terms of reference. This, at least, was what the Secretary of the MRC, the organisation mainly involved in funding molecular biology, emphatically felt. In his eyes, the CSP had got itself into the awkward position 'whereby they had lent themselves to giving cover to a committee of Council employees to pass comments and recommendations on how the Council should do its business'. According to Himsworth's angry view, 'there would be a chaos if this kind of thing were allowed to pass'. The matter was so grave that he felt the Treasury should be informed. Regarding the specific recommendations made in the report, he felt that some of them would backfire by compromising the ability of the Research Councils to take independent action. This applied in particular to the suggestion made in the report of a closer collaboration between universities and the MRC in staff appointments. 'If this system had been in force in Cambridge twenty years ago', he pointed out acidly, 'molecular biology would never have existed in this country. It was only because we were able to stand outside the University and take an independent line that we were able to back it.'[40]

Following Himsworth's angry reaction, the more detailed specifications regarding the administration of the proposed 'focal centres' were relegated to an appendix. They were, however, published with the report. As Swann reported to Himsworth, Kendrew 'had shown a disposition to stand by his principles'.[41]

In this debate Himsworth vehemently defended the independence of his Council, which was directly responsible to government (this would

---

[40] H. Himsworth to J. Kendrew, 29 June 1967; H. Himsworth, 'Interview with Professor M. M. Swann' (notes of 25 July 1967); H. Himsworth to J. Gowans, 3 August 1967; H. Himsworth to J. Gowans, 31 January 1968; H. Himsworth to M. Swann, 31 January 1968; H. Himsworth to H. Massey, 5 February 1968; file D747/1, MRC Archives. Himsworth was not alone in regarding the LMB as the leading example of the report and as synonymous with molecular biology in Britain (see below).

[41] H. Himsworth, 'Interview with Professor M. M. Swann' (notes of 25 July 1967); file D747/1, MRC Archives. Himsworth repeatedly suggested taking the 'policy recommendation' out of the report and submitting it separately to a joint committee of the CSP, the University Grants Committee and the Research Councils which was currently discussing the relationship between research councils and universities.

change a few years later). Kendrew, the main mover behind the work of the committee, for his part, was determined to use the report and science policy as a tool to require action regarding the promotion of molecular biology. A mere assessment of current teaching and research in the field was not sufficient. In resonance with the current political rhetoric, molecular biology was presented as 'modern' biology, as well as 'good value for money' ([Kendrew Report] 1968, 2 and 9). To allow the 'modern approach to biology' to permeate the whole of biology, planning and state intervention were necessary. Rather than an issue of university politics, the establishment of molecular biology became an issue of science policy and science policy a tool scientists embraced to further their case. In this, French molecular biologists who had been instrumental in forging a five-year plan of concerted government action for their science may well have shown their British colleagues the way (see above, chapter 8). Indeed, the documents regarding the French *action concertée* in molecular biology were circulated as CSP papers.

In drawing up his recommendations, Kendrew quite clearly drew on the experience of the resistances met to setting up the Laboratory of Molecular Biology in the precinct of Cambridge University. But the anomalous position of the LMB *vis-à-vis* the university was now taken as the model according to which research and teaching in the universities were to be reformed. The LMB experience, albeit in a negative key, was also decisive where suggestions deviated from the model, as for instance in the call for a close integration of the focal centres into the fold of universities. Readers certainly associated the 'Kendrew Report' with the LMB. The list of four unnamed examples of 'some existing molecular biology laboratories' – here the original denomination slipped in – did but reinforce this connection. It led with certainty only to the LMB, the only institution which actually carried that name, allowing ample speculation regarding the other 'model institutions'. *Nature* guessed that '"existing centres" must mean the MRC laboratory at Cambridge and perhaps the National Institute of Medical Research', while biochemists harboured no doubt that the LMB represented the 'prototype' for the suggested 'focal centres' (Anonymous 1968b, 107; [Krebs Report] 1969, 24).[42]

---

[42] In response to a later inquiry by a social sciences researcher, the anonymous laboratories mentioned in the report were disclosed as: MRC Biophysics Research Unit, King's College London; MRC Experimental Virus Research Unit, Institute of Virology, London; MRC Laboratory of Molecular Biology, Cambridge; MRC Microbiological Genetics Research Unit, Hammersmith Hospital, London; S. S. Blume (Department of Education and Science) to L. Doughty (Manchester), October 1971, Kendrew Papers, MS Eng. c.2539, J.41, Bodl. Lib. All four MRC establishments mentioned had representatives in the working party.

The working group was worried whether its arguments in favour of giving more support to molecular biology would have the right impact, especially 'in the present difficult financial position'. Wilkins, who went through Kendrew's draft, felt that they had to 'face the fact that many important medical advances such as dealing with polio, partial dealing with cancer and with tissue transplantation have had little help from molecular biology'. Most progress had come from 'empirical approaches'. Falling back on a postwar consensus which was losing its grip, he wondered whether 'the main case of supporting molecular biology is that it is really fundamental science and that in *any* really fundamental field there is a great likelihood that important applications will follow eventually'. Wilkins also suggested trying the draft report on 'science journalists, scientists in other fields, politicians or civil servants'.[43] The final report drew on the fundamental character of research of 'biology at the molecular level' and its 'deep intellectual significance' as well as on the 'very important advances in many applied fields' with 'social and economic dividends of inestimable value', which would most likely flow from molecular research. The biomedical as well as the agricultural field were mentioned ([Kendrew Report] 1968, 1).

The CSP decided to publish the report of the working group, on the basis that 'much of what is said . . . about the distribution and variable quality of biology at the molecular level is equally applicable to other fields of science' ([Kendrew Report] 1968, ii). It recommended that the Working Group on the Support of Scientific Research in Universities, initiated by the CSP, together with the University Grants Committee, should pursue the issues involved and make suggestions for their implementation. The government paper received widespread attention in both the scientific and the political press, with *Nature*, the *New Scientist*, *The Economist*, and the *Guardian* and other daily papers reporting on it, often in their editorials. Discussion was mainly linked to the 'brain drain' or to more general issues regarding the support of British science. Even the *New Scientist*, although denouncing the report as 'arrogant', concluded: 'An investment in such centres, while probably enhancing the molecular biologists' sense of divinity even further, would be a sound outlay of scientific capital' (Anonymous 1968d; see also Anonymous 1968a; 1968b; 1968c; Tucker 1968). Response also came from individuals who found their own frustrating experiences in setting up interdisciplinary research groups in universities confirmed by the report (the case of Professor Baddiley at the University of Newcastle

---

[43] M. Wilkins to J. Kendrew, 24 November 1967 [italics in original], Kendrew Papers, MS Eng. c.2539, J.38, Bodl. Lib.

was widely cited). Concerted negative response, however, came from bio-chemists who contended their own rights on the territory of molecular biology. Their strong reaction was the most direct indication of the impact of the report.

Signs of protest gathered early. On reading of the setting up of Kendrew's working group in the 1966 CSP Report of Scientific Policy, James N. Davidson, Professor of Biochemistry at the University of Glasgow and himself the author of an authoritative textbook on nucleic acids, expressed surprise at the affirmation that teaching in molecular biology was not ac-tive enough in universities in Britain. What was molecular biology if not biochemistry, which was widely taught in British universities? In a special twist Davidson expressly excluded protein crystallography, Kendrew's own field of specialisation, from molecular biology or biochemistry, thus under-mining his correspondent's authority to speak for either.[44] On publication of the CSP report on molecular biology, Davidson's colleague at Glasgow, Robin M. S. Smellie, urged the Secretary of the Biochemical Society to take action against the report, which he described as inaccurate and misleading and seriously damaging for biochemistry (Olby 1991). As a result, the Soci-ety set up a subcommittee with Nobel Laureate Hans Krebs as chairman, and, less than a year later, published a counter-report ([Krebs Report] 1969). Without much ado, the subcommittee substituted the term 'biology at the molecular level' with 'biochemistry'. While recognising 'an urgent need for great expansion of biochemistry', the subcommittee saw 'no case for setting up separate departments of "molecular biology"' which was not seen as more than a 'subdivision of biochemistry', dealing with nucleic acids and proteins instead of with the whole range of biological molecules ([Krebs Report] 1969, 8). For the rest the report largely agreed with the Kendrew Report, except that it urged more funding through the Research Councils rather than setting up focal centres to ensure a more competitive research climate.[45]

---

[44] J. N. Davidson to J. Kendrew, 3 June 1966. See also J. Kendrew to J. N. Davidson, 4 July 1966; J. N. Davidson to J. Kendrew, 11 July 1966; Kendrew Papers, MS Eng. c.2539, J.37, Bodl. Lib. Kendrew circulated his correspondence with Davidson to the working group members. Kornberg, himself a biochemist, expressed his 'amusement' about Davidson's tirades; H. Kornberg to J. Kendrew, 15 September 1966, Kendrew Papers, MS Eng. c.2539, J.37, Bodl. Lib. Kornberg worked at the University of Leicester which, like other new universities, operated an integrated 'School of Biology', as recommended in the 1962 Royal Society Report discussed above and in many points taken up by the molecular biology working group. On a sequel to the argument between Davidson and Kendrew on the occasion of the Biochemical Society's five-hundredth meeting see Abir-Am (1992a, 226–32).

[45] The Society's Yearbook for 1969 reported: 'Wisely the subcommittee concentrated on suggesting steps for the implementation of the original document, rather than throwing up destructive criticism' (*The Biochemical Society Yearbook* 1970, 10).

The dispute between biochemists and molecular biologists in Britain stretched far beyond the two rival reports and has been the subject of detailed historical studies (Olby 1983, 1991; Abir-Am 1992a; de Chadarevian and Gaudillière 1996).[46] In the context of this chapter, it serves to highlight the particular use molecular biologists made of the science policy arena. While molecular biologists used science policy bodies sanctioned by government to promote their subject, biochemists relied on the more traditional channels of their professional society (channels which were not available to molecular biologists) to address their concerns.[47] Indeed, by publishing the report the Society went beyond its traditional role of facilitating scientific communication and exchange, even if it carefully explained that the report contained 'no more, and no less, than the opinions...of a small group of British biochemists' and the expression of a 'corporate opinion' was categorically dismissed ([Krebs Report] 1969, 3). Similarly, while molecular biologists linked their demands to current and more general issues of 'modernisation', competitiveness, national performance, assessment and planning and used the respective rhetoric, biochemists showed more preoccupation with defending their disciplinary authority in academia, though this included a reform of secondary school curricula as well as the strengthening of contacts with industry. While molecular biologists saw a need for 'national planning and central direction',[48] biochemists expressed themselves against 'administrative measures' imposed from above and preferred to rely on independent peer review as the approved mechanism to maintain high standards of research. The suggestion of the Krebs Report to increase the responsibilities of the Reseach Councils for financing scientific research in the universities was based on the observation that the councils were more successful in applying these quality controls and making it the basis for their financial support. In contrast, universities found it 'difficult, even unpleasant, to differentiate in terms of finance between one individual and another, or between one department and another' ([Krebs Report] 1969, 19). The suggestion for Research Councils to assume more responsibility clearly rested on the assumption that these remained free of government interventions,

---

[46] On the particular tensions between molecular biologists and biochemists at Cambridge see above (chapter 7). American and French biochemists jumped much more readily on the molecular biology bandwagon. For an attempt to explain these national differences in the response of biochemists to molecular biology see Abir-Am (1992a, 225–6).

[47] On this point see also Abir-Am 1992a. Apart from the fact that molecular biologists did not have any weight in university politics, the integration of the biological disciplines under the banner of molecular biology could hardly have been proposed in a faculty meeting.

[48] Paper by S. Brenner on proposals for the improvement of graduate education in biology presented to the fourth meeting of the Working Group on Molecular Biology, 20 January 1967 [no title, no date], Kendrew Papers, MS Eng. c.2539, J.33, Bodl. Lib.

a point which was very much up for discussion. Biochemists also appealed to the 'morale' of university staff, mainly understood as a strength of character, to keep up standards ([Krebs Report] 1969, 26–7). As an official report (a command paper) presented to Parliament by the Secretary of State for Education and Science, the Kendrew Report was assured circulation in political circles. In contrast, the first addressees of the Krebs Report were the members of the Biochemical Society (4700 at that time), all of whom received a copy.[49] Although biochemists apparently welcomed the Society's report, *Nature* announced it under the title 'Dog wags tail' as a 'depressingly predictable little tract' (Anonymous 1969b, 969).

Before trying to assess the impact of these interventions on policy level for the new science of molecular biology, I will return to the negotiations regarding EMBL. Here molecular biologists carried their case one step further and negotiated at the intergovernmental level. As we have seen, the international activities formed the background to Kendrew's working group. In the official report no reference was made to the discussions regarding Britain's support of the international project, but those responsible did not fail to make the connection. As Himsworth, a prominent opponent of the EMBL project, did not fail to point out, the 'desirable expansion' of molecular biology in Britain was 'likely to be remote' if the country was to contribute 'the best part of a million a year from the budget available for biological research... in order to support an international laboratory in this field'.[50] To understand the political and scientific implication of the project and the role played by Cambridge molecular biologists some background information on the early history of the project is necessary.

A Europe of biology

So far only brief outlines of the history of EMBL exist. These were mainly written for celebratory occasions by scientists involved in the creation and administration of the laboratory. This is in stark contrast to other European scientific projects like CERN or ESRO (European Space Research Organisation) which have been the subject of extensive historical research

---

[49] I am grateful to Mike Whitnall, Associate Director of the Society, for making this information available to me. Since the report was freely distributed, it is difficult to assess how many actually read it. None of the biochemists I interviewed remembered the report (though this is also true of the Kendrew Report) and one of them described the actions of the Society regarding teaching 'as always non-events'. This contrasts with Davidson's description of the report as a 'best seller' (Davidson 1970, 3).

[50] H. Himsworth to M. Swann, 31 January 1968, file D747/1, MRC Archives.

(Hermann *et al.* 1987–90; Krige and Russo 1995; Krige 1996).[51] Here I will focus on the early history of the laboratory, and more specifically on how Cambridge molecular biologists used EMBL to affirm their position, both at home and abroad.

EMBL was officially opened in Heidelberg in 1978, after a decade and a half of intense negotiations in view of stark opposition to the plan not only by politicians but also by members of the biological community. These protracted negotiations offer important clues to how the laboratory plan was made to fit political requirements and politics was used to establish the discipline. The early history of EMBL also shows that, despite all rhetoric of the inherent international character of science, even the establishment of an international laboratory is mediated by local negotiations conducted by a small network of people (scientists and administrators) who gain the power to define the project and thus consolidate their claim to authority. I have already pointed to Kendrew's leading role in defining the laboratory project and in (more or less overtly) steering the negotiations for Britain's participation in the project. In this he was helped by other colleagues from the LMB who counted among the strongest supporters of the international laboratory and generally dominated the British delegations. Such was their influence that some voiced the fear that EMBL would just be a second 'MRC Laboratory of Molecular Biology at Heidelberg'.[52] The early dominance of the project by British molecular biologists did not exclude other countries from using their participation in the laboratory project for their own national and local politico-scientific aims and interests. A full history of EMBL would have to take these national and local histories into account.[53]

In an earlier chapter I have already referred to a first plan for a European Institute of Molecular Biology discussed between an American delegation

---

[51] For an excellent overview of the politics of European scientific collaboration see Krige (1997).

[52] A. E. Evangelopoulos to J. Kendrew, 30 April 1973, Kendrew Papers, MS Eng. c.2428, F.113, Bodl. Lib. The EMBO member from Greece was also concerned that after some time the laboratory would become German dominated. He suggested that the number of German scientists should not exceed 20 per cent, if the international character of the laboratory was to be preserved. Kendrew responded that competence, not nationality, would be of importance for acceptance to EMBL and that there would be no minimal or maximal quotas, though efforts would be made to avoid predominance by any country. At an earlier stage of the negotiations it had indeed been considered whether to create the EMBO laboratory at the LMB or even to expand the LMB into a European laboratory; see for instance H. Himsworth, 'Laboratory of Molecular Biology and the University of Cambridge. Interview with the Vice-Chancellor of Cambridge, 9 May 1966', file E243/109, vol. 2, MRC Archives. The growth of the LMB itself made this option look increasingly unrealistic. In the final negotiations about the site, Cambridge was not included.

[53] On the impact of EMBO and EMBL to help the depressed state of molecular biology in Spain see Santesmases and Muñoz (1997a; 1997b).

from Washington and a group of molecular biologists from the Pasteur Institute in Paris in November 1958 (see pp. 254–7). The scheme, devised in the aftermath of the Sputnik launch, was to encourage peaceful collaboration between East and West in the biomedical field. Though the initiative was enthusiastically received by the Pasteurians and plans were drawn up and discussed, in the end funding was not forthcoming.

Another international event triggered new discussions for a European Laboratory of Molecular Biology, this time under quite different circumstances. The Cuban crisis had induced Leo Szilard, an ex-European émigré, temporarily to leave the United States and settle in Geneva where, together with another American of European origin, CERN director Victor Weisskopf, and Italian physicist Gilberto Bernardini, he conceived the idea of a European laboratory for the biological sciences on the model of CERN. They proceeded to get a number of molecular biologists interested. Famously, Kendrew and Watson were invited to Geneva on their return from the Nobel ceremony in Stockholm to discuss the project, with Kendrew soon emerging as possible director of the proposed laboratory.[54] Significantly the first acronym coined for the project was CERB (Centre Européenne de Recherches Biologiques), in analogy to CERN, and the main rationale for the new laboratory was once more to stop the 'brain drain' of European scientists to the better-equipped laboratories across the Atlantic.[55] Despite the apparent 'anti-Americanism', American scientists and science administrators supported the project, which was in line with the United States' Cold War concerns to strengthen European science. In contrast to CERN, there was not a single, large and very expensive machine which justified the creation of an international laboratory, but Kendrew's proposal (the one he circulated among members of the CSP in 1965) argued that the multiplicity of technologies and multidisciplinary approaches which made up molecular biology required a 'substantial capital of men and equipment', putting its realisation beyond the reach of many European countries (Kendrew 1968, 841).

---

[54] On this last point see L. Szilard to V. Weisskopf, 21 January 1963, and J. Kendrew to V. Weisskopf, 13 February 1963, Kendrew Papers, MS Eng. c.2418, F.1, Bodl. Lib. Regarding the beginnings of the talks on a European molecular biology laboratory, I have relied on the standard account. Based on new archival research, John Krige has recently gone some way to revise this account by pointing to a number of earlier initiatives, supported by Weisskopf, of creating an international biological laboratory on the model of CERN. Of special importance for the history of EMBL was Adriano Buzzati-Traverso's plan to establish an International Laboratory of Genetics and Biophysics at Naples. See Krige (forthcoming). On the Italian initiative see also Capocci and Corbellini (forthcoming).

[55] The same acronym had been used before to denominate a Swiss initiative to establish a European Centre for Research in Radiobiology. It was one of the projects Weisskopf was engaged in before supporting the European molecular biology project.

The project soon took form through a number of meetings and the creation of EMBO, a private non-profit organisation registered in Switzerland; 140 molecular biologists from twelve West European countries and Israel were invited to join.[56] Perutz became its first chairman. In addition to the plan for a central laboratory, EMBO embraced the project of a European fellowship, training and travel programme. To some extent this represented the boiled down version of what had started as a counter-proposal to a central laboratory: Waddington's project of a 'federation' of biological laboratories closely connected to the universities. While favouring European action in the field of 'fundamental biology', Waddington considered the idea of integrating by centralising as 'old-fashioned . . . dating from before the age of telephones, airlines and motor-roads', and in contrast to the way the cultural and economic integration of Europe was going to proceed. 'There can never be one European "capital city", playing the role for the continent which Paris has played for France, or London for Britain', he expanded. 'There will rather be a network of centres, between which communication is easy.'[57] On the particular point of a closer integration of the European project in the university system favoured by Waddington, he wrote to Kendrew: 'Your special, and I should hope abnormal, experience of Cambridge probably gives you a biased slant on this – if all British universities were as self-satisfiedly imbecile as Cambridge, there would be nothing to be done with them, but this isn't really so.'[58]

---

[56] Membership grew continuously and reached 700 by 1989. The inclusion of Israel among the original member states depended on Kendrew's personal links with the Israeli scientific community, especially the Weizmann Institute, and on a generous financial gift by the Israeli government to the initial fund for EMBO. On the history of EMBO see Tooze (1981; 1986) and European Molecular Biology Organisation (1989).

[57] C. H. Waddington, 'Proposal for an European Biological Organisation (EBO)' (document prepared for a discussion on molecular biology held at CERN on 28 June 1963). With regard to this meeting see also J. C. Kendrew, 'The proposal for an international laboratory of molecular biology in Europe' (dated 28 May 1963) and 'Record of discussion on molecular biology held at CERN on 28 June 1963'; Kendrew Papers, MS Eng. c.2418, F.3, Bodl. Lib. After what was generally viewed as a 'confused' and inconclusive discussion on Waddington's and Kendrew's proposals at the meeting in Geneva, Kendrew was asked to produce a single paper combining the two projects. This document became the basis for discussions for a larger gathering of self-appointed molecular biologists (or fundamental biologists – the designation kept shifting) in Ravello (Italy) in September 1963 at which the creation of EMBO was decided. See 'Proposal for a European Organization of Fundamental Biology', Kendrew Papers, MS Eng. c.2421, F.34, Bodl. Lib.

[58] C. H. Waddington to J. Kendrew, 11 March 1963, Kendrew Papers, MS Eng. c.2418, F.1, Bodl. Lib. Waddington's and Kendrew's proposals (EBO and CERB respectively) were discussed at a meeting in Geneva in June 1963. Waddington remained active with regard to the fellowship programme, but completely retreated from the laboratory plan, which for Britain was now exclusively represented by members of the LMB.

Through a grant from the newly created Volkswagen Foundation the more modest exchange and training programme was soon successfully launched.[59] In contrast, the plan for a European laboratory continued to meet with a 'lukewarm and even hostile' reception, by politicians because of the high and continuing costs involved, and by scientists because they disliked and saw no necessity for a large international institution which might cream off the best researchers from their national laboratories (Alton 1989, 149).

The creation, in 1969, of a more formal intergovernmental organisation, the European Molecular Biology Conference (EMBC), brought about with the active help of the Swiss government and through intensive lobbying behind the scenes by EMBO members in their national governments, gave the EMBO fellowship programme a more secure financial basis. Initial signs to the contrary, the new organisation also maintained the laboratory project.

The turning point for the controversial project was marked by a conference in Constance in November 1969 where, under Kendrew's chairmanship and after animated debate, a more modest and much revised plan for the labororatory was drawn up. There was a decisive cut in the proposed size of the laboratory: from an eventual size of about 260 scientists, an equivalent number of technical staff and an annual budget of £3.5 million to just about 60 scientists and an annual budget of £850,000. According to the new plan, the laboratory would be strongly orientated towards the development of advanced instrumentation – the finally found substitute for the single, large and very expensive machine which justified the creation of CERN. The wide-ranging scientific programme of the earlier proposals comprising all imaginable subject areas from biological structure determinations to immunology, embryology and cancer research was narrowed down to two main fields of research: cell genetics and subcellular structures. Both these fields were seen as lying 'outside what is classically known as molecular biology' but as 'the areas into which a molecular approach is likely to spread'. They both strongly depended on the development of new instrumentation.[60] The change in the scientific programme was justified

[59] On the role of the Volkswagen Foundation in the creation of German molecular biology see Gottweis (1998a, 48–9).

[60] J. C. Kendrew, 'Summary of EMBO's new proposals for an international laboratory', Kendrew Papers, MS Eng. c.2427, F.102, Bodl. Lib. The ambitious project of a 'complete solution' of *E. coli*, proposed by Crick for the European laboratory and included in the 1967 laboratory document, was also dropped from the programme; see EMBC, 'Proposal for a European Laboratory of Molecular Biology prepared by the Council of the European Molecular Biology Organisation (November 1967)', Kendrew Papers, MS Eng. c.2434, F.182, Bodl. Lib., and 'JCK draft', Kendrew Papers, MS Eng. c.2435, F.189, Bodl. Lib. See also Crick (1973) and above pp. 294–5.

10.3 Building work starts at the site of the future European Molecular Biology Laboratory at Heidelberg (March 1975). *Source*: Bodleian Library, MS Add. Kendrew Papers, R.176.

by the rapid development of the field. But changing political considerations and the necessity to trim down the programme to give the project a chance were as crucial. Indeed, a difficult balance needed to be struck between too small scale a project to warrant international support and too ambitious a proposal to find governments prepared to join.

After further negotiations and revisions of the project, the laboratory agreement was finally signed by ten of the thirteen EMBC states, establishing EMBL. Kendrew, who had led most of the negotiations, became its first Director-General. By that time, in a politically motivated move aimed at compensating France's recent acquisition of 'CERN 2' (the site of the 300 million electron volts accelerator), Heidelberg (not Nice as originally planned) had been selected as the site for the main laboratory, with DESY, the electron synchroton and high energy X-ray source at Hamburg, and the Institut Laue Langevin in Grenoble, for the use of neutron beams, as outstations (Fig. 10.3).[61]

Britain was among the original signatories of the laboratory agreement. In view of the earlier general hostility to the plan this may come as a surprise.[62] Indeed the decision was all but straightforward. As mentioned

[61] On the politics of a site for international scientific institutions see Krige (1997).
[62] More generally Morange has argued that the development of 'simple, ready-made' molecular biological techniques in the 1970s made many of the arguments supporting the

earlier, the CSP had on several occasions considered the EMBO proposal for a European laboratory. Since December 1968 a working party under the chairmanship of Massey had been examining the problem in detail. Kendrew participated as 'assessor'. Before making its report to the Secretary of State, the working party solicited wide discussion of the issues involved. *Nature* pirated a report on a meeting at the Royal Society, intended to be private. Participants included stout defenders as well as strong opponents of the laboratory plan. The question to be considered was 'whether substantial funds be applied to the expansion of molecular biology, on the understanding that they would come from UK science budget; and on whether this additional support, if forthcoming, be used as the British contribution to an EMBO laboratory, or laboratories, or alternatively for support of UK national laboratories'. The question was clearly one of a redistribution of funds, not one of new funds being made available, and thus was destined to stir up opposition.

Blackett, as President of the Royal Society, was in the chair.[63] Papers were presented by Kendrew to explain the EMBO proposal, by M. G. P. Stoker to report on the CSP working party, and by Brenner who talked about the experience of working at a laboratory, the LMB, which was not integrated in the structure of the university. From a chart Kendrew compiled during the meeting on a small notebook page (his famous filofax system) it appears that only five of the eighteen invitees spoke in favour of a central European laboratory.[64] The report in *Nature* confirmed the general hostility to the project and revealed the stark tones used

---

creation of a central European laboratory redundant. It follows that the foundation of a European laboratory should have appeared much less of a priority in the mid-1970s than ten years earlier, when it was first proposed (Morange 1995). That EMBL was none the less realised in the 1970s and not in the 1960s confirms that scientific arguments were not the determinant factor.

[63] A few weeks before the meeting Blackett had insisted to Waddington that he 'must try to sort out the deep split which . . . is present in British biologists between those who want the laboratory and those who won't have it at any price'. Waddington sent invitations to M. Swann, N. A. Mitchison, P. Walker, M. Pollock, W. Hayes and M. Birnstil to discuss the matter (letter C. H. Waddington to M. Swann, 6 October 1969, Kendrew Papers, MS Eng. c.2435, F.189, Bodl. Lib.). Mitchison and Pollock also attended the Royal Society meeting, the first speaking against, the second in favour of the EMBL plan.

[64] According to Kendrew's chart, those who spoke for EMBL were M. Pollock (Edinburgh), G. Pontecorvo (Glasgow), M. Bretscher (LMB), M. Wilkins (London) and R. Appleyard (EMBO Executive Secretary); F. Crick (LMB) and J. Gurdon (LMB) spoke for multiple laboratories while E. Chain (London), B. [Hartley] (LMB), N. A. Mitchison (London), L. Wolpert (London), D. Phillips (Oxford), J. Z. Young (London), A. P. Mathias (London), B. Katz (London), [R. H. S.?] Thompson (London), J. A. V. Butler (London) and H. Himsworth (MRC) spoke against the project; [note in J. Kendrew's handwriting], Kendrew Papers, MS Eng. c.2435, F.189, Bodl. Lib. H. Harris (Oxford) and H. L. Kornberg (Leicester) do not appear on Kendrew's list but, according to the report in *Nature*, also strongly opposed the project (Anonymous 1969c).

especially by the opponents of a central laboratory which they described as 'chimera, a glittering idealistic concept', the 'thin end of a white elephant' and nothing less than 'a disaster for molecular biology' (Anonymous 1969c).

In spite of such heavy opposition the CSP recommended UK participation in the international laboratory project, subject to further examination of the scientific programme, costing and site. A second CSP working group, set up to consider the revised plan, confirmed the recommendation which the Council submitted to the Secretary of State. Was this all the making of an influential clique of scientists? This was certainly a view which circulated widely, despite Kendrew's determined efforts to dispel it.[65] However, rather than just marking the success of a pressure group, the decision rested on a convergence of interests between scientists and politicians. As Stoker had made clear in his presentation at the Royal Society, 'there may be political arguments for U.K. support of an EMBO laboratory which are independent of scientific considerations, but which may substantially influence the ultimate decision'.[66] Molecular biologists supporting the plan recognised and eagerly seized the opportunity.

In 1969 Britain was negotiating its entrance into the European Community, a step made possible by the more flexible attitude of George Pompidou, De Gaulle's successor as President of France earlier that year. Not joining EMBL could give a negative political signal. This was true even if EMBL (like CERN) was not an official project of the European Community.[67]

Scientists, both for and against the EMBL project, made reference to this political background. Kendrew, especially, lamented the lack of a 'European attitude' in Britain and explicitly viewed the EMBL plan in the context of European collaboration more generally.[68] At the other end

---

[65] See for instance 'JCK draft' [typescript of speech prepared by J. Kendrew for the Royal Society meeting of 21 October 1969], Kendrew Papers, MS Eng. c.2435, F.189, p. 6, Bodl. Lib.

[66] M. G. P. Stoker, 'The United Kindgom and the proposed EMBO laboratory', paper prepared for the Royal Society Discussion on UK participation in the European Molecular Biology Organization, 21 October 1969, Kendrew Papers, MS Eng. c.2435, F.189, Bodl. Lib.

[67] Similar considerations played a role in Britain's suffered decision to participate in the new 300 GeV particle accelerator at CERN in 1970.

[68] 'JCK draft' [typescript of speech prepared by J. Kendrew for the Royal Society meeting of 21 October 1969], Kendrew Papers, MS Eng. c.2435, F.189, Bodl. Lib. Besides his influence in the British discussions regarding EMBL, Kendrew also used his role as chairman of the CSP Standing Committee on International Scientific Relations and multiple representational functions in international meetings, coming with this position, to engage foreign colleagues in discussions on EMBL. Between 1971 and 1972, for

of the spectrum Bernard Katz, biophysicist at University College London, warned his colleagues that they were being 'used by the politicians to gain a foothold in Europe' and that they should not allow themselves to be so used.[69] In the event, the CSP formulated its final recommendation to join the laboratory project in the same month (January 1972) in which Britain signed the treatise of accession to European Community membership. The laboratory agreement, with Britain among the original signatories, was signed in February 1973, one month after Britain's admission to the European Community became operative. Both decisions remained controversial in the country.

Not only on the planning but also on the organisational level, the new laboratory owed much to the input of Cambridge molecular biologists.[70] Indeed, EMBL shared many features with the LMB. Like the LMB, EMBL had no formal links with the university and counted a small number of permanent staff in relation to the number of visitors. The strong workshop tradition which the LMB had taken over from the Cavendish became a main feature of EMBL. Structural studies were as strong in Heidelberg as in Cambridge, though protein crystallography was at first not included, on the assumption that the problem of protein structure was already 'solved' – which represents but another example of the impact of the Cambridge laboratory, if in inverse key. From the beginning, however, Cambridge molecular biologists were keen users of the strong electron and neutron sources of the EMBL outstations, driving technological developments at the sites (Huxley and Holmes 1997). Finally, just as many of the LMB publications appeared in the *Journal of Molecular Biology*, EMBL scientists before long published their papers in the *EMBO Journal*. Questioned to what extent the LMB functioned as a model for EMBL, Kendrew cunningly responded: 'The LMB was never consciously adopted as a model. But I hope it did resemble it, because I was doing it.'[71]

Despite the export of much of the 'LMB culture' to the European level, not everyone in the Cambridge laboratory welcomed the creation of EMBL. Perutz, for one, reckoned that the European laboratory was an expensive and ineffective way of conducting research. The early history

instance, Kendrew as Chairman of the CSP Committee attended a series of informal meetings between France, Germany and Britain aimed at discussing fields suitable for collaborative efforts, including EMBL, at a time which was crucial for the final decision on the project (Alton 1989, 287).

[69] Anonymous (1969c) and correction to the *Nature* report (Katz 1969).

[70] This is true also in geographical terms. The LMB was Kendrew's institutional base until 1974, and much of the work of EMBO, EMBC and EMBL emanated from there. Brenner, Huxley and other members of the LMB (but not Perutz – see below) also supported the laboratory project.

[71] Interview with J. Kendrew, 14 July 1992.

of EMBL none the less shows that, despite all the rhetoric, international laboratories also, in the end, are made by a few people and, welcomed or not, bear the imprint of local traditions which, through this process of translation (from local to international), gain further recognition. This is also true of molecular biology for which the international space has been seen as 'constitutive' (Abir-Am 1992b).

An assessment of the scientific achievements of EMBL is outside the scope of this book.[72] However, with the negotiations for EMBO and EMBL, molecular biologists certainly acquired important political visibility. In the mid-1960s, the Council of Europe, an interministerial organisation comprising nineteen European countries, expressed the view that 'the development of research in the field of molecular biology is a matter of the highest importance'.[73] With EMBC, an intergovernmental organisation specifically dedicated to the advancement of molecular biology was created. Its principal role was to agree a budget paid by the member states, which was distributed by EMBO, a private organisation of individual scientist members.[74] Through this mechanism a direct channel of communication between molecular biologists and politicians was established. The EMBC agreement, and with it the special support of a fellowship, travel and course programme in molecular biology, has been periodically renewed until the present day. A second intergovernmental organisation, the Laboratory Council, is exclusively responsible for the much larger budget of the central European laboratory.

An early example of the decisive impact of EMBO and EMBL on governmental politics occurred in the context of the recombinant DNA debate (Jackson and Stich 1979; Watson and Tooze 1981; Krimsky 1982; Dickson 1988, 243–60; Wright 1994; Gottweis 1998a). In 1974, following earlier debates, Paul Berg, one of the pioneers of recombinant DNA experiments, published a letter in *Nature*, signed by eleven other leading scientists, calling for a voluntary moratorium on research with the new technologies (Berg *et al.* 1974). In February 1975, 140 scientists met at an international conference in Asilomar (California) to discuss the implications of the new technologies. To regain control over a debate which was stirring up increasing public concern and to allow research to resume, the scientists at the Asilomar conference decided the drafting of guidelines which would regulate experimentation with recombinant DNA. Shortly after these events, the EMBO Council decided to establish a Standing

---

[72] On this point see Morange (1995).

[73] 'Note on Council of Europe action in the field of molecular biology' [CEBM 67/7E], Kendrew Papers, MS Eng. c.2429, F.32, Bodl. Lib.

[74] For figures on the fast-growing EMBC budget between 1970 and 1981 see Tooze (1981).

Advisory Committee on Recombinant DNA Research. Its role was described as to advise governments, other organisations and individual scientists on scientific and technical questions.[75] EMBO thus assumed the role of a central 'expert body' with the clear intent of influencing policy decision. In its first report, published in 1976, the Committee played down the risks of recombinant DNA experiments, especially those connected with the unintentional production and spread of genes into the environment (Watson and Tooze 1981, 213). Later it discussed the recombinant DNA guidelines drawn up by the National Institutes of Health (NIH) in the USA which introduced several levels of physical and biological containment for work with recombinant DNA. These guidelines were criticised by many – concerned citizens, environmentalists and some scientists – for being too lax and only concerning work safety, not long-term effects on the environment. In contrast, the EMBO committee expressed its concern over the increasing regulation of recombinant DNA work and reported its opinion to EMBC. A joint EMBO–NIH workshop on the issue led to the recommendation of a massive reduction of the levels of containment laid down in the NIH guidelines issued only a few years before. The argument was that recombinant DNA techniques made experiments with the genomes of dangerous pathogens safer, not less safe. The recommendation was endorsed by a national committee set up by the NIH. It marked a turning point in the history of recombinant DNA regulation, introducing a series of relaxations of the safety procedures, in the American guidelines, as well as in the guidelines discussed in European countries (Watson and Tooze 1981, 291–303, 337–429; Wright 1994).

The efforts of the EMBO Committee notwithstanding, and indeed quite in line with its philosophy, EMBL was the first European laboratory to offer a high-level containment unit for experiments with recombinant DNA technologies. With the laboratory still in course of construction when the first guidelines were published, the plans were altered to make room for the high-cost unit, which seemed to offer additional justification for EMBL.[76]

---

[75] For the following see European Molecular Biology Organisation (1989, 17–18). Brenner, who was to play a key role in the British policy discussions on genetic engineering, was a member of the EMBO committee, as was Lennart Philipson, EMBL's second director. From the beginning Brenner was a strong advocate of regulating genetic engineering as a means to protect research.

[76] In Britain the first containment unit for work with recombinant DNA was built at Imperial College London. The LMB received a category III containment facility in 1978, after substantial opposition by the University Committee for Safety which was reluctant to agree that such a facility should be introduced on the top floor of a building on a major hospital site. The plan also caused substantial public concern. The only high-containment unit in Britain was established at the Chemical and Biological Defence Establishment at Porton Down in the mid-1980s (Carter 1992, 83).

Throughout its history a basic rationale for EMBL was to allow European molecular biologists to pursue fundamental research, and to keep its independence *vis-à-vis* industrial investments and interests. Facilitating research with recombinant DNA technologies and pushing for the relaxation of regulations imposed on experiments with recombinant DNA, however, EMBO helped prepare the ground for massive government and industrial investment into and exploitation of the new technology. This guaranteed new funds for molecular biologists as well as a key place in government policies. Both molecular biologists and governments could see this as a justification of their earlier claims regarding the promises of and their investments in the new field. By the 1980s, an increasing number of molecular biologists themselves turned into enterpreneurs of new biotech firms, fundamentally altering what being molecular biologists was about. In this, Heidelberg, which in the late 1970s became the hub of biotech industries in Germany, preceded Cambridge.[77]

In this chapter I have shown how molecular biologists used both national and international political channels to establish their science. The point is not only that molecular biologists moved 'willy-nilly into politics and diplomacy' (Alton 1989, 186), but that molecular biology itself, like nuclear physics before, though under different socio-economic conditions and propelling different expectations, became a political issue and an item on the governmental agenda. Molecular biologists were actively engaged in achieving this aim. Molecular biology, then, was constructed as much in political committees as at the bench, and science policy was as much a tool in the hands of the scientists to further their projects as it became a tool for politicians to 'manage' science. If, from the late 1970s onwards, molecular technologies became key sites of economic and health policies, the (political) groundwork for these interventions was certainly laid before.

The final chapter will investigate the role and place of the LMB in the changing political and scientific landscape of the 1970s.

---

[77] On biotechnology policy in Germany compared to France and Britain see Gottweis (1998a). On the development of biotech industries in Cambridge see Segal Quince & Partners (1985) and Gonzales-Benito, Reid and Garnsey (1997). See also below (chapter 11).

11

...

The end of an era

In the mid-1970s, the MRC initiated a major review of molecular biology. The initiative was prompted by the imminent retirement of Max Perutz. The exercise, however, differed quite markedly from the standard procedure all MRC establishments had to undergo when their directors reached retirement age (not unusally with the effect of the establishment closing down). Most importantly, even before a review committee (the MRC Cell Board Subcommittee on Molecular Biology) was set up, the MRC not only had extended Perutz's tenure for five years, but had also agreed that the laboratory represented a 'national asset' and that it was 'inconceivable' that it should not continue after the chairman's retirement.[1] As a consequence, the review committee's terms of reference were at the same time more restricted and larger, giving rise to some ambiguity. The aim of the general review was to formulate a Council policy for the future support of molecular biology 'in universities and research institutes, both in the United Kingdom and abroad'.[2] However, once again the problem was to define the term and determine the boundaries of the field. The subcommittee decided that in the present case this could best be done 'in terms of the historical relationship of the Cambridge Laboratory to the subject's development'. 'Molecular biology', then, comprised 'all the content of the programmes of the Cambridge Laboratory, and, by extension, the programmes of EMBLab', which was seen as a direct filiation of the LMB. The general review became a review of the LMB after all.[3]

[1] 'MRC Laboratory of Molecular Biology. Future of the Laboratory after Dr Perutz's retirement: recommendation from the Cell Board', MRC 74/909, file E243/261, MRC Archives. The recommendation was based on separate reviews of the three divisions conducted in the early 1970s which had all highly praised the work of the laboratory.

[2] MRC, UGTA note, number 57, 3 October 1975, file A602/1, vol. 4, MRC Archives. Though somewhat ambiguous in its terms (see below), this was not the first special review of a field initiated by the MRC. In 1969, for instance, a committee reported on the whole field of radiobiology (Anonymous 1969a).

[3] 'Cell Board Subcommittee set up to review molecular biology. Unconfirmed minutes of first meeting, 21 July 1975', and 'Note for file', by J. Dowman, 14 November 1975; both in file A147/14, vol. 1, MRC Archives. An implicit criticism of this restriction was that the

After six meetings and consideration of written and oral submissions by seventy-five individuals as well as ten group submissions from LMB members and other grant holders or teaching and research institutes in Britain the subcommittee, chaired by David Phillips, Professor of Molecular Biophysics at Oxford, issued a report which suggested a 25 per cent cut of the LMB staff and budget and a thorough revision of the managerial and administrative structure. For the first time it also formulated precise terms of reference for the laboratory.[4]

As can be expected, the report as well as the whole exercise preceding it and the mechanisms put in place to enforce it, had a demoralising effect on the researchers at the LMB, even if the committee wistfully suggested that 'morale was probably no lower in the Laboratory than elsewhere' in the current situation of financial stringency following the oil crisis and general recession.[5] The cuts were remembered as the 'Phillips cuts' and the report is generally taken to mark the 'end of an era' which especially the older members of the LMB identified with the retirement of Himsworth, the Secretary of the MRC who had overseen the fast growth of the establishment from the original two-man unit in the late 1940s to its current size of 200 employees including scientific, technical and administrative staff.[6] In reality, the Phillips Report reflected and responded to much more wide-ranging changes which had started to affect the laboratory for some time. They concerned the political, economic and administrative culture, including the approach to science policy as well as developments in molecular biology to which the very success of the LMB had contributed decisively. The request for accountability of scientific expenditure on the side of government coincided with

LMB had failed to 'seed' the subject in university laboratories (see *ibid*. and this chapter, below). The 'EMBLab people' were considered as 'indistinguishable' from those at the LMB; J. Dowman [MRC], Note for file, 14 November 1975, file A147/14, MRC Archives. The fact that the MRC, which provided the funds for the LMB, also administered the UK contribution to EMBL, the creation of which remained highly controversial, made this connection seem even closer. The discussion on the future support of the two laboratories inevitably became linked.

[4] 'Subcommittee to Review Molecular Biology. Report to Cell Board', MRC 76/610, file S1202/1, vol. 2, MRC Archives (in the following referred to as [Phillips Report]). Other members of the subcommittee were: W. F. Bodmer from the Genetics Laboratory at Oxford; K. Burton from the Biochemistry Department of Newcastle upon Tyne and member of the Science Research Council; R. Dulbecco from the Imperial Cancer Research Fund Laboratories; H. Harris from the Dunn School of Pathology at Oxford; A. Huxley from the Department of Physiology at University College London; H. Kornberg from the Department of Biochemistry at Cambridge; B. R. Rabin from the Department of Biochemistry at University College London; M. H. F. Wilkins from the Department of Biophysics at King's College London.

[5] 'Subcommittee to Review Molecular Biology. Unconfirmed minutes of the 2nd Meeting, held on 6 January 1976', file A147/14, vol. 2, MRC Archives.

[6] Interviews with M. Perutz, Cambridge, 25 June 1992, and S. Brenner, Cambridge, 17 July 1992.

a crisis of scientific authority and public demand for greater social responsibility of scientists. Together these changes marked the end of a postwar bonanza for the sciences (Ravetz 1971; also Dickson 1988). For molecular biologists the new climate became manifest in the recombinant DNA debate, started by concerned scientists and then picked up by the public. The work of the Phillips Committee closely overlapped in time with the ongoing negotiations regarding the drafting of guidelines for recombinant DNA research. The report of the committee, based on widespread discussions among biomedical scientists, thus represents a unique window to assess the transformations regarding the role and place of the LMB and of molecular biology more generally in the changing politico-economic and scientific climate.

By the mid-1970s, science policy had become an integral part of policy-making generally (Brickman and Rip 1979). Having acquired political visibility, molecular biology became increasingly the subject of policy decisions. As a government-funded institution, the LMB, even if enjoying a privileged position on account of its record of achievements, was not immune to these changes. After a brief discussion of the new government policies towards science and their reflection in the Phillips Report with respect to the future of the LMB, I will focus on two issues which were repeatedly addressed in the review exercise and gained increasing importance in the following years. One concerns the growing expectation of medical pay-offs of molecular biology; the other the question of industrial applications and commercial exploitations of molecular biology research. The 'scandal' which developed around the apparent failure to patent the procedure to produce monoclonal antibodies developed at the LMB in the mid-1970s (Tansey *et al.* 1997, 4) points especially clearly to dramatic changes in the attitude of governments towards research and to subsequent changes in the support, direction and practice of the biomedical sciences which persist to the present. Georges Köhler, a young visitor from Germany, and César Milstein, an Argentinian émigré and senior staff member in Sanger's division, first showed that antibody-producing cells isolated from immunised laboratory animals could be fused with cancer cells, retaining the capacity both to produce antibodies and to continue growing *in vitro*, a characteristic of cancer cells (Köhler and Milstein 1975). Through cloning of fused cells one particular kind of antibody could be produced in large amounts. The debate around the commercial exploitation of this technology was fuelled by the growth of new biotechnology (especially genetic engineering) firms in the United States and fears that Britain would fall behind in what looked set to become an important new economic sector. However, extensive discussions on the economic and medical pay-offs of molecular biology had preceded these developments. They created the expectations put into the new technologies.[7]

'The party is over'[8]

Swept to power in an unexpected victory in the 1970 elections, Edward Heath, Britain's new Conservative Prime Minister, vowed to reorganise the machinery of central government with the aim of ensuring higher accountability and efficiency. The review included the organisation of civil research and development, which had grown into a sizeable apparatus. Recommendations regarding this sector were formulated in two distinct reports published together in a Green Paper (a discussion document for Parliament): the report by Lord Rothschild, head of the Central Policy Review Staff, Heath's 'think tank', on *The Organisation and Management of Government R&D* and the report *On the Future of the Research Council System*, presented by Sir Frederick Dainton, chairman of the Council for Scientific Policy, to the then Secretary of State for Education and Science, Margaret Thatcher ([Rothschild Report] 1971 and [Dainton Report] 1971). After extensive public and parliamentary discussions, the main recommendations were taken up in a White Paper which laid down a new framework for government R&D. The salient points were that, as proposed by Rothschild, applied research and development commissioned by the government was to be organised according to the 'customer/contractor principle'. This meant extending the application of the principle from the military sector, where it was already in operation, to the civil sector, where there was no commissioning or control of research work by a customer. The Research Council System, which had come under attack in the new drive towards general accountability, was to be preserved – as recommended by the Dainton Report – but part of the Science budget devoted to the Councils was to be transferred to the Service Departments from where it could be recouped by the Councils through contract work (*Framework* 1972; Wilkie 1991, 68–95).

With the Rothschild Report, government spending in civil R&D shifted from 'fundamental' science which had attracted large government funds after World War II to 'applied' research with more direct accountability. Arguing for these controversial changes, Rothschild, a Cambridge scientist who had moved to industry, chose a polemical tone. He vehemently

---

[7] See Gottweis' thesis on the invention, by policy discourse, of biotechnology as a new high-tech industry for Europe (Gottweis 1998a).

[8] The phrase was coined by Shirley Williams, Labour MP and Secretary of State for Education and Science in the following Labour Government (1974–9). It denoted the end of the era of rising science budgets (Williams 1971).

denied that 'basic' (or 'fundamental') and 'applied' research were difficult to distinguish, as especially the Research Councils liked to argue[9] – probably, he suspected, with the intent to protect themselves 'from the imaginary ravages of applied R&D users'. The end-product of basic research, he argued, was an 'increase in knowledge' or 'the discovery of rational correlations and principles', while the end-product of applied research was a product, a process or a method of operation. Of course this did not exclude 'that the results of pure research may sometimes be of applied or practical value, and that applied research may produce results of "pure" interest and importance'. But according to Rothschild, the 'country's needs were not so trivial as to be left to the mercies of a form of scientific roulette with many more than the conventional 37 numbers on which the ball may land'. And if, as the Councils argued, much of their research was effectively 'applied', there was to be a customer to 'commission and approve it'. 'However distinguished, intelligent and practical scientists may be', the report concluded, 'they cannot be so well qualified to decide what the needs of the nation are, and their priorities, as those responsible for ensuring that those needs are met' ([Rothschild Report] 1971, 3–4, 10). At least in principle, the new directions also opened up science to public scrutiny.

While introduced to guarantee greater efficiency and more direct accountability in the vastly extended sector of government research under tighter budget conditions, the control functions instituted by the new policy framework required new bureaucratic structures. As a commentator critically put it, 'scarcely a decade after Lord Hailsham [the first Minister of Science in the preceding government] had been content with but a busload of bureaucrats, Lord Rothschild would have had them by the trainload' (Wilkie 1991, 81).

The new government policies for the administration of science directly affected the Medical Research Council. According to the new policy, a quarter of its budget was to be transferred to the Department of Health and Social Security in the first instance.[10] While this money could in principle be recouped, it none the less meant a squeeze on the existing programmes. The situation was exacerbated in the mid-1970s, when, following general

---

[9] See for instance Landsborough Thomson (1973, 95–7) and above, p. 60.

[10] The new policy abolished the 'Haldane Principle' on which the Councils were founded and which guaranteed their independence from the Service Departments ([Rothschild Report] 1971, 17–19). The cuts in the MRC budget following the Rothschild Report were less severe than those affecting the Agricultural and Natural Environment Research Council which lost nearly half and a third of their budgets respectively. However, in the following years, with the Health Department failing to commisssion new research, the MRC fell short of its expected appropriation. As a consequence, by 1981, the MRC (in contrast to the other Research Councils) managed to reclaim its full budget (Wilkie 1991, 88).

economic recession, the growth of the science budget (if only at a modest 4 per cent as compared to the over 13 per cent of the mid-1960s) was replaced by decline.

The new framework with its need for tighter selection and control further swelled the administrative structure of the MRC. By that time the budget of the MRC had grown from £1.3 million in the late 1940s to £23 million. With the money the administration also grew. An increasing number of boards were set up to deal with different subject fields. They referred back to Council.[11] Later special grants committees were set up to deal with the incoming applications. John Gray, Secretary of the MRC at the time of the Phillips Report, described the changes thus: 'In Himsworth's time there were few people in the MRC. Everything was done by the Council, there were no Boards. Himsworth did not delegate...He had a tremendous impact, he dominated the Council...But do not generalise, it was not true for my time.'[12]

If the LMB researchers wished back Himsworth's days, it was also, and maybe above all, a time when paperwork was minimal and dealings were more informal and personal. Indeed, many of the stories relating to the founding years of the institution circulating in the laboratory, even if obviously told with hindsight, relate to the informal way things were dealt with at the time. Of course these were also the days when much hinged on 'old boy networks' and when the science budget continued to grow, leaving space for new developments. The setting up of the unit, the story goes, was decided by Bragg, the Cavendish Professor housing the protein crystallographers, and Mellanby, the then Secretary of the MRC, over lunch at the Athenaeum, the exclusive London club next to the Royal Society. Kendrew adds: 'The fascinating thing about it was that we then proceeded to have their money for ten years without producing any results at all. – I don't think it could happen today, but it did happen then.'[13] And regarding the decision to hire Brenner as new staff member in the mid-1950s, Perutz recalls: 'No panel, no referees, no interview, no lengthy report, just a few men with good judgement at the top' (Perutz 1987, 41).

The Phillips Report in many respects reflected the new policy framework and aimed at bringing the LMB in line with current administrative

---

[11] Each board covered all aspects from basic to applied research. The Clinical Board was set up in the early 1950s to deal with the changes in the provision of research following the introduction of the National Health System. The Cell Board, whose responsibility included radiobiology, cancer and molecular biology, the Board for Tropical Medicine and the Biological Research Board followed suit.

[12] Interview with J. Gray, Cambridge, 1 December 1995. For similar developments at the National Institutes of Health in America after J. Shannon's retirement in 1968 see Strickland (1989).

[13] Interviews with J. Kendrew, Cambridge, 14 July 1992, and Linton, 18 March 1993.

requirements. Hence the importance attached to laying down, for the first time, formal terms of reference for the laboratory, and to an overhaul of the 'managerial and administrative constitution' of the laboratory; hence also the need for cutting down the size of the laboratory and for the introduction of not only cost-efficiency but cost-benefit considerations.

Whilst most members of the LMB favoured retention of the existing informal organisation of the laboratory which avoided 'unnecessary paperwork' and dealt 'promptly and efficiently' with administrative problems, the subcommittee had 'no confidence' that the existing system, even if it had worked in the past, would work in the future. In any case it would 'not measure up to the demands of increased financial stringency and consequent competition for resources'.[14] The report, therefore, went into much detail suggesting a revision of the confederative system to a more dirigistic system with clearly identified responsibilities at the top. Such a structure was regarded as necessary in view of harsh decisions the future would require with respect to staff reductions and scientific programme.

The suggested cut of 25 per cent, while in line with a general squeeze on the Council's resources, was justified in the report on scientific grounds. Here X-ray crystallography, Phillips' own speciality, was singled out for substantial cuts. The laboratory, it was suggested, should focus on large molecular structures (i.e. molecules above 150,000 Daltons), the analysis of which lay beyond the competence of other laboratories (a proposal dismissed as nonsensical by the crystallographic division).[15] The cuts were further justified by the need to spread support for molecular biology. Here the failure of the LMB to interact and collaborate with other institutions in Britain, as well as Britain's commitment to EMBL, came under scrutiny. The LMB, the report acknowledged, had made a significant impact abroad, where many centres of molecular biology 'have been colonised by people who have had a substantial part of their early training in the Laboratory and have become imbued by the style of thinking and work that has been developed there'. However, the same was not true at national level ([Phillips Report], paragraph 2.8). According to the committee, the MRC policy to concentrate support for molecular biology in four major

---

[14] 'Report to the Phillips Committee from the members of the ASTMS [Association of Scientific, Technical and Managerial Staff] in the Laboratory', 12 January 1976, file A147/14, vol. 2, MRC Archives, and [Phillips Report], paragraph 2.10.

[15] LMB crystallographers interpreted the proposed cuts in their division as the revenge of a competitor. Phillips, then working under Bragg in the Royal Institution, had collaborated with Kendrew on the myoglobin project. Later a rivalry between the two groups developed. However, it should also be noted that, since his move to Oxford, Phillips was working in a university department which might have given him another perspective.

centres (Cambridge, London, Oxford and Edinburgh) had exacerbated the problem. The effect was that the financial allocation to the laboratory represented as much as 30 per cent of MRC support for molecular biology as a whole (including the UK contribution to EMBO and EMBL) and 5 per cent of the Council's total budget, a situation which the committee felt needed to be redressed.[16] Regarding EMBO, one suggestion discussed was that, instead of being supported by EMBC, it should be supported by the European Science Foundation where it would compete for funds with other sciences.[17]

Against the general assertion that the LMB was particularly lavishly funded, the subcommittee was 'impressed and intrigued by the economy with which the LMB had been run hitherto'.[18] This was apparently achieved by a low technician-to-scientist ratio as well as by the presence of a high number of postdoctoral researchers with foreign grants. As noted earlier, this represented a distinctive and crucial feature of the laboratory, but was regarded here under strictly economic terms. Besides cost-efficiency, the subcommitte considered cost-benefits of research expenditures. Here the subcommittee took the view that molecular biology, 'potentially at least, could find application in any field of biomedical research'. The report listed a number of areas in which the molecular understanding of the etiology or, in some cases, the causation of disease, was offering 'hope for rational therapy'. It none the less suggested that, in future, the Council might identify medical fields which were ready for a 'molecular attack' and consider what benefits would accrue from such an approach.

The Phillips Report dealt with molecular biology mainly as an administrative problem. Of course the MRC had always had to administer its funds and account for them. But the new legislative framework for science which requested higher accountability, and the general squeeze on public spending which also affected the science budget, represented a new

---

[16] 'Subcommittee to Review Molecular Biology. Unconfirmed summary record of a visit to the Laboratory of Molecular Biology on Monday, 12 January 1976', file E243/23, vol. 1, MRC Archives. The attempt of the committee to trace the original decision to focus support for molecular biology in four major centres revealed that such a policy had never been explicitly formulated. For the MRC the concentration of funds was not the result of a 'geographically planned strategy' but the 'result of the skills and facilities available in particular universities'; The Select Committee on Science and Technology (Science Sub-Committee), 'Abstract from Minutes of Evidence, Wednesday 30 April 1975', file E243/23, vol. 1, MRC Archives.

[17] Extract from Minutes of Council Meeting held 25 November 1976, minute 11.41, file S1202/1, vol. 2, MRC Archives.

[18] 'Subcommittee to Review Molecular Biology. Unconfirmed minutes of the 2nd meeting, held on 6 January 1976', paragraph 13, file A147/14, vol. 2, MRC Archives. See also [Phillips Report], paragraph 4.3.

situation. Though the MRC resisted the changes introduced by the new legislative framework for government R&D, it had to comply.

The change of government policy was in important ways an effect of the growth of government expenditure in civil science and thus of the very success of the sector. Grown to represent a sizeable part of the total budget and still set to grow, it had become an object of policy interventions. Molecular biology as a research field had undergone a similar development. Among those who submitted evidence to the subcommittee, no one, not even the strongest critic, questioned the very existence of the field, as was still usual a decade earlier. The question was not if, but how much funding should be spent on molecular studies. Here the answers varied widely. Some argued angrily that the importance of molecular biology, especially in the medical field, had been 'greatly overemphasized' while other fields, notably functional biochemical research, had suffered correspondingly. Others found that there was 'a good case for switching resources from areas like astronomy and nuclear physics into molecular biology'.[19]

The problem for Perutz's successor – after much consultation the choice fell on Brenner who succeeded Perutz as director, not chairman of the laboratory and thus with more decisionary power – was not only to deal with the new administrative measures. He also had to ensure that research in the laboratory remained 'innovative' (a new political buzzword), despite the imposed cuts and a new career scheme which guaranteed tenured posts in the laboratory to much more junior researchers.

We have already identified some of the practices and strategies developed in answer to the new challenges of the 1970s. In the worm project, work on an organism was used to construct increasingly ambitious research goals. These were designed to attract a growing number of researchers as well as new funds and actively expand the field of molecular biology, scientifically and institutionally. As we have seen, the very success of this strategy affected the central role of the LMB.

Other developments contributed to changes in research practices and in the place and role of the LMB. These hinged on two key issues which already figured largely in the review process: the growing expectation of, in particular, medical pay-offs of molecular biology research and, connected to it, industrial interest in the field.

---

[19] [Phillips Report], Annex 3: E. Chain to T. Vickers (MRC), 9 February 1976, MRC 76/1092, file S1202/1, vol. 2, MRC Archives, and W. Ferdinand, P. Banks, P. C. Engel, P. M. Harrison and C. B. Taylor (Department of Biochemistry, Sheffield), 'MRC research policy in molecular biology', file A147/14, vol. 1, MRC Archives.

From basic to applied?

The question of the relations, and indeed the relevance, of molecular biological research to medical practice was not new, but in the depressed economic climate of the mid-1970s, connected to the increasing need for a social accountability of science, it gained new urgency, with important effects not only for the justification but for the very direction of research work. Here I will briefly review some observations made in earlier chapters on this question in order to point to the most salient changes between the immediate postwar years and the mid-1970s.

When Bragg in 1947 applied to the MRC for funding for Perutz and Kendrew's work in protein crystallography, he stressed the 'fundamental' character of their research. This was at a time when scientists, pointing to the momentous wartime achievements of radar, the atomic bomb and penicillin, had successfully argued that 'fundamental' research produced the pool of knowledge which was essential for useful applications, especially in emergency situations as presented by wars. Successive governments agreed to a growing science budget (see chapter 2). The MRC, from its foundation devoted to the promotion of 'basic' medical research, fully espoused the official political line, taking advantage to strengthen its position as main patron of biomedical research.[20] Research into the structure of 'living molecules', mostly subsumed under 'biophysics' at the time, was 'fundamental' both in the sense that it promised new 'discoveries' and insight into the *basic* mechanisms of life processes. Such knowledge was expected to yield 'unforeseen applications' in the medical field. Again, penicillin and its 'chance discovery' served as example. In postwar America, where support of basic research took the place of general health care provision, the term 'biomedicine' was coined in justification of such a politics (above, chapter 3).

In the application to the MRC for a new Laboratory of Molecular Biology ten years after the founding of the Cambridge unit, references to pathology and medicine were added only in a second draft, and were cosmetic rather than substantial (see above, p. 211, note 33). Others at the time, however, developed more ambitious visions of the integration of basic biological research into medical practice. Among these was Joseph Mitchell, Regius Professor of Physic at Cambridge, who stepped in to rescue the

---

[20] With the increased provision for clinical research following the National Insurance Act of 1946, the MRC also established itself as main patron of clinical research in Britain (see above, p. 217, note 47). Support of biomedical research in Britain which, unlike America, saw the introduction of a national healthcare system straight after World War II, always remained far below American allocations.

LMB plan from foundering, by suggesting that the new laboratory should be built next to his own Department of Radiotherapeutics on the new hospital site at the outskirts of the city. For the molecular biologists, who were keen to keep up their connections with the natural sciences laboratories in the town centre, this was but a last resort. Mitchell's grand plan for the new Postgraduate Medical School, however, saw hospital and research laboratories closely integrated. Clinical and research staff were to work side by side, with hospital staff having access to laboratory facilities, and research laboratories (including those for fundamental research) being located in close proximity to the hospital departments. Bridges, underground corridors and lifts would make it possible to move patients into any of the university laboratories and lecture rooms (what this could mean for patients was not reflected upon).[21]

The reality, at least at the beginning, looked markedly different. The plans for the postgraduate medical school never materialised, and for a long time the buildings housing the Department of Radiotherapeutics and the LMB stood quite lonely on the site (see Fig. 7.6). Eventually the hospital and the Clinical School of the university did move to the new site, but the LMB remained quite separate. In confirmation of this fact, in the recent history of Addenbrooke's Hospital there is only a fleeting reference to the LMB (Rook, Carlton and Cannon 1991).[22]

From the late 1960s, the medical applications of what had come to be established as molecular biology none the less became an issue and molecular biologists felt increasingly obliged to address it. 'Has classical molecular biology already had important medical applications?', Crick asked an audience of medical scientists and clinicians. 'There has been nothing as spectacular or as useful as, say, penicillin', he conceded. But according to Crick, it was only a question of time. 'Because of the very fundamental discoveries which are going to be made', he prepared his audience, 'you are going to have a change in the nature of medicine' (Crick 1969, 187).

Around the same time, Perutz, together with Hermann Lehmann, clinical biochemist at Cambridge, in a paper entitled 'Molecular pathology of human haemoglobins', related clinical symptoms of patients with abnormal haemoglobins to structural changes of the haemoglobin molecule, represented by Perutz's atomic model (Fig. 11.1). In the Phillips Report,

---

[21] 'University of Cambridge Postgraduate Medical School. Memorandum submitted by Dr J. S. Mitchell, Regius Professor of Physic, to the Faculty Board of Medicine and to the General Board of the Faculties on the need for integration between the University laboratory and the Hospital Wards' (1959), General Board Paper 5178, University Archives, CUL.

[22] On more recent plans for a closer integration of laboratory and hospital see below.

11.1 Atomic model of horse oxyhaemoglobin used by Max Perutz and Hermann Lehmann for their work on abnormal haemoglobins. Courtesy of Max Perutz.

Perutz and Lehmann's collaboration stood out as a prime example of a fruitful interaction between clinical research staff and LMB scientists. The collaboration reached back to the mid-1950s, when Lehmann provided Ingram, working in Perutz's research group, with probes of sickle cell and other haemoglobins from his rich collection, gifts which Ingram reciprocated by introducing Lehmann to the techniques of protein fingerprinting for characterising haemoglobins (de Chadarevian 1998a). The collaboration with Lehmann on the abnormal haemoglobins gave Perutz deep satisfaction. He rejoiced that his work had become medically important.[23] But Perutz and Lehmann's joint paper closed on a more sober note. 'The data presented here', the authors concluded, 'do not hold out any hope that the lesions in mutant protein molecules could be repaired directly.' According to the authors, genetic repair was 'imaginable, but still utopian'.

[23] Interview of M. F. Perutz by H. F. Judson, November 1987. Video recording. The Biochemical Society, London.

Less remote seemed the possibility of replacing the organ which produced
the abnormal haemoglobin. For the time being, however, the only hope
lay in preventing the distribution of genes carrying strongly pathogenic
mutations (Lehmann and Perutz 1968, 909). Knowledge of the molecular
structure of haemoglobin was not necessary to this end.[24] It should also be
noted that for the particular case of sickle cell haemoglobin which, since
its description by Pauling in 1949, has been used to herald the promises
of a molecular approach to medicine, the mechanism proposed by Perutz
and Lehmann proved far too simple to explain the dramatic physiological
effects of the altered molecule (Pauling *et al.* 1949; Gray 1951; Pauling
1952). The exact molecular mechanism of the sickling of red blood cells
is still debated today and fifty years of molecular research have not brought
a cure for the disease.

Title and conclusions of Lehmann and Perutz's paper responded to the
increased need to justify research in molecular biology with respect to
its medical usefulness.[25] The expectations rested on the view that basic
research is 'applied' to the clinic where it leads to therapeutic breakthroughs. However, the view of a unidirectional flow of knowledge from
the laboratory to the clinic conceals the fact that projects like Perutz's,
from the beginning, heavily relied on material resources and functional
knowledge from the clinic. While his collaborative paper with Lehmann
announced that clinical symptoms could be explained in molecular terms,
the reverse was also true: being able to relate clinical symptoms to atomic
structure, Perutz gained decisive clues for the normal mechanics of the
molecule. More importantly, perhaps, the expectation of medical breakthroughs from basic knowledge ignores the work necessary to translate
laboratory practices into clinical routines. Figures like Lehmann and material resources like his collection of haemoglobins played a crucial role

[24] Screening programmes for sickle cell anaemia, the most widespread haemoglobin
disease, affecting especially black populations, were introduced in America in the 1970s.
The experience was generally regarded as disastrous (Michaelson 1972; Kevles 1985,
255ff; Duster 1990). Similar programmes were also introduced for thalassemia, a
haemoglobin condition common in Mediterranean countries.

[25] Discussions among molecular biologists preceding the presentation of the EMBL project
to governments confirm the growing pressure to justify research in medical terms. One
participant put it most clearly: 'Typical criticisms raised by enemies of molecular biology
and of EMBO are that the code being solved, the era of molecular biology is over and its
proper place is already in the science museum . . . that in spite of its brilliant discoveries,
molecular biology has not helped to cure any disease or to improve crops etc. . . . When
such remarks are made to government officials, they are disastrous to the EMBO scheme.
We should . . . in our presentation of the case for the Laboratory insist that molecular
biology is really important for the world', H. Chantrenne to J. Kendrew, 15 August 1967,
file EMBO/Nice, fonds Monod, Paris. See also J. Wyman to J. Kendrew, 12 April 1967,
file EMBO/Nice, fonds Monod, Paris.

in this mediation. Lehmann relied on 'molecular' technologies (first electrophoresis, later protein finger printing) for the characterisation of the abnormal haemoglobins. The collection served both as research tool in the laboratory and as reference for the classification and diagnosis of diseases. This double function of the collection indicates the continued importance of natural history type approaches for 'analytical' work in molecular biology (Pickstone 1993a; 1993b; 1994). It also shows that clinicians and laboratory scientists shared a 'molecular culture' before this necessarily translated into new therapeutic measures (de Chadarevian 1998a).

While the expected medical applications were slow to develop, molecular biologists participated in a resuscitation of eugenic ideas. A forum to air such ideas was the CIBA Foundation meeting on *Man and His Future* held in London in 1963. Among the participants called together to discuss the potentials of the new biology were old-time eugenic advocates like Julian Huxley and J. B. S. Haldane; proponent of 'germinal choice theory', Hermann Muller; designer of the pill, Gregory Pincus; and latter-day molecular biologists like Joshua Lederberg and Francis Crick. Leading a discussion on 'Eugenics and genetics', Crick agreed with Lederberg that, despite recent momentous advances in molecular genetics, 'the practical possibilities of synthesizing or modifying the germinal material are very far in the future'. However, this was not a call for inertia. 'I think we all agree', he prompted his audience, 'that on the long-term basis we do have to do something... and because public opinion on this subject is so far behind, we should start to do something about now' (Wolstenholme 1963, 274–5).[26] The ensuing discussion turned on a wide range of measures aimed at counteracting the perceived deterioration of the human gene pool – variously attributed to population explosion and to increased radioactive fall-out – as well as positively enhancing the genetic make-up of individuals. The proposals ranged from licensing child-birth to using frozen sperm of selected donors (Nobel laureates always being a good choice). Among the participants, the enthusiasts for such measures certainly outnumbered the sceptics (Wolstenholme 1963). Pauling's notorious suggestion that carriers of the sickle cell gene should have a symbol tatooed on the forehead dates from around the same time (Pauling 1968, 269).[27]

---

[26] Expecting less 'clumsy' methods of interventions 'within a few generations', Lederberg in his presentation had suggested to focus on 'euphenics', the engineering of human development, as intermediary goal. Possible areas of intervention according to Lederberg comprised pre- and postnatal regulation of human brain size (and thus intelligence) by aid of hormones and tissue and organ replacements (including the development of artificial organs) (Lederberg 1963).

[27] The measure was intended to help young people carrying the gene for sickle cell anaemia not to fall in love with one another and not to create children. Pauling argued that such a

It has been suggested that an early and continuing resonance existed between the programme of molecular biology based on technological interventions at the subcellular level and eugenic agendas of technocratic control and improvement of the human race (Kay 1993a, especially 9, 45–6, 280–2). The eugenic measures discussed by molecular biologists in the 1960s, however, only vaguely, if at all, related to new molecular technologies. What most strikes the reader of these discussions today are the presumption of molecular biologists in defining the problems of society and the naivety of the proposed solutions.

While eugenic questions are again in the centre of debates surrounding the human genome project, policy interventions in the 1970s shifted the attention of biomedical researchers into other directions. The attempt by the Nixon administration to improve its image for public health spending by injecting huge funds for cancer research reverberated in Britain in substantive criticism, in scientific journals and the national press, of the support for cancer research in the country.[28] Rising political pressure and the need to formulate a 'quotable policy' moved government and MRC into action. Nothing comparable to the American 'war on cancer' campaign developed in Britain. Indeed the emphasis of the British response lay on a 'more effective use of present resources rather than in increased financial allocations'.[29] The discussions on the topic which coincided with the review of government R&D and the drafting of the Rothschild Report, however, did introduce an increased pressure for clearer justification and control of government spending in medical research. These pressures also affected the LMB.

Even before the final drafting of the American National Cancer Act, the MRC initiated a review aimed at establishing how much of the fundamental research that it was supporting, especially in molecular biology and immunology, even if it 'had not hitherto been considered as "cancer research"', could effectively be classified as such. The same review committee identified the study of tumour antigens and of chromosome structure and function as most promising areas of cancer research and resolved that they merited increased support.[30]

measure would help to eliminate a source of suffering in the world and was therefore not only morally justified but called for. On the eugenic discussions of the 1960s see also Kevles (1985).

[28] On the background of the American Cancer Act see Strickland (1972) and Rettig (1977). On the reactions in Britain see Austoker (1989, 296–9).

[29] Extracts from Council minutes for meeting of November 1971. Minute 160: Research on cancer; file S18/4, vol. 2, MRC Archives.

[30] 'Review of some fundamental research that may be of relevance to cancer problems', MRC 71/1284, Annex 1 (MRC 71/970), MRC Archives. On the earlier history of the MRC support for cancer research see Landsborough Thomson (1975, 10–12).

The report failed to impress researchers in the field. As a prominent spokesman put it:

There is nothing fundamentally wrong with the report, but it is so frightfully conventional . . . It is, I expect, common ground that in a small and relatively poor country, as we now are, with a tiny budget for medical research in comparison with, say, the United States, we should concentrate on doing original things and not simply follow the trends being pursued with enormous vigour across the Atlantic. But your report, with the exception of one or two specific items, gives me the impression of being a 'me too' exercise: it simply reveals that the MRC is also supporting activity in the big 'band-wagon' areas of world cancer research.[31]

MRC officials held a more positive view. An internal note read: 'With suitable editing, [the report] could serve a number of purposes including counteracting the arguments of those who say that basic research has no relevance to practical problems and [for] general education of the public who clamour for more cancer research.'[32]

The government commissioned Zuckerman who, after his retirement as Chief Scientific Adviser, continued to act as government consultant, to undertake an independent inquiry into the state of cancer research in Britain. Visiting all major research centres in the country, Zuckerman did not find 'a single good idea which was not being followed up for lack of money'. He concluded that government was spending enough money on research. If more funds were to be allocated, these should go into the care of cancer patients (Zuckerman 1972; 1988, 456–7). Predictably, Zuckerman's conclusions were not well received by the major biomedical funding bodies, especially the MRC and the Imperial Cancer Reserch Fund, the biggest charity in the field. In the end, government agreed to make a special contribution of $100,000 to the International Agency for Research on Cancer, while the MRC included a sum of around £800,000 for recurrent costs of new developments in the cancer field in its Forward Estimates for the years 1973–7 (in comparison the American government allocated 7 to 8 billion dollars to its National Cancer Program between 1972 and 1981).

The LMB was included in the MRC survey. Much of the work performed in the three divisions was listed as 'relevant to cancer research'. The list comprised all studies relating to viruses, if concerning their physical and

---

[31] H. Harris to J. Neale (MRC), 30 June 1971, file S18/4, vol. 1, MRC Archives. Harris was head of the Sir William Dunn School of Pathology in Oxford, honorary director of the Cancer Research Campaign, Cell Biology Unit, and a member of the Scientific Advisory Committee of the Cancer Research Campaign.

[32] Handwritten file note, J. Faulkner to J. Neale [4 June 1971], file S18/4, vol. 1, MRC Archives. A revised version of the report was published in *The Lancet* (Walker 1972).

chemical structure, their genetic make-up or the biochemical processes involved in virus infection; studies on the control of protein synthesis and the structure of proteins and nucleic acids involved; membrane studies and studies on the formation and structure of antibodies and proteins. Only a handful of projects were listed as 'projects not of direct relevance to cancer research'.[33] As just noted, this 're-naming' exercise proved politically useful, although discussions about whether the balance of cancer research was 'fair to the patient' persisted (Ford 1974; Austoker 1988, 298–301).

Grasping the mood of the time, Brenner, greatly encouraged by the MRC, submitted an application for a new Tumour Biology Group in the laboratory. Research was to focus on the structure and characterisation of the surface antigens of tumour cells. This corresponded to one of the two areas the cancer research survey had singled out for further funding. Brenner did not ask for new posts since the group would be staffed from within his and Crick's division. However, the plan did require funds for the provision of new laboratory space. In his application Brenner stressed that the research was 'likely to have practical consequences' and that he was 'anxious' that anything the group might develop could be rapidly applied to 'human material'. He had secured the supply of human tumour material from the Department of Pathology, and hoped to develop closer links with clinical cancer immunologists.[34] In the course of a few years the MRC created a new Clinical Oncology Unit under Norman Bleehan and a Mechanisms in Tumour Immunity Group under Peter Lachmann, an immunologist from Hammersmith Hospital, as part of its 'cancer effort' at Cambridge. Both groups were explicitly linked to Brenner's initiative, with Lachmann's immunological work expected to provide the 'middle ground between the molecular biologists and the clinicians'.[35]

The Phillips Report praised the new collaboration between Brenner's, Bleehan's and Lachmann's groups in the field of cancer research. But neither Brenner's own group nor the interaction with the other groups produced the expected results. Soon after taking over the directorship of

[33] 'Work being undertaken at the Laboratory of Molecular Biology of relevance to cancer research' [attached to letter by M. Perutz to J. Neale, 3 April 1970], file S18/4, vol. 1, MRC Archives.

[34] 'Proposal to set up a Tumour Biology Group at the MRC Laboratory of Molecular Biology', Annex 1 – proposal of Dr Brenner; MRC 72/1132, file BRB, December 1972, MRC Archives.

[35] P. J. Lachmann, 'Proposal to join the tumour biology enterprise in Cambridge', MRC 75/595, Annex 1, file E243/24, vol. 1, MRC Archives. Parallel to the efforts at Cambridge, the MRC expanded a pre-existing Mammalian Genome Group under P. M. B. Walker at the University of Edinburgh into a fully fledged Unit of Fundamental Chromosome Studies in Relation to Cancer, which covered the second area identified for special support in Walker's own report.

the LMB, Brenner disbanded the group he had helped create. The blame for the failure of the venture went to the clinicians. 'The original conceived collaboration with Bleehan', he bluntly stated, 'has just not worked out and it is clear that while we remain on gentlemanly good terms with him, there is plainly not going to be any scientific collaboration of any magnitude.' Some years later, questioned on the prospects of 'molecular medicine' in Cambridge, Brenner elaborated his position, explaining that, while the LMB played an active role in applying molecular biology, collaborations were up to individuals and 'could only be encouraged, not decreed'. According to Brenner, much depended on the attitude of the clinicians, 'who could not expect the Laboratory staff to do their work for them'.[36] By that time, the hopes placed in tumour immunology for rapid advances in the diagnosis and therapy of cancer were waning. At the same time, the development of new technologies and the prospect of commercial ventures in biotechnology opened other (and more rewarding) prospects for bridging the gap between the molecular biology laboratory and the clinic.[37]

The monoclonal antibody scandal

To the authors of the Phillips Report the application of the techniques of DNA recombination still seemed 'conjectural'. In easier reach seemed the use of knowledge regarding the structure of enzymes and ligands for the rational design of drugs. The subcommittee was convinced that 'the strategic investment in the whole field [of molecular biology] is now at a stage where substantial dividends will be yielded' and intimated that 'a loss of nerve at this stage could prejudice that investment' ([Phillips Report], paragraph 3.14). However, the rapidity and the extent to which industrial investments and especially new venture capital (created in the United States through changes in capital gains tax and security laws) would flow into the field were hardly foreseeable.

The first 'new' biotechnology companies were all American.[38] However, by the late 1970s, the American developments raised concerns in

---

[36] S. Brenner to A. V. Harrison (MRC), 22 October 1980, and 'MRC Laboratory of Molecular Biology – Appointment of Director. Minutes of 1st meeting of Council Selection Committee held on 19 June 1986', file E243/23, vol. 2, MRC Archives. For at least partly more successful examples of collaborations between research laboratories and clinical establishments in the field of cancer immunology set up in the 1970s in Britain see Austoker (1988, 301–13). On the interaction of scientists and clinicians in the field of cancer immunology see also Löwy (1996).

[37] On the role of industry in the molecularisation of biology and medicine see de Chadarevian and Kamminga (1998b, 10–12).

[38] Recently, Rasmussen has suggested that there was nothing new in the 'new biotechnology', which marked but the late arrival of genetics to the commercialisation of life technologies (Rasmussen 1999). While stressing the continuities with the 'century-old

Britain about the state of biotechnology in the country. When in 1978 American researchers from the Wistar Institute in Philadelphia filed – and obtained – two patents for the production of monoclonal antibodies, a technology pioneered at the LMB in Cambridge, the spectre of Britain's missed opportunity to secure its rights on the commercialisation of penicillin was conjured up (Dickson 1979).[39] By the time Margaret Thatcher became Prime Minister, the 'failure' to apply for a patent of the hybridoma technique in 1975 had grown into a scandal which pushed the new government into action. In 1980, the state-owned National Enterprise Board announced the formation of a new biotechnology firm. Celltech had the backing of the government and the City and an agreement guaranteed the new firm preferential access to the commercialisation of research funded by the MRC, with the LMB as its prime target (Fig. 11.2).

The plan to create a biotechnology firm in Britain had been around for some time. Especially influential was the publication of a report on biotechnology, known as the Spinks Report after the chairman of the reporting committee, former director of research for ICI, Alfred Spinks. The committee, initiated by the second Wilson government, was set up by the Advisory Council for Applied Research and Development (a new advisory body set up in the wake of the Rothschild reforms), the Royal Society and the Advisory Board of the Research Councils. The report confirmed the promises of biotechnology (including the new recombinant DNA technologies) for the innovation of British industry and called for a coordinated effort in its support. In view of 'the shortage of venture capital and other resources of financial support for innovation in the United Kingdom' it specifically recommended the creation, 'with some public funds', of

tradition of biotechnology', Bud holds that the 'wedding with genetics' in the 1970s gave biotechnology new prominence (as it raised new fears) (Bud 1993). From a more epistemological point of view, Rheinberger has argued that genetic manipulations marked a radical shift from the extra-cellular representation of intra-cellular processes to the intra-cellular representation of an extra-cellular project, or the 're-writing' of life itself (Rheinberger 1995, 33–4). For further discussion on the question of continuities and discontinuities between the old and new biotechnologies see the collection of essays in Thackray (1998). On the rise of biotechnology firms linked to recombinant DNA technologies see Krimsky (1991) and Wright (1994, esp. 79–109); for a more journalistic account of the role of Genentech, the first new biotechnology firm, in the production of human insulin see Hall (1988). On the court case surrounding the foundation of Genentech see Gillis (1999).

[39] A similar patent application in Britain by the same scientists was not successful. On the construction of Köhler and Milstein's original experiments into the invention of a new 'technology' see Cambrosio and Keating (1992). On the patent issue in the field of monoclonal antibody research and more general considerations regarding the impact of patenting on the political economy of science and technology see Mackenzie, Keating and Cambrosio (1990).

11.2 Margaret Thatcher on a visit to the LMB in 1980. On her right Denis Thatcher and Max Perutz. Courtesy of Medical Research Council Laboratory of Molecular Biology, Cambridge.

a 'research-orientated biotechnology company of the kind taking shape elsewhere' ([Spinks Report] 1980, 41). But the report alone was not enough to convince the new Thatcher government to intervene. Not the concern with biotechnology itself, but the concern to improve the process of technology transfer in Britain in order to avoid further débâcles like the missed opportunities to patent the technology to produce monoclonal antibodies, moved the Prime Minister to make the investment, despite her anti-interventionist philosophy (Dodgson 1990, 17–24).[40]

[40] On the Spinks Report see also Wright (1994, 409–10). Among those serving on the committee was Hartley, who had moved on from the LMB to become Professor of Biochemistry at Imperial College London. In his new position he strongly pushed the development of biotechnology. Like other members of the committee he also had close links with industry, being a founding member of Biogen, a Swiss biotechnology firm. In addition to supporting the industrial development of biotechnology in Britain, the Spinks Report called for new funds for research in the field. In this context the customer–contractor principle was criticised for putting up an 'arbitrary and unreal divide between fundamental and applied research' ([Spinks Report] 1980, 22). In his critical appreciation of the new biotechnology Yoxen has described the creation of Celltech as 'a kind of compromise between the free market buccaneering in America and the bureaucracy of the NRDC' (Yoxen 1983, 132). Celltech's exclusive rights for MRC research, however, were soon lifted. For a critical perspective on Britain's biotechnology

By providing the cause for action, as well as staunch support for the initiative, the LMB played a crucial role in the formation of the new firm.[41] When, just a year after its official launch, the company took the decision to pursue recombinant DNA research (in addition to research on monoclonal antibodies), Brenner and Milstein, who sat on the Science Council of the firm, felt that this was a move away from the distinctive strength of the LMB which had supplied the research on which Celltech was founded.[42] They resigned in protest. This, however, was hardly the end of the collaboration between Celltech and the LMB. On the contrary, Celltech continued to rely on LMB research and, soon, also on the transfer of personnel.[43]

For LMB researchers this was not their first contact with industry. Starting with the rotating anode tube developed by Anthony Broad in 1952 and later, in an improved design, commercially produced by Elliott Bros, various technological developments in the laboratory were licensed, by the National Research Development Corporation (NRDC), to outside manufacturers. From the late 1960s onwards, an increasing number of LMB scientists had also taken on industrial consultancies, a step facilitated by

policy, based on state intervention, see Gottweis (1998b). In putting the hopes for industrial renewal in biotechnology (broadly construed), Britain followed Japan and especially Germany who had made this choice in the early 1970s (before the invention of recombinant DNA technologies). An important attraction of biotechnology in these countries was that it could be shown to offer a response to environmental concerns, an issue less central in political discussions in Britain at the time (Bud 1993, 141–62; also Gottweiss 1998a, esp. 181–209).

[41] In the year preceding the creation of Celltech, Brenner, LMB's newly installed director, held talks with the NRDC, the National Enterprise Board and Wellcome regarding commercial prospects for LMB research and the possibility of contract work being undertaken by the laboratory. Thanking Brenner for his efforts, an MRC officer ventured to comment: 'This exercise is going very well. If pushed to a successful end [it] could be important politically way beyond any simple question of money changing hands. I do hope you will encourage your people to play along. The whole question of the exploitation, development and marketing of MRC innovation is one the Secretary is very engaged with, as you know', T. Vickers (MRC) to S. Brenner, 24 August 1979, file E243/55, MRC Archives.

[42] The decision was based on the strong expertise in recombinant DNA technologies of one of the first Celltech employees, Norman Carey, who had joined the firm from GD Searle. Several researchers from Searle followed his move. The proximity to Searle's laboratories was one of the reasons which spoke for Slough, west of London (instead of Cambridge as originally planned) as the site for the new firm. Not locating the firm at Cambridge, however, also meant displaying independence from the LMB (Dodgson 1990, 25–6). The decision was taken at a time when Cambridge was gaining rising importance as a new high-tech centre with many start-up companies, though initially mainly in the sector of information technology and scientific instrumentation. On the 'Cambridge Phenomenon' of the 1980s see Segal Quince & Partners (1985), Garnsey and Cannon-Brookes (1993) and Gonzales-Benito, Reid and Garnsey (1997).

[43] In 1983 E. Lennox and in 1986 D. Secher from the LMB joined Celltech as project managers.

the redrafting of the MRC policy in this matter.[44] Though often quite unrelated to their actual research work, this side-activity not only was lucrative but also accustomed scientists to the business world.

Why, then, was the hybridoma technology not patented in 1975? And why did it become a 'scandal' only in 1980?[45] In their first publication on what came to be known as hybridoma technology, Köhler and Milstein did not fail to point out the possible commercial implications of their work. Their paper ended with the much cited sentence: 'Such cultures [of hybridised antibody-producing cells] could be valuable for medical and industrial use' (Köhler and Milstein 1975, 497).[46] Also the patent question was discussed in 1975. Hearing Milstein present his results at an internal MRC meeting, an officer of the MRC headquarters noted the possible commercial implications of the work, and – at least informally – contacted the NRDC. MRC researchers could only apply for patents through this body, set up in the postwar years to remedy what was already perceived as a British weakness, namely the capacity to turn scientific inventions into commercial products. It appears that a formal approach to the NRDC was not made until a year later (Tansey *et al.* 1997, 26). After considering the case, the NRDC took the position that there were not 'any immediate practical applications which could be pursued as commercial venture, even assuming that publication had not already occurred' (Tansey *et al.* 1997, 8–9). Not surprisingly, in the ensuing debates the NRDC bureaucrats were blamed for the missed opportunity.[47] However, it is also important to note that the first exchange between Milstein and the MRC headquarter happened only *after* the paper describing the experiments was already

---

[44] See 'Collaboration with industry. Proposed scheme of industrial consultancies', MRC 69/1261, 21 January 1970, file A804/1, MRC Archives. Among those engaged in consultancies in the 1970s were M. Perutz (from 1969 with Schering), U. Arndt (from 1970 with Enraf Nonius DELFT), B. Hartley (from 1971), S. Brenner (by 1970 with Du Pont; from 1980 with Rothschild's Biotechnology Investments Ltd), F. Crick (from 1970 with Depenidam), A. R. Faruqi (1978 with Marconi Avionics), R. Sheppard (from 1980 with ICI); A804, Industrial consultants [file index], MRC Archives.

[45] This question has been amply debated by participants and analysts. For the following see especially Mackenzie, Keating and Cambrosio (1990) and Tansey *et al.* (1997).

[46] The sentence recalls Watson and Crick's equally laconic and much cited final statement in their 1953 paper in *Nature* announcing the double helical structure of DNA: 'It has not escaped our notice that the specific pairing we have postulated immediately suggests a possible copying mechanism for the genetic material' (Watson and Crick 1953a, 737). For a discussion see Cambrosio and Keating (1992).

[47] In contrast to this later perception, the Spinks Report attributed much of the blame for the missed opportunity to seek early patent protection for monoclonal antibodies to the scientists. The passage much irritated Milstein and contributed to his decision to pursue more vigorously the industrial potentials of his research for some years thereafter (Tansey *el al.* 1977, 7–9).

accepted for publication by *Nature*. Yet according to British law a patent can only be granted prior to publication. This stands in contrast to American law, which allows one year of lapse between publication and patent application (Mackenzie, Keating and Cambrosio 1990, 80 n.29). Thus, while not alien to industry connections and commercial exploitation of their work, LMB researchers were more anxious to secure the publication of their results than to secure patent rights – that is, provided that Köhler and Milstein's research actually qualified for patent application, which remains an open question (Cambrosio and Keating 1992, 220). As one participant later summed up the situation, 'the LMB was not in a patent culture in 1975' (Tansey *et al.* 1997, 27). The same was true for British biomedical research laboratories more generally. It should be added that for government-funded laboratories in particular there was no incentive for a different attitude. Neither MRC researchers nor the MRC itself had a share in income from patents.

While under the effect of recombinant DNA technologies and the monoclonal antibody scandal this situation was quickly to change, LMB researchers were keen to keep up 'fundamental' research and defended this as the main mission of the laboratory.[48] A case in point is Milstein's own research trajectory. In view of claims by other groups, for some years he actively pursued the potentials of the new fusion technique. As he himself recalled: 'it dawned on me that it was up to us to demonstrate that the exploitation of our newly-acquired ability to produce monoclonal antibodies "à la carte" was of more importance than our original purpose... For several years I shelved the antibody diversity problem to demonstrate the practical importance of monoclonal antibodies in other areas of basic research and clinical diagnosis' (Milstein 1985, 203–4). Around this time, though apparently nearly against his will, we also find Milstein sitting on the scientific advisory board of Celltech. 'I was asked to be on the Council of Celltech and didn't want to do it', he later explained, 'but thought I ought to' (Tansey *et al.* 1997, 29). However, only a few years later he deliberately returned to more 'fundamental' problems of immunology. When he and Köhler were presented with the Nobel Prize for their achievements in immunology, Milstein ended his speech with a

---

[48] It has been argued that only the euphoria surrounding the potentials of the new genetic technologies made people appreciate the commercial potentials of monoclonal antibodies. On the share market, dividends in the field of monoclonal antibodies quickly outstripped the value of investments in recombinant DNA kits and products. For figures see Mackenzie, Cambrosio and Keating (1988, 157) and recent market analysis reports which predict dramatic growth rates for investments in the field of monoclonal antibodies (Silverman and Wright Marino 1998; Kim and van den Broeck 1999).

11.3 César Milstein in his laboratory at
the LMB. Photograph by Pete Addis.
From *New Scientist*, 21 May 1987, p. 54.
Reprinted with permission.

passionate defence of 'basic research'. The hybridoma technology, he declared, was a

clear-cut example of the enormous practical impact of an investment in research
which might not have been considered commercially worthwhile, or of immediate medical relevance. It resulted from esoteric speculations, for curiosity's sake;
only motivated by a desire to understand nature. It is to the credit of the MRC
in Britain to have fully appreciated the importance of basic research to advance
medicine. We are delighted to belong to the small, lucky group of those who are
at the window-dressing end of the justification for the wisdom of that policy.
(Milstein 1985, 213–14) (Fig. 11.3)

By 1980, LMB researchers had agreed to get involved in 'work of potential commercial interest', but on the condition that 'applied research
was not carried out at the expense of fundamental research to which the
LMB must continue to have a unique contribution to make'. However,
the rise of commercial interests around molecules and techniques developed by molecular biologists affected work practices more generally and
LMB researchers also had to concede that 'restrictions on publication
may sometimes have to be imposed – either because the national interest

transcends scientific interests or because commercial agencies may impose such restrictions as part of a contract'. The distribution of cell lines could fall under similar restrictions.[49] This concession notwithstanding, even researchers keen to develop their results commercially resisted giving up publication rights and control over the applications. The interactions between Celltech and LMB researcher Greg Winter regarding the commercialisation of techniques for 'humanising' mouse monoclonal antibodies, involving both LMB and Celltech patents, were marred by these difficulties (Dodgson 1990, 78). Winter eventually started his own biotech company, Cambridge Antibody Technology, exploiting further developments of the technology. A few years later, he resigned from his position in the company and resumed his LMB research position, though he continued to benefit from shares in the company. Despite the initial difficulties and some setbacks early on, the commercialisation of humanised monoclonal antibodies became the most successful venture for both Celltech and LMB. Revenues from patents and products are expected to grow dramatically in the next few years. Winter most openly describes himself as a scientist-entrepreneur.[50]

The case of the hybridoma technology shows that changes in the attitude towards research – in government circles as well as in industrial and certain scientific quarters – rather than the maturation of technological breakthroughs *per se* initiated the new commercial era of molecular biology. The change was supported by a new economic climate and the attempt to define new 'key technologies' for an industrial policy aimed at innovation as the basis for competitiveness and economic growth (see also Wright 1994, 108; Gottweis 1998a, especially 195–209). Molecular biologists who since the 1960s had sought and won political support for their 'modern' approach to biology had prepared the ground for the new expectations placed in their latest technologies. The new technologies, they now promised, would lead to important applications in the medical and agricultural fields and thus deliver the long-awaited economic pay-offs.

By positioning itself as a 'fundamental' research laboratory in this new landscape, the LMB kept something of a special position, continuing to profit from large government funds (after a period of restriction these began to grow again). But not only did the new commercial ventures offer

---

[49] P. B. Loder (MRC), 'Note for file', 11 March 1980 [relating to a discussion between S. Brenner, C. Milstein and the Secretary of the MRC], file E 243/55, MRC Archives.

[50] See his intervention at the 25th Anniversary of the Invention of Monoclonal Antibodies, London, 13 July 2000. Significantly, Winter belongs to a younger generation of scientists. His elder colleagues, who had worked on mission-orientated projects during the war but had built their careers by engaging in 'fundamental' research after the war, on the whole were more reluctant to embrace fully the new commercial trends.

new and to some people more attractive opportunities in the field of molecular biology; in the long run also the LMB could not resist the new trend. By 1985 the LMB had a patent liaison officer, and since the 1990s a small but rising percentage of the LMB income has come from royalties. With twelve new start-up companies based on developments from the laboratory, the LMB proudly takes its share in the growth of new high-tech companies around the university. While biotechnology companies were late-comers in what has been dubbed the 'Cambridge Phenomenon', in the early 1990s they represented the fastest-growing sector (Segal Quince & Partners 1985; Gonzales-Benito, Reid and Garnsey 1997). The extent to which the research culture has changed since 1975 was clearly visible at the 25th anniversary of the invention of monoclonal antibodies. This event, which was celebrated in style at the Queen Elizabeth II Conference Centre in the heart of London in July 2000, saw the participation of scientists, industrial entrepreneurs, financial analysts, health care managers, patent agents, high-level science administrators and the media. They all joined in celebrating the success and in devising ways for the future scientific *and* industrial development of the technology.

As already noted, other changes concurred with the push to commercialisation. In the early 1990s, a group of researchers moved out of the laboratory into the newly built Sanger Centre, where DNA sequencing was performed along industrial lines (see chapter 9).[51] It is too early to tell whether the expensively equipped and highly automated sequencing establishments will become 'service centres' to the research laboratories or if they will develop into the research centres of the future. But one observation can be made with certainty: if in the 1960s and 1970s molecular biology in Britain was taken by many to be synonymous with the LMB (with even the MRC getting confounded with it), in the vastly expanded and diversified landscape of the 1980s and 1990s, the LMB, though still a high-ranking laboratory and recipient of various Nobel Prizes, has become one among many sites where molecular biology is practised. At the same time, however, recent developments and especially the excitement surrounding the human genome project heighten the historical role of the LMB. It is celebrated as the place where the double helical structure of DNA was first proposed and where, in the 1970s, Fred Sanger

---

[51] Continuing the tradition of the LMB, the Sanger Centre, funded by the Wellcome Trust, the biggest charity for biomedical research in Britain (and world-wide), together with other publicly funded centres in several countries, competed fiercely with Celera Genomics of Rockeville, Maryland, headed by scientist-entrepreneur Craig Venter, to finalise the human genome sequence and secure that the information remained in the public domain. See the special issues of *Nature*, 15 February 2001 (409, no. 6822), and *Science*, 16 February 2001 (291, no. 5507).

developed the techniques which are at the basis of current approaches to DNA sequencing. The LMB also figures largely in plans to develop the hospital site at Cambridge into a 'biomedical campus', able to take up the challenges set by new developments in the health sector. The ambitious plans, drawn up jointly by the Addenbrooke's Hospital NHS Trust, the University of Cambridge and the MRC, comprise largely expanded facilities for general and specialised hospital care, education and research as well as for commercial settlements in the biomedical field.[52]

In considering the push to commercialisation we have moved far beyond the time period set for this book. The 1970s are generally taken to mark a shift in the history of molecular biology. This shift is attributed to the development of new technologies, especially recombinant DNA technologies, and the rise of commercial interests in the field, as well as to rising public awareness of the potentialities and dangers of the new technologies. Studying the political background relating to the general review of the LMB, the debates regarding the support of fundamental research and the history of the monoclonal antibody scandal, I have argued that the changes affecting the field in that period cannot be interpreted merely in terms of the application of new technologies. The scientific, political and economic uses of these technologies (recombinant DNA as well as hybridoma technologies) rather represented the response to much more wide-ranging changes regarding the practice, place and funding of molecular biology. Changes in government policies for science and growing competition among molecular biologists were as decisive as long-nurtured expectations of practical, especially medical pay-offs of basic biological research. The ground for the government move into biotechnology, finally, was well prepared by molecular biologists' tenacious negotiations, from the 1960s onwards, aimed at putting the promotion of their science on the governmental agenda.

[52] *Addenbrooke's: The 2020 Vision*, brochure issued by the Addenbrooke's NHS Trust, September 1999. Also 'Your local NHS', consultation document produced by Cambridgeshire Health Authority, September 1999. The diversification of research sites for molecular biology might fit the model of a 'distributed knowledge system' postulated by some sociologists as an emerging new mode of knowledge production (Gibbons *et al.* 1994). The thesis, however, lacks historical specificity. To what extent it is really a 'new mode' remains an open question.

12

...

Conclusions

The Laboratory of Molecular Biology serves as a vantage point to study the making of molecular biology. The history of the laboratory is instructive, not because of the sequence of Nobel Prize winning achievements, but because it offers the opportunity to examine in detail the complex web of activities and cultural resources used to build molecular biology in the decades following World War II. The episodes in the study are chosen so as to illuminate major transformations. The solutions found have been local, but they are symptomatic of wider changes.

The history of the laboratory started in the immediate postwar years, when reconstruction fuelled the hopes and offered the opportunities to build a new science of life, most often captured under the umbrella term of 'biophysics'. The different approaches gathered under this term mostly built on prewar research traditions, but their scope vastly increased after the war. Skills, networks and status acquired in the war, new government funds for basic biomedical research, technological developments, and the political and cultural attractions of a 'physics of life' were used to introduce new ways of producing knowledge of biological structures and processes in the laboratory. Recycled radar gears and electronic digital computers, together with new recruits and more secure financial backing, helped protein crystallographers in the Cambridge laboratory refine their techniques and produce intricate models of molecular structures which travelled widely. However, their approach was only one of the research directions assembled under the term biophysics.

Despite attempts to amalgamate the different research traditions which made up biophysics after the war, by the late 1950s centrifugal forces prevailed. When the protein crystallographers, together with Sanger's protein sequencing group, applied for new funds for the creation of an independent laboratory, they settled on 'molecular biology', a term which was around but not in frequent use, as the name for their new enterprise. Besides protein crystallography and protein sequencing, a structural approach to phage genetics established by Brenner and Crick in the laboratory became part of molecular biology as practised in Cambridge. As I have stressed, these moves were not inevitable but contingent.

The laboratory grew in a time of expanding science budgets. But money and scientific output are not enough to understand the extraordinary expansion of the laboratory and its role in the establishment of the new science. Cambridge molecular biologists did not only work at the bench. They created a journal, participated in recruiting tours, presented their work in the media and sat on governmental committees. They teamed up with other molecular biologists to create a European organisation for the promotion of molecular biology. They left the LMB to build up and head new institutions like EMBL or the Sanger Centre and, although initially somewhat reluctantly, established commercial links. Technologies created in the in-house workshops were commercially produced and used by laboratories around the world; postdoctoral researchers, attracted to and hosted by the laboratory in large numbers, exported the local way of doing things; the atomic models of the molecular structures established in the laboratory did not fail to impress visitors and played a crucial role in the public presentation of the science, in lecture rooms, exhibitions halls and on television. The laboratory itself became the pivot around which these activities turned. To understand the institution, its growing size as well as its role in the establishment of the field, requires an appreciation of this two-way traffic between the laboratory and the world which supported this kind of knowledge.

From the 1950s to the 1970s the laboratory profited from its status as a government-funded research institution, at arm's length from the university, while still benefiting from Cambridge's reputation as a place of academic excellence. By the 1980s this same status set limitations to the role of the LMB in the commercialisation and institutional diversification of molecular biology.

While offering glimpses of later developments, the study presented here has consciously focused on what I described as the 'long' postwar period, that is, the time of rising science budgets and of high public esteem for the sciences following the scientific mobilisation of World War II. This period coincided with the history of the establishment of molecular biology.

Despite the focus on this unique historical period, the study offers tools for a critical evaluation of current trends. Arguably, genomics, the study of whole genomes, linked to ever faster and automated ways of producing data in the laboratory, the influx of large amounts of venture capital and vested agricultural and pharmaceutical interests, is introducing a transformation of the biomedical sciences comparable in scale to that after World War II. These transformations help sharpen our historical understanding of the time in which molecular biology was instituted. Then government funds were available for fundamental research, the pressure of practical outcomes was low, the pace of research slower, and scientists overall enjoyed unbroken public recognition. Molecular biologists are the first to underline the changes their field is undergoing. They present tales of a 'world lost' similar to those of nuclear physicists after World War II.

There are all the same important continuities between the postwar era and the situation today. The private funds invested into biomedical research are matched by public funds, with governments investing unprecedented amounts of money into targeted areas, including especially genomics and electronic technologies. This public support of the sciences, and of the biomedical sciences in particular, builds as much on the legacies of World War II and the later efforts of molecular biologists to interest governments in their science, as the hopes of medical breakthroughs from the laboratory, the focus on molecules and the introduction of computers and data banks in biological research. The wartime success of penicillin and other medical mobilisation programmes gave the laboratory a new valency for medical research and provided tools and incentives for the investigation of molecules and subcellular structures, while the first genetic data banks were built with information gathered in studies of survivors of the first atomic bombs dropped on Japan. The studies on the effects of radiation fell under the same heading of biophysics as the structure analyses of the protein crystallographers who, as we have seen, were among the first to use electronic computers to handle their calculations. The media play a crucial role in the making of public perceptions about the new technologies and, indirectly, for keeping share prices high. Audiences have become sophisticated (and more diffident) in analysing some of the messages. But several of the popular scientific periodicals still in circulation were started in the late 1940s or 1950s. The first science programmes on television also date from that period.

These admittedly sketchy observations may suffice to confirm the continuing impact of practices and structures set in place after the war, and as an effect of the war. This might lead us to revise some of our views on postwar sciences. So far the picture has been dominated by accounts of nuclear physics, 'big' machines and large technological projects. The strategic role of nuclear physics in the Cold War era is hard to dispute. Biophysicists and molecular biologists in many ways depended on – and profited from – the resources, the prestige and the menace of nuclear physics. The postwar attraction of biophysics rested on the hope of turning some of its technologies to peaceful, medical uses as well as studying and preventing the destructive effects of nuclear radiation in the fabrication and testing of home-grown bombs or in the feared scenario of a nuclear war. Molecular biologists chose physicists as their founding figures, they made use of some of their most sophisticated technologies (above all computers and ever more powerful sources of radiation for the study of molecular structures) and relied on their example for institutional development, notably in the creation of EMBL. However, the work of molecular biologists was never centred on big machines in the way physics was, and it did not receive the military funding physics did, at least not in Britain, even if military funding did play a role. Although expanding fast, molecular biology was 'small science'. With the massive electronic databases and the industrial and financial alliances created in recent years, it may have entered the status of 'big' science, but arguably this happened only once molecular biologists began relying on biological mechanisms of replication and multiplication and under the changed political conditions of the post-Cold War era. In order to understand how molecular biology achieved its dominant position today, we need to make place for its 'small' beginnings.

Besides highlighting continuities and discontinuities with the situation today, the history of postwar molecular biology can make us appreciate that the current transformation of the field, if such a transformation is indeed taking place, will be institutional, political and cultural as much as narrowly cognitive or technical. It will be tied to changes of values extending to how we understand ourselves as well as life and reproduction more generally. To grasp these changes, the course of which is neither inevitable nor fixed, this study can only mark a beginning.

Bibliography

Aaserud, F. 1990. *Redirecting Science: Niels Bohr, Philanthropy, and the Rise of Nuclear Physics* (Cambridge: Cambridge University Press).

Abir-Am, P. G. 1980. 'From biochemistry to molecular biology. DNA and the acculturated journey of the critic of science Erwin Chargaff', *History and Philosophy of the Life Sciences* 2: 3–60.

1982a. 'The discourse of physical power and biological knowledge in the 1930s. A reappraisal of the Rockefeller Foundation's policy in molecular biology', *Social Studies of Science* 12: 341–82 and the 'Responses and replies', *Social Studies of Science* 14 (1984): 225–63.

1982b. 'Essay review. How scientists view their heroes: some remarks on the mechanism of myth construction', *Journal of the History of Biology* 15: 281–315.

1985. 'Themes, genres and orders of legitimation in the consolidation of new scientific disciplines: deconstructing the historiography of molecular biology', *History of Science* 23: 73–117.

1987. 'The biotheoretical gathering, trans-disciplinary authority and the incipient legitimation of molecular biology in the 1930s: new perspective on the historical sociology of science', *History of Science* 25: 1–70.

1989. 'Synergy or clash: disciplinary and marital strategies in the career of mathematical biologist Dorothy Wrinch', in P. Abir-Am and D. Outram (eds.), *Uneasy Careers and Intimate Lives: Women in Science 1789–1979* (New Brunswick: Rutgers University Press), pp. 239–80.

1991. 'Essay review. Noblesse oblige: lives of molecular biologists', *Isis* 82: 326–43.

1992a. 'The politics of macromolecules. Molecular biologists, biochemists, and rhetoric', *Osiris* 7: 210–37.

1992b. 'From multidisciplinary collaboration to transnational objectivity: international space as constitutive of molecular biology', in E. Crawford, T. Shinn and S. Sörlin (eds.), *Denationalizing Science* (Dordrecht: Kluwer), pp. 153–85.

1992c. 'A historical ethnography of a scientific anniversary in molecular biology: the first protein X-ray photograph (1984, 1934)', *Social Epistemology* 6: 323–54 (including the following comments and responses, pp. 355–87).

1997. 'The molecular transformation of twentieth-century biology', in J. Krige and D. Pestre (eds.), *Science in the Twentieth Century* (London: Harwood), pp. 495–524.

Abir-Am, P. G. (ed.) 1998. *La mise en mémoire de la science: pour une éthnographie historique des rites commemoratifs* (Amsterdam: Editions des Archives Contemporaines).

Abir-Am, P. G. and Elliott, C. A. (eds.) 1999. *Commemorative Practices in Science: Historical Perspectives on the Politics of Collective Memory*, Osiris 14.

Agar, J. 1998a. 'Introduction. History of computing: approaches, new directions and the possibility of informatic history', *History and Technology* 15: 1–5.

1998b. 'Digital patina. Texts, spirit and the first computer', *History and Technology* 15: 121–35.

1998c. *Science and Spectacle: The Work of Jodrell Bank in Post-War British Culture* (London: Harwood).

Agar, J. and Balmer, B. 1998. 'British scientists and the Cold War: the Defence Research Policy Committee and information networks, 1947–1963', *Historical Studies in the Physical and Biological Sciences* 28: 209–52.

Agar, J. and Hughes, J. 1995. 'Between government and industry: academic scientists and the reconfiguration of research practice at TRE', paper presented at Bournemouth University, December.

[Air Ministry]1963. *The Origins and Development of Operational Research in the R.A.F.* Air Publication 3368 (London: Her Majesty's Stationary Office).

Aldhous, P. 1993. 'Europe's genomes come home to roost', *Science* 260: 1741.

Allen, G. 1978. *Life Science in the Twentieth Century* (Cambridge: Cambridge University Press).

Alperovitz, G. 1995. *The Decision to Use the Atomic Bomb and the Architecture of an American Myth* (London: HarperCollins).

Alton, J. 1989. *Catalogue of the Papers and Correspondence of Sir John C. Kendrew*. Vols. I–II (Oxford: National Cataloguing Unit for the Archives of Contemporary Scientists).

Amis, K. 1954. *Lucky Jim* (London: Gollancz).

Ankeny, R. A. 1997. 'The conqueror worm: an historical and philosophical examination of the use of the nematode *C. elegans* as a model organism'. Ph.D. dissertation, University of Pittsburgh.

Annan, N. 1978. '"Our age". Reflections on three generations in England', *Daedalus* (Fall): 81–109.

Anonymous 1940. *Science in War* (Harmondsworth: Penguin).

1953a. 'The coronation', *Nature* 171: 943–4.

1953b. 'X-ray discovery', *Varsity* (Cambridge) 30 May: 1.

1956. 'This is our policy', *New Scientist* 1: 5.

1968a. 'Slipping through our fingers', *The Economist* 228 (13 July): 59.

1968b. 'Too little, too thin', *Nature* 219: 107–8.

1968c. 'Centres of excellence and despair', *Nature* 219: 212–13.

1968d. 'Lords of creation' [editorial], *New Scientist* 39: 117.

1969a. 'New look for MRC', *Nature* 223: 437–8.

1969b. 'Dog wags tail', *Nature* 223: 969.

1969c. 'The EMBO question debated', *Nature* 224: 406–7.

1974. 'Science broadcasting statistics', *Nature* 250: 362.

Appel, T. B. 2000. *The National Science Foundation and American Biological Research, 1945–1975* (Baltimore: The Johns Hopkins University Press).

Archer, P. 1997. 'A history of the Medical Artists' Association of Great Britain, 1949–1997'. Ph.D. dissertation, University College London.

Arndt, U. W., Crowther, R. A. and Mallett, J. F. W. 1968. 'A computer-linked cathode-ray tube microdensitometer for X-ray crystallography', *Journal of Scientific Instruments* (Series 2) 1: 510–16.

Ash, M. G. and Söllner, A. (eds.) 1996. *Forced Migration and Scientific Change: Emigré German-Speaking Scientists and Scholars after 1933* (Washington, DC: German Historical Institute).

[Association of Scientific Workers] 1947. *Science and the Nation* (Harmondsworth: Penguin Books).

Astbury, W. T. 1939. 'X-ray studies of the structure of compounds of biological interest', *Annual Review of Biochemistry* 8: 113–32.

1950–1. 'Adventures in molecular biology', *Harvey Lectures* 46: 3–44.

1961. 'Molecular biology or ultrastructural biology?', *Nature* 190: 1124.

Austoker, J. 1988. *A History of the Imperial Cancer Research Fund 1902–1986* (Oxford: Oxford University Press).

1989. 'Walter Morley Fletcher and the origins of a basic biomedical research policy', in J. Austoker and L. Bryder (eds.), *Historical Perspectives on the Role of the MRC: Essays in the History of the Medical Research Council of the United Kingdom and Its Predecessor, the Medical Research Committee, 1913–1953* (Oxford: Oxford University Press), pp. 23–33.

Balchin, N. 1985. *The Small Back Room* (Oxford: Oxford University Press).

Banham, M. and Hillier, B. (eds.) 1976. *A Tonic to the Nation: The Festival of Britain 1951* (London: Thames and Hudson).

[Barlow Report] 1946. *Scientific Manpower: Report of a Committee Appointed by the Lord President of the Council under the Chairmanship of Sir Alan Barlow* (London: His Majesty's Stationery Office), Cmnd. 6824.

Beatty, M. S. 1991. 'Genetics in the atomic age: the Atomic Bomb Casualty Commission, 1947–1956', in K. B. Benson, J. Maienschein and R. Rainger (eds.), *The Expansion of American Biology* (New Brunswick: Rutgers University Press), pp. 284–324.

Bennett, J. M. and Kendrew, J. C. 1952. 'The computation of Fourier synthesis with a digital electronic calculating machine', *Acta Crystallographica* 5: 109–16.

Berg, P. *et al.* 1974. 'Potential biohazards of "recombinant" DNA molecules', *Science* 185: 303.

Bernal, J. D. 1937–9. 'Structure of proteins', *Proceedings of the Royal Institution of Great Britain* 30: 541–57.

1939a. *The Social Function of Science* (London: Routledge and Kegan Paul).

1939b. 'X-ray evidence for the structure of protein molecules', in T. Svedberg *et al.*, 'A discussion on the protein molecule', *Proceedings of the Royal Society of London B* 127: 36–9.

1968. 'The pattern of Linus Pauling's work in relation of molecular biology', in A. Rich and N. Davidson (eds.), *Structural Chemistry and Molecular Biology: A Volume Dedicated to Linus Pauling by His Students, Colleagues and Friends* (San Francisco and London: Freeman and Co.), pp. 370–9.

1975. 'Lessons of the war for science', *Proceedings of the Royal Society of London A* 342: 555–74 [originally published in *Reports on Progress in Physics* 10 (1944–45): 418–36].

Bernal, J. D. and Crowfoot, D. 1934. 'X-ray photographs of crystalline pepsin', *Nature* 133: 794–5.

The Biochemical Society Yearbook and Annual Reports 1970. (London: Biochemical Society).

Birse, R. M. 1994. *Science at the University of Edinburgh 1583–1993: An Illustrated History to Mark the Centenary of the Faculty of Science and Engineering, 1893–1993* (Edinburgh: Faculty of Science and Engineering, University of Edinburgh).

Blow, D. M. 1997. 'The tortuous story of Asp . . . His . . . Ser: structural analysis of α-chymotrypsin', *Trends in the Biochemical Sciences* 22: 405–8.

Blow, D. M., Birktoft, J. J. and Hartley, B. S. 1969. 'Role of a buried acid group in the mechanism of action of chymotrypsin', *Nature* 221: 337–40.

Bodo, G., Dintzis, H. M., Kendrew, J. C. and Wyckoff, H. W. 1959. 'The crystal structure of myoglobin. V. A low resolution three-dimensional Fourier synthesis of sperm-whale myoglobin in crystals', *Proceedings of the Royal Society A* 253: 70–102.

Booth, C. 1989. 'Clinical research', in J. Austoker and L. Bryder (eds.), *Historical Perspectives on the MRC: Essays in the History of the Medical Research Council of the United Kingdom and Its Predecessor, the Medical Research Committee, 1913–1953* (Oxford: Oxford University Press), pp. 205–41.

Boyer, P. 1985. *By the Bomb's Early Light: American Thought and Culture at the Dawn of the Atomic Age* (New York: Pantheon).

Bragg, W. L. 1929. 'The determination of parameters in crystal structures by means of Fourier series', *Proceedings of the Royal Society of London A* 123: 537–59.

1941–3. 'Physicists after the War', *Proceedings of the Royal Institution* 32: 253–72.

1959. 'Introduction au Palais International de la Science', in M. Lambilliotte (ed.), *Le Mémorial officiel de l'Exposition universelle et internationale de Bruxelles 1958. Vol. VI: Les Sciences* (Brussels: Etablissements généraux d'Imprimerie), pp. 17–21.

1965. 'First stages in the X-ray analysis of proteins', *Reports on Progress in Physics* 28: 1–16.

1967. 'Introduction', *Proceedings of the Royal Society B* 167: 349.

Bragg, L. and Perutz, M. F. 1954. 'The structure of haemoglobin. VI: Fourier projections on the 010 plane', *Proceedings of the Royal Society of London A* 225: 315–29.

Brannigan, A. 1981. *The Social Basis of Scientific Discoveries* (Cambridge: Cambridge University Press).

Brenner, S. 1954. 'The physical chemistry of cell processes: a study of bacteriophage resistance in *Escherichia coli*, strain B.' Ph.D. dissertation, University of Oxford.

1973. 'The genetics of behaviour', *British Medical Bulletin* 29: 269–71.

1974a. 'New directions in molecular biology', *Nature* 248: 785–7.

1974b. 'The genetics of *Caenorhabditis elegans*', *Genetics* 77: 71–94.

1984. 'Nematode research', *Trends in Biochemical Sciences* 9: 172.

1997. *A Life in Science Told to Lewis Wolpert: Recorded on 10 April–21 May 1994* (London: Science Archive).

Brenner, S. and Barnett, L. 1959. 'Genetic and chemical studies on the head protein of bacteriophages T2 and T4', in *Structure and Function of Genetic Elements. Report of Symposium Held June 1–3, 1959.* Brookhaven Symposia 12 (Upton, NY: Brookhaven National Laboratory).

Brenner, S., Barnett, L., Crick, F. H. C. and Orgel, A. 1961. 'The theory of mutagenesis', *Journal of Molecular Biology* 3: 121–4.

Brenner, S., Benzer, S. and Barnett, L. 1958. 'Distribution of proflavin-induced mutations in the genetic fine structure', *Nature* 182: 983–5.

Brenner, S. and Horne, R. W. 1959. 'A negative staining method for high resolution electron microscopy of viruses', *Biochimica and Biophysica Acta* 34: 103–10.

Brenner, S., Streisinger, G., Horne, R. W., Champe, S. P., Barnett, L., Benzer, S. and Rees, M. W. 1959. 'Structural components of bacteriophage', *Journal of Molecular Biology* 1: 281–92.

Brickman, R. and Rip, A. 1979. 'Science policy advisory councils in France, the Netherlands and the United States, 1957–77', *Social Studies of Science* 9: 167–98.

Briggs, A. 1979. *The History of Broadcasting in the United Kingdom. Vol. IV: Sound and Vision* (Oxford: Oxford University Press).

Brooke, C. N. 1993. *A History of the University of Cambridge. Vol. IV: 1870–1990* (Cambridge: Cambridge University Press).

Bud, R. 1993. *The Uses of Life: A History of Biotechnology* (Cambridge: Cambridge University Press).

1998. 'Penicillin and the new Elizabethans', *British Journal for the History of Science* 31: 305–33.

Bud, R. and Gummett, P. (eds.) 1999. *Cold War, Hot Science: Applied Research in Britain's Defence Laboratories 1945–1990* (London: Harwood Academic Publishers).

Bud, R., Niziol, S., Boon, T. and Nahum, A. 2000. *Inventing the Modern World. Technology since 1750* (London: Dorling Kindersley).

Bude, H. 1988. 'Der Fall und die Theorie. Zum erkenntnislogischen Charakter der Fallstudien', *Gruppendynamik* 19: 421–7.

Burian, R. M. 1996. 'Underappreciated pathways towards molecular genetics as illustrated by Jean Brachet's cytochemical embryology', in S. Sarkar (ed.), *The Philosophy and History of Molecular Biology: New Perspectives* (Dordrecht and London: Kluwer Academic Publishers), pp. 67–83.

Burian, R. and Thieffry, D. (eds.) 1997. 'Research programs at the Rouge-Cloître Group', special issue, *History and Philosophy of the Life Sciences* 19, no. 1: 5–142.

Butler, J. A. V. and Huxley, H. E. 1963. 'Preface', *Progress in Biophysics and Molecular Biology* 13: [v].

Butler, J. A. V. and Katz, B. 1957. 'Preface', *Progress in Biophysics and Biophysical Chemistry* 8: vii.

Butler, J. A. V. and Randall, J. T. 1950. 'Preface', *Progress in Biophysics and Biophysical Chemistry* 1: vii.

C. elegans Sequencing Consortium 1998. 'Genome sequence of the nematode *C. elegans*: a platform for investigating biology', *Science* 282: 2012–18.

Cairns, J. 1992. 'Preface to the expanded edition', in J. Cairns, G. S. Stent, J. D. Watson (eds.), *Phage and the Origins of Molecular Biology* (Cold Spring Harbor, NY: Cold Spring Harbor Press), pp. v–viii.

Cairns, J., Stent, G. S. and Watson, J. D. 1966. *Phage and the Origins of Molecular Biology* (Cold Spring Harbor, NY: Cold Spring Harbor Laboratory of Quantitative Biology).

Calder, R. 1953. 'Why you are you: nearer the secret of life', *News Chronicle*, 12 May, p. 1.

1955. *Science Makes Sense* (London: George Allen and Unwin).

Cambrosio, A. and Keating, P. 1992. 'Between fact and technique: the beginnings of hybridoma technology', *Journal of the History of Biology* 25: 175–230.

1995. *Exquisite Specificity: The Monoclonal Antibody Revolution* (New York and Oxford: Oxford University Press).

Campbell-Kelly, M. 1992. 'The airy tape: an early chapter in the history of debugging', *IEEE Annals of the History of Computing* 14, no. 4: 16–26.

Capocci, M. and Corbellini, G. forthcoming. 'The rise of molecular biology in Italy', in S. de Chadarevian and B. Strasser (eds.), *Molecular Biology in Postwar Europe*, special issue, *Studies in History and Philosophy of Biological and Biomedical Sciences*.

Carswell, J. 1985. *Government and the Universities in Britain, 1960–1980* (Cambridge: Cambridge University Press).

Carter, G. B. 1992. *Porton Down: 75 Years of Chemical and Biological Research* (London: Her Majesty's Stationery Office).

Chain, E. B. 1964. 'The chemical structure of penicillins. Nobel Lecture, March 20, 1946', in *Nobel Lectures including Presentation Speeches and*

Laureates' Biographies. Physiology or Medicine, 1942–1962 (Amsterdam: Elsevier Publishing Company), pp. 110–43.

Chargaff, E., 1963. *Essays on Nucleic Acids* (Amsterdam: Elsevier Publishing Company).

1974. 'The tower of Babel', *Nature* 248: 776–97.

Chibnall, A. C. 1966. 'The road to Cambridge', *Annual Review of Biochemistry* 35: 1–22.

Chomet, S. (ed.) 1995. *D.N.A.: Genesis of a Discovery* (London: Newman-Hemisphere).

Clark, R. W. 1962. *The Rise of the Boffins* (London: Phoenix House).

1965. *Tizard* (London: Methuen).

1972. *A Biography of the Nuffield Foundation* (London: Longman).

Clarke, A. E. and Fujimura, J. H. (eds.) 1992. *The Right Tools for the Job: At Work in Twentieth-Century Life Sciences* (Princeton: Princeton University Press).

Clarke, P. 1996. *Hope and Glory: Britain 1900–1990* (Harmondsworth: Penguin).

Cohen, G. N. 1986. 'Four decades of Franco-American collaboration in biochemistry and molecular biology', *Perspectives in Biology and Medicine* 29: S141–S148.

Cohen, I. B. 1988. 'The computer: a case study of support by government, especially the military, of a new science and technology', in E. Mendelsohn, M. R. Smith and P. Weingart (eds.), *Science, Technology and the Military*. Sociology of the Sciences Yearbook 12, no. 1 (Dordrecht: Reidel Publishing Company), pp. 119–54.

Cohen, S. S. 1975. 'The origins of molecular biology', *Science* 187: 827–30.

1984. 'The biochemical origins of molecular biology. Introduction', *Trends in the Biochemical Sciences* 9: 334–6.

Committee of the Privy Council for Medical Research 1947. *Medical Research in War: Report of the Medical Research Council for the Years 1939–1945* (London: Her Majesty's Stationery Office), Cmnd. 7335.

Committee on Manpower Resources for Science and Technology 1967. *The Brain Drain: Report of the Working Group on Migration* (London: Her Majesty's Stationery Office).

Cook-Deegan, R. 1994. *The Gene Wars: Science, Politics, and the Human Genome* (New York and London: Norton & Co.).

Corbellini, G. 1999. *Le grammatiche del vivente: storia della biologia e della medicina molecolare* (Rome: Laterza).

Cox, E. G., Gross, L. and Jeffrey, G. A. 1947. 'Use of punched card tabulating machines for crystallographic Fourier syntheses', *Nature* 159: 433–4.

Crawford, E. 1984. *The Beginnings of the Nobel Institution: The Science Prizes, 1901–1915* (Cambridge: Cambridge University Press and Paris: Editions de la Maison des Sciences de l'Homme).

1992. *Nationalism and Internationalism in Science, 1880–1939: Four Studies of the Nobel Population* (Cambridge: Cambridge University Press).

Creager, A. N. H. 1993. 'Schlieren patterns, sedimentation coefficients, and sub-units: how the analytic ultracentrifuge structured proteins', paper presented at the ISHPSSB meeting at Brandeis University, July 1993.

—— 1996. 'Wendell Stanley's dream of a free-standing biochemistry department at the University of California, Berkeley', *Journal of the History of Biology* 29: 331–60.

—— 1998. 'Producing molecular therapeutics from human blood: Edwin Cohn's wartime enterprise', in S. de Chadarevian and H. Kamminga (eds.), *Molecularizing Biology and Medicine: New Practices and Alliances, 1910s–1970s* (London: Harwood), pp. 107–38.

—— In press. *The Life of a Virus: Tobacco Mosaic Virus as an Experimental Model, 1930–1965* (Chicago: University of Chicago Press).

Creager, A. N. H. and Gaudillière, J.-P. 1996. 'Meanings in search of experiments or vice-versa: the invention of allosteric regulation in Paris and Berkeley (1959–1967)', *Historical Studies in the Physical and Biological Sciences* 27: 1–89.

Crick, F. H. C. 1953. 'Polypeptides and Proteins. X-Ray Studies'. Ph.D. dissertation, University of Cambridge.

—— 1958. 'On protein synthesis', in *The Biological Replication of Macromolecules*. Symposia of the Society of Experimental Biology 12 (Cambridge: Cambridge University Press), pp. 138–63.

—— 1964. 'On the genetic code. Nobel lecture, December 11, 1962', in *Nobel Lectures Including Presentation Speeches and Laureates' Biographies. Physiology or Medicine, 1942–1962* (Amsterdam: Elsevier Publishing Company), pp. 811–19.

—— 1965. 'Recent research in molecular biology: introduction', *British Medical Bulletin* 21: 183–6.

—— 1969. 'Molecular biology and medical research', *Journal of the Mount Sinai Hospital* 36: 178–88.

—— 1970. 'Molecular biology in the year 2000', *Nature* 228: 613–15.

—— 1973. 'Project K: "The complete solution of *E. coli*"', *Perspectives in Biology and Medicine* 17: 67–70.

—— 1987. 'Ruthless research in a cupboard', *New Scientist* 114: 66–8.

—— 1990. *What Mad Pursuit: A Personal View of Scientific Discovery* (Harmondsworth: Penguin).

Crick, F. H. C., Barnett, L., Brenner, S. and Watts-Tobin, R. J. 1961. 'General nature of the genetic code for proteins', *Nature* 192: 1227–32.

Crick, F. H. C. and Orgel, L. 1981. *Life Itself* (New York: Simon and Schuster).

Crick, F. H. C. and Watson, J. D. 1956. 'Structure of small viruses', *Nature* 177: 473–6.

Croarken, M. 1990. *Early Scientific Computing in Britain* (Oxford: Clarendon Press).

—— 1992. 'The emergence of computing science research and teaching at Cambridge, 1936–1949', *IEEE Annals of the History of Computing* 14: 10–15.

Crowther, J. G. 1974. *The Cavendish Laboratory 1874–1974* (London: Macmillan).

Crowther, J. G., Howarth, O. J. R. and Riley, D. P. 1942. *Science and the World Order* (Harmondsworth: Penguin Books).

Crowther, J. G. and Whiddington, R. 1947. *Science at War* (London: His Majesty's Stationery Office).

Cullis, A. F., Muirhead, H., Perutz, M. F., Rossmann, M. G. and North, A. C. T. 1962. 'The structure of haemoglobin VIII. A three-dimensional Fourier synthesis at 5.5 Å resolution: description of the structure', *Proceedings of the Royal Society A* 265: 161–87.

[Dainton Report] 1971. *The Future of the Research Council System*, published as part of the Green Paper *A Framework for Government Research and Development* (London: Her Majesty's Stationery Office), Cmnd. 4814.

Dale, H. 1955. 'Edward Mellanby 1884–1955', *Biographical Memoirs of the Fellows of the Royal Society* 1: 193–222.

Davidson, J. N. 1970. 'Chairman's introduction', in T. W. Goodwin (ed.), *British Biochemistry Past and Present*. Biochemical Society Symposia 30 (London and New York: Academic Press), pp. 3–4.

de Chadarevian, S. 1994. 'Architektur der Proteine. Strukturforschung am Laboratory of Molecular Biology in Cambridge', in M. Hagner, H.-J. Rheinberger and B. Wahrig-Schmidt (eds.), *Objekte, Differenzen und Konjunkturen: Experimentalsysteme im historischen Kontext* (Berlin: Akademie Verlag), pp. 181–200.

1996a. 'Sequences, conformation, information: biochemists and molecular biologists in the 1950s', *Journal of the History of Biology* 29: 361–86.

1996b. 'Memoirs of a scientist-historian' (essay review), *Isis* 87: 507–10.

1997a. 'Using interviews for writing the history of science', in T. Söderqvist (ed.), *The Historiography of Contemporary Science and Technology* (London: Harwood Academic Publishers), pp. 51–70.

1997b. 'Internationale Wissenschaft und die Ökonomie wissenschaftlicher Kommunikation – Überlegungen im Anschluss an Mareys "La méthode graphique" (1978) und Bernals "The Social Function of Science" (1939)', in C. Meinel (ed.), *Fachschrifttum, Bibliothek und Naturwissenschaft im 19. und 20. Jahrhundert* (Wiesbaden: Harrassowitz Verlag), pp. 125–35.

1998a. 'Following molecules: hemoglobin between the clinic and the laboratory', in S. de Chadarevian and H. Kamminga (eds.), *Molecularizing Biology and Medicine: New Practices and Alliances, 1910s–1970s* (London: Harwood), pp. 171–201.

1998b. 'Of worms and programmes: *Caenorhabditis elegans* and the study of development', *Studies in the History and Philosophy of the Biological and Biomedical Sciences* 29: 81–105.

1999. 'Protein sequencing and the making of molecular genetics', *Trends in the Biochemical Sciences* 24: 163–207.

2000. 'Mapping development or how molecular is molecular biology?', *Journal for the History ond Philosophy of the Life Sciences* 22: 335–50.

Forthcoming. 'Models and the making of molecular biology', in S. de Chadarevian and N. Hopwood (eds.), *Displaying the Third Dimension: Models in the Sciences, Technology and Medicine* (Stanford: Stanford University Press).

de Chadarevian, S. and Gaudillière, J.-P. 1996. 'The tools of the discipline: biochemists and molecular biologists', *Journal of the History of Biology* 29: 327–30 [introduction to the special issue dedicated to this theme with contributions by A. N. H. Creager, S. de Chadarevian, H.-J. Rheinberger, J.-P. Gaudillière, L. E. Kay and R. M. Burian].

de Chadarevian, S. and Hopwood, N. (eds.) forthcoming. *Displaying the Third Dimension: Models in the Sciences, Technology and Medicine* (Stanford: Stanford University Press).

de Chadarevian, S. and Kamminga, H. (eds.) 1998a. *Molecularizing Biology and Medicine: New Practices and Alliances, 1910s–1970s* (London: Harwood).

de Chadarevian, S. and Kamminga, H. 1998b. 'Introduction', in de Chadarevian and Kamminga (eds.), *Molecularizing Biology and Medicine: New Practices and Alliances, 1910s–1970s* (London: Harwood), pp. 1–16.

de Chadarevian, S. and Strasser, B. (eds.) forthcoming. *Molecular Biology in Postwar Europe*, special issue, *Studies in History and Philosophy of Biological and Biomedical Sciences*.

de Maria, M., Grilli, M. and Sebastiani, F. (eds.) 1989. *Proceedings of the International Conference on the Restructuring of Physical Sciences in Europe and the United States, 1945–1960. Università 'La Sapienza', Rome, Italy, 19–23 September 1988* (Singapore: World Scientific), pp. 96–104.

[Deer Report] 1965. 'Report to the General Board of the Committee of the Board on the Long-Term Needs of Scientific Departments', *Cambridge University Reporter* 96: 545–616.

Deichmann, U. 1996. *Biologists under Hitler* (Cambridge, MA: Harvard University Press.)

Dennis, M. A. 1994. 'Our first line of defense: two university laboratories in the postwar American state', *Isis* 85: 427–55.

Dickerson, R. E. 1992. 'A little ancient history', *Protein Science* 1: 182–6.
    1997a. 'Irving Geis, molecular artist, 1908–1997', *Protein Science* 6: 2483.
    1997b. 'Irving Geis, 1908–1997', *Structure* 5: 1247–9.
    1997c. 'Molecular artistry', *Current Biology* 7: R740–R741.

Dickerson, R. E. and Geis, I. 1969. *The Structure and Action of Proteins* (New York: Harper and Row).

Dickson, D. 1979. 'California set to cash in on British discovery', *Nature* 279: 663–4.
    1988. *The New Politics of Science* (Chicago: Chicago University Press).

Dixon, B. 1971. 'D-Day for New Scientist and Science Journal', *New Scientist and Science Journal* 49: 99.

Dodgson, M. 1990. *Celltech: The First Ten Years of a Biotechnology Company*. Science Policy Research Unit. Discussion Paper Series (Brighton: University of Sussex).

Dore, S. 1996. 'Britain and the European Payments Union. British policy and American influence', in F. H. Heller and J. R. Gillingham (eds.), *The United States and the Integration of Europe: Legacies of the Postwar Era* (Basingstoke: Macmillan), pp. 167–97.

Dounce, A. 1953. 'Nucleic acids and template hypothesis', *Nature* 172: 541–2.

Doyle, R. 1994. 'Dislocating knowledge, thinking out of joint: rhizomatics, *Caenorhabditis elegans*, and the importance of being multiple', *Configurations* 2: 47–58.

1997. *On Beyond Living: Rhetorical Transformations of the Life Sciences* (Stanford: Stanford University Press).

Dubos, R. J. 1976. *The Professor, the Institute, and DNA* (New York: Rockefeller University Press).

Duster, T. 1990. *Backdoor to Eugenics* (London: Routledge).

Dyson, F. J. 1970. 'The future of physics', *Physics Today* 23: 23–8.

Edelson, E. 1998. *Francis Crick and James Watson and the Building Blocks of Life* (New York and Oxford: Oxford University Press).

Edge, D. O. and Mulkay, M. J. 1976. *Astronomy Transformed: The Emergence of Radio Astronomy in Britian* (New York: John Wiley & Sons).

Edgerton, D. E. H. 1991. *England and the Aeroplane: An Essay on a Militant and Technological Nation* (London: Macmillan).

1992. 'Whatever happened to the British warfare state? The Ministry of Supply, 1945–1951', in H. Mercer, N. Rollings and J. D. Tomlinson (eds.), *Labour Governments and Private Industry: The Experience of 1945–1951* (Edinburgh: Edinburgh University Press), pp. 91–116.

1996a. 'British scientific intellectuals and the relations of science, technology and war', in P. Forman and J. M. Sánchez-Ron (eds.), *National Military Establishments and the Advancement of Science and Technology: Studies in 20th Century History* (Dordrecht: Kluwer Academic Publishers), pp. 1–35.

1996b. *Science, Technology and the British Industrial 'Decline', 1870–1970* (Cambridge: Cambridge University Press).

1996c. 'The "White Heat" revisited: the British government and technology in the 1960s', *Twentieth Century British History* 7: 53–82.

1997. 'C. P. Snow as historian of British culture', paper presented at the The British Society for the History of Science 50th Anniversary Conference: The History of Science as Public Culture, at the British Association Festival of Britain, University of Leeds, 9–11 September.

Edsall, J. T. *et al.* 1996. 'A proposal of standard conventions and nomenclature for the description of polypeptide conformations', *Journal of Molecular Biology* 15: 399–407.

Edwards, P. N. 1996. *The Closed World: Computers and the Politics of Discourse in Cold War America* (Cambridge, MA: MIT Press).

Ellwood, D. W. 1992. *Rebuilding Europe: Western Europe, America and Postwar Reconstruction* (London: Longman).

Elzen, B. 1986. 'Two ultracentrifuges: a comparative study of the social construction of artefacts', *Social Studies of Science* 16: 621–62.

The Emigration of Scientists from the United Kingdom. Report of a Committee Appointed by the Royal Society 1963 (London: The Royal Society).

Ephrussi, B., Leopold, U., Watson, J. D. and Weigle, J. J. 1953. 'Terminology in bacterial genetics', *Nature* 171: 701.

European Molecular Biology Organisation: 25th Anniversary 1964–1989 (Heidelberg: EMBO).

Ewald, P. P. 1969–70. 'The myth of myths. Comments on P. Forman's paper on "The discovery of the diffraction of X-rays in crystals" ', *Archive for History of Exact Sciences* 6: 72–81.

Ewald, P. P., (ed.) 1962. *Fifty Years of X-Ray Diffraction. Dedicated to the International Union of Crystallography on the Occasion of the Commemoration Meeting in Munich, July 1962* (Utrecht: Oosthoek's Uitgeversmaatschapp).

Feinstein, A. R. 1995. 'The crisis in clinical research', *Bulletin of the History of Medicine* 69: 288–91.

Fenton, B. 1996. 'Day they dropped a nuclear bomb on Birmingham', *The Daily Telegraph*, Monday, 23 December, p. 7.

Fermi, G. and Perutz, M. 1981. *Haemoglobin and Myoglobin* (vol. 2 of *Atlas of Molecular Structures in Biology*, edited by D. C. Phillips and F. M. Richards) (Oxford: Clarendon Press).

Ferry, G. 1998. *Dorothy Hodgkin: A Life* (London: Granta Books).

Fincham, J. R. S. 1993. 'Genetics in the United Kingdom – the last half century', *Heredity* 71: 111–18.

Fischer, P. and Lipson, C. 1988. *Thinking about Science: Max Delbrück and the Origins of Molecular Biology* (New York: Norton).

Fleming, D. 1969. 'Emigré physicists and the biological revolution', in D. Fleming and B. Bailyn (eds.), *The Intellectual Migration: Europe and America, 1930–1960* (Cambridge, MA: Belknap Press of Harvard University Press), pp. 152–89.

Ford, B. 1974. 'Does cancer research have realistic aims?', *Nature* 249: 299–300.

Forgan, S. 1997. 'Native genius and landmarks of progress: science and national identity at the Festival of Britain 1951', paper presented at the British Society 50th Anniversary Conference *The History of Science as Public Culture?* at the British Association Festival of Science, University of Leeds, 9–11 September.

1998. 'Festivals of science and the two cultures: science, design and display in the Festival of Britain 1951', *British Journal of the History of Science* 31: 217–40.

Forman, P. 1969–70. 'The discovery of the diffraction of X-rays by crystals: a critique of the myths', *Archive for History of Exact Sciences* 6: 38–81.

1988. 'Behind quantum electronics: national security as basis for physical research in the United States, 1940–1960', *Historical Studies in the Physical and Biological Sciences* 18: 149–229.

1989. 'Social niche and self-image of the American physicist', in M. de Maria, M. Grilli and F. Sebastiani (eds.), *Proceedings of the International*

Conference on the Restructuring of Physical Sciences in Europe and the United States, 1945–1960. Università 'La Sapienza', Rome, Italy, 19–23 September 1988 (Singapore: World Scientific), pp. 96–104.

1996. 'Into quantum electronics: the maser as "gadget" of cold-war America', in P. Forman and J. M. Sanchez-Ron (eds.), *National Military Establishments and the Advancement of Science and Technology* (Dordrecht: Kluwer Academic Publishers), pp. 261–326.

Fortun, M. A. 1993. 'Mapping and making genes and histories: the genomics project in the United States, 1980–1990'. Ph.D. dissertation, Harvard University.

1998. 'The Human Genome Project and the acceleration of biotechnology', in A. Thackray (ed.), *Private Science: Biotechnology and the Rise of the Molecular Sciences* (Philadelphia: University of Pennsylvania Press), pp. 182–201.

1999. 'Projecting speed genomics', in M. Fortun and E. Mendelsohn (eds.), *The Practices of Human Genetics*. Sociology of the Sciences Yearbook 21 (Dordrecht: Kluwer), pp. 25–48.

Fortun, M. and Schweber, S. S. 1993. 'Scientists and the legacy of World War II: the case of operations research', *Social Studies of Sciences* 23: 595–642.

Forty, A. 1976. 'Festival politics', in M. Banham, and B. Hillier (eds.), *A Tonic to the Nation: The Festival of Britain 1951* (London: Thames and Hudson), pp. 26–38.

Fox Keller, E. 1983. *A Feeling for the Organism: The Life and Work of Barbara McClintock* (New York: Freeman and Company).

1990. 'Physics and the emergence of molecular biology: a history of cognitive and political synergy', *Journal for the History of Biology* 23: 389–409.

1995a. *Reconfiguring Life: Metaphors of Twentieth-Century Biology* (New York: Columbia University Press).

1995b. 'The body of a new machine: situating the organism between telegraphs and computers', in *Reconfiguring Life: Metaphors of Twentieth-Century Biology* (New York: Columbia University Press), pp. 79–118.

1996. 'Drosophila embryos as transitional objects. The work of Donald Poulson and Christiane Nüsslein-Volhard', *Historical Studies in the Physical and Biological Sciences* 26: 313–46.

2000. *The Century of the Gene* (Cambridge, MA: Harvard University Press).

A Framework for Government Research and Development 1971. (London: Her Majesty's Stationery Office), Cmnd. 4814.

Framework for Government Research and Development 1972. (London: Her Majesty's Stationery Office), Cmnd. 5046.

Francoeur, E. 1997. 'The forgotten tool: the design and use of molecular models', *Social Studies of Science* 27: 7–40.

Francoeur, E. and Segal, J. forthcoming. 'From model kits to interactive computer graphics', in S. de Chadarevian and N. Hopwood (eds.), *Displaying* the Third Dimension: Models in the Sciences, Technology and Medicine (Stanford: Stanford University Press).

Franklin, S. Brooks 1988. 'Life Story. The gene as fetish object on TV', *Science as Culture* 3: 92–100.

Frayn, M. 1986. 'Festival', in M. Sissons and P. French (eds.), *Age of Austerity* (Oxford: Oxford University Press), pp. 305–26.

Fritsch, O. R. 1980. *What Little I Remember* (Cambridge: Cambridge University Press).

Fruton, J. S. 1992. *A Skeptical Biochemist* (Cambridge, MA: Harvard University Press).

    1999. *Proteins, Enzymes, Genes: The Interplay of Chemistry and Biology* (New Haven: Yale University Press).

Fujimura, J. H. and Fortun, M. 1996. 'Constructing knowledge across social worlds: the case of DNA sequence database in molecular biology', in L. Nader (ed.), *Naked Science: Anthropological Inquiry into Boundaries, Power and Knowledge* (New York and London: Routledge), pp. 160–73.

Gaber, B. P. and Goodsell, D. S. 1997. 'The art of molecular graphics. Irving Geis: dean of molecular illustration', *Journal of Molecular Graphics and Modelling* 15: 57–9.

Galison, P. 1987. *How Experiments End* (Chicago: Chicago University Press).

    1988. 'Physics between war and peace', in E. Mendelsohn, M. R. Smith and P. Weingart (eds.), *Science, Technology and the Military*, vol. 1 (Dordrecht: Reidel Publishing Company), pp. 47–86.

    1994. 'The ontology of the enemy. Norbert Wiener and the cybernetic vision', *Critical Inquiry* 21: 228–66.

    1997. *Image and Logic: A Material Culture of Microphysics* (Chicago: University of Chicago Press).

Galison, P. and Hevly, B. (eds.) 1992. *Big Science: The Growth of Large-Scale Research* (Stanford: Stanford University Press).

Gamow, G. 1954a. 'Possible relation between deoxyribonucleic acid and protein structure', *Nature* 173: 318.

    1954b. 'Possible mathematical relation between deoxyribonucleic acid and protein', *Biologiske meddelelser, kongelige Danske Videnskabernes Selskab* 22: 1–13.

Gardiner, J. 1999. *From the Bomb to the Beatles* (London: Collins & Brown).

Garnsey, E. and Cannon-Brookes, A. 1993. *The 'Cambridge Phenomenon' Revisited: Aggregate Changes among Cambridge High Technology Companies since 1985*. Research Paper 1992–1993, No. 3, Judge Institute of Management Studies, Cambridge.

Gaudillière, J.-P. 1991. 'Biologie moléculaire et biologistes dans les années soixante: la naissance d'une discipline. Le cas français'. Ph.D dissertation, Université Paris VII.

1993. 'Molecular biology in the French tradition? Redefining local traditions and disciplinary patterns', *Journal of the History of Biology* 26: 473–98.

1996. 'Molecular biologists, biochemists, and messenger RNA: the birth of a scientific network', *Journal of the History of Biology* 29: 417–45.

1998. 'The molecularization of cancer etiology in the postwar United States: instruments, politics and management', in S. de Chadarevian and H. Kamminga (eds.), *Molecularizing Biology and Medicine: New Practices and Alliances, 1910s–1970s* (London: Harwood), pp. 139–70.

In press. *L' invention de la biomédicine: la construction des sciences de la vie et de la maladie entre France et Amérique, 1945–1965* (Paris: La Découverte).

Geison, G. L. 1978. *Michael Foster and the Cambridge School of Physiology: The Scientific Enterprise in Late Victorian Society* (Princeton, NJ: Princeton University Press).

Geison, G. L. and Holmes, F. L. (eds.) 1993. *Research Schools. Historical Reappraisals*, Osiris 8.

Gibbons, M., Limoges, C., Nowotny, H., Schwartzman, S., Scott, P. and Trow, M. 1994. *The New Production of Knowledge: The Dynamics of Science and Research in Contemporary Societies* (London: Sage Publications).

Gillis, J. 1999. '20 years later, stolen gene haunts a biotech pioneer', *Washington Post*, 17 May, p. A01.

Glusker, J. P., Patterson, B. K. and Rossi, M. (eds.) 1987. *Patterson and Pattersons: Fifty Years of the Patterson Function* (New York and Oxford: Oxford University Press).

Glusker, J. P. and Trueblood, K. N. 1985. *Crystal Structure Analysis: A Primer* (New York and Oxford: Oxford University Press).

Glynn, J. 1996. 'Rosalind Franklin, 1920–1958', in E. Shils and C. Blacker (eds.), *Cambridge Women: Twelve Portraits* (Cambridge: Cambridge University Press), pp. 267–82.

Goldsmith, M. 1980. *Sage: A Life of J. D. Bernal* (London: Hutchinson).

Golinski, J. 1998. *Making Natural Knowledge: Constructivism and the History of Science* (Cambridge: Cambridge University Press).

Gonzales-Benito, J., Reid, S. and Garnsey, E. 1997. *The Cambridge Phenomenon Comes of Age*. Research Paper in Management Studies WP 22/9 (Cambridge: Judge Institute of Management Studies).

Gooding, D., Pinch, T. and Schaffer, S. 1989. *The Uses of Experiments: Studies in the Natural Sciences* (Cambridge: Cambridge University Press).

Gosling, R. G. 1995. 'X-ray diffraction studies of NaDNA with Rosalind Franklin', in S. Chomet (ed.), *D.N.A.: Genesis of a Discovery* (London: Newman-Hemispere), pp. 43–73.

Gossling, T. H. and Mallett, W. F. 1968. 'Use of computers in a molecular biology laboratory', Fall Joint Computer Conference, 9–11 December 1968, San Francisco, CA, *American Federation of Information Processing Societies Conference Proceedings* 33(2): 1089–98.

Gottweis, H. 1995. 'Genetic engineering, democracy, and the politics of identity', *Social Text* 42: 127–52.

1998a. *Governing Molecules: The Discursive Politics of Genetic Engineering in Europe and the United States* (Cambridge, MA: MIT Press).

1998b. 'The political economy of British biotechnology', in A. Thackray (ed.), *Private Science: Biotechnology and the Rise of the Molecular Sciences* (Philadelphia: University of Pennsylvania Press), pp. 105–30.

Gould, P. 1997. 'Women and the culture of university physics in late nineteenth-century Cambridge', *British Journal for the History of Science* 30: 127–49.

Gowing, M. 1964. *Britain and Atomic Energy, 1939–1945* (London: Macmillan).

1974a. *Independence and Deterrence: Britain and Atomic Energy, 1945–1952. Vol. I: Policy Making* (London: Macmillan).

1974b. *Independence and Deterrence: Britain and Atomic Energy, 1945–1952. Vol. II: Policy Execution* (London: Macmillan).

Gray, G. W. 1951. 'Sickle cell anemia', *Scientific American* 185 (August): 56–9.

Green, D. W., Ingram, V. M. and Perutz, M. 1954. 'The structure of haemoglobin. IV: Sign determination by the isomorphous replacement method', *Proceedings of the Royal Society A* 225: 287–307.

Gross, M. A. G. 1990. *The Rhetoric of Science* (Cambridge, MA: Harvard University Press).

Gummett, P. 1980. *Scientists in Whitehall* (Manchester: Manchester University Press).

Hacker, B. 1987. *The Dragon's Tail: Radiation Protection in the Manhattan Project, 1942–1946* (Berkeley: University of California Press).

Hacking, I. 1986. 'Weapons research and the form of scientific knowledge', *Canadian Journal of Philosophy* Suppl. 12: 237–60.

Hägg, G. 1964. 'Chemistry 1962. Presentation speech', in *Nobel Lectures Including Presentation Speeches and Laureates' Biographies. Chemistry 1942–1962* (Amsterdam: Elsevier Publishing Company), pp. 649–52.

Hall, S. S. 1988. *Invisible Frontiers: The Race to Synthesize a Human Gene* (London: Sidgwick & Jackson).

Halsey, A. H. 1994. 'Oxford and the British universities', in B. Harrison (ed.), *The History of the University of Oxford. Vol. VIII: The Twentieth Century* (Oxford: Clarendon Press), pp. 577–606.

Hardy, W. B. 1928. 'Living matter', *Colloid Symposium Monograph* 6: 7–16.

Hargittai, I. 1999. 'Sidney Altman. Interview', *The Chemical Intelligencer* 5(2): 12–16.

Hartley, B. S. 1962. 'On the structure of chymotrypsin', *Brookhaven Symposia in Biology* 15: 85–100.

1964. 'Amino-acid sequence of bovine chymotrypsinogen-A', *Nature* 201: 1284–7.

1970. 'The primary structure of proteins', in T. N. Goodwin (ed.), *British Biochemistry Past and Present*. Biochemical Society Symposium 30 (London and New York: Academic Press), pp. 29–41.

Hartree, E. F. 1968. 'Cambridge: molecular biology', *Science* 160: 252.

Hauptman, H. A. 1998. 'David Harker', *Biographical Memoirs National Academy of Sciences of the U.S.A.* 74: 127–43.

Hayes, W. 1966. 'Sexual differentiation in bacteria', in J. Cairns, G. S. Stent and J. D. Watson (eds.), *Phage and the Origins of Molecular Biology* (Cold Spring Harbor, NY: Cold Spring Harbor Laboratory of Quantitative Biology), pp. 201–15.

Hecht, G. 1998. *The Radiance of France: Nuclear Power and National Identity after World War II* (Cambridge, MA: MIT Press).

Heilbron, J. L. 1974. *H. G. J. Moseley: The Life and Letters of an English Physicist, 1887–1915* (Berkeley: University of California Press).

Hennessy, P. 1992. *Never Again: Britain 1945–1951* (London: Jonathan Cape).

Hermann, A., Krige, J., Mersits, U. and Pestre, D. 1987–90. *History of CERN*, vols. 1–2 (Amsterdam: North Holland).

Hilgartner, S. 1998. 'Data access policy in genome research', in A. Thackray (ed.), *Private Science: Biotechnology and the Rise of the Molecular Sciences* (Philadelphia: University of Pennsylvania Press), pp. 201–18.

Hill, A. V. 1931. *Adventures in Biophysics* (Oxford: Oxford University Press).
1956. 'Why biophysics?', *Science* 124: 1233–7; originally published in Postgraduate Medical Federation (ed.), *Lectures on the Scientific Basis of Medicine. Vol. IV: 1954–55* (London: Athlone Press), pp. 1–17.

Hinshelwood, C. N. 1946. *The Chemical Kinetics of the Bacterial Cell* (Oxford: Clarendon Press).

Hinsley, F. H. and Stripp, A. (eds.) 1994. *Codebreakers: The Inside Story of Bletchley Park* (Oxford: Oxford University Press).

Hoagland, M. B. 1990. *Toward the Habit of Truth* (New York: W. W. Norton & Company).

Hobsbawm, E. 1994. *Age of Extremes: The Short Twentieth Century, 1914–1991* (London: Michael Joseph).

Hodges, A. 1992. *Alan Turing: The Enigma* (London: Vintage).

Hodgkin, A. 1992. *Chance and Design: Reminiscences of Science in Peace and War* (Cambridge: Cambridge University Press).

Hodgkin, D. Crowfoot 1949. 'The X-ray analysis of the structure of penicillin', *Advancement of Science* 6: 85–9.
1980. 'John Desmond Bernal, 10 May 1901–15 September 1971', *Biographical Memoirs of the Fellows of the Royal Society* 26: 17–84.

Hodgkin, D. Crowfoot and Riley, D. P. 1968. 'Some ancient history of protein X-ray analysis', in A. Rich and N. Davidson (eds.), *Structural Chemistry and Molecular Biology* (San Francisco and London: Freeman and Co.) pp. 15–29.

Hodgkin, J. 1989. 'Early worms', *Genetics* 121: 1–3.
1991. 'The interactive worm', *The New Biologist* 3: 951–4.

Hodgson, M. L., Clews, C. J. B. and Cochran, W. 1949. 'A punched-card modificaton of the Beevers-Lipson method of Fourier synthesis', *Acta Crystallographica* 2: 113–16.

Holland, R. 1991. *The Pursuit of Greatness: Britain and the World Role, 1900–1970* (London: Fontana Press).

Holmes, A. 1993. 'Duke honours DNA Nobel duo', *Cambridge Evening News*, 9 June, p. 3.

Holmes, F. L. 2000. 'Seymour Benzer and the definition of the gene', in P. Beurton, R. Falk and H.-J. Rheinberger (eds.), *The Concept of the Gene in Development and Evolution: Historical and Epistemological Perspectives* (Cambridge: Cambridge University Press), pp. 115–55.

Hopwood, N. 1999. ' "Giving body" to embryos: modeling, mechanism, and the microtome in late nineteenth-century anatomy', *Isis* 90: 462–96.

Forthcoming. 'Embryology', in P. J. Bowler and J. V. Pickstone (eds.), *The Cambridge History of Science. Vol. VI: Life and the Earth Sciences since 1800* (New York and Cambridge: Cambridge University Press).

Horner, D. 1993. 'The road to Scarborough: Wilson, Labour and the scientific revolution', in R. Coopey, S. Fielding and N. Tiratsoo (eds.), *The Wilson Governments 1964–1970* (London and New York: Pinter Publishers), pp. 48–71.

Horvitz, H. R. 1977. 'A uniform genetic nomenclature for the nematode *Caenorhabditis elegans*', *The Worm Breeder's Gazette* 2, no. 3, part B: 1–14.

Hubbard, R. 1990. *The Politics of Women's Biology* (New Brunswick, NJ: Rutgers University Press).

Hunt, F. and Barker, C. 1998. *Women at Cambridge: A Brief History* (Cambridge: The Press and Publications Office, University of Cambridge).

Huxley, A. F. 1972. 'The quantitative analysis of excitation and conduction in nerve. Nobel lecture, December 11, 1963', in *Nobel Lectures Including Presentation Speeches and Laureates' Biographies. Physiology or Medicine 1963–1970* (Amsterdam: Elsevier Publishing Company), pp. 52–69.

Huxley, H. E. 1987. 'Double vision reveals the structure of muscle', *New Scientist* 114: 42–5.

1990. 'An early adventure in crystallographic computing', in J. M. Thomas and D. Phillips (eds.), *Selections and Reflections: The Legacy of Sir L. Bragg* (Northwood: Science Reviews), pp. 133–4.

1996. 'A personal view of muscle and motility mechanisms', *Annual Review of Physiology* 58: 1–19.

Huxley, H. E. and Holmes, K. C. 1997. 'Development of synchroton radiation as a high-intensity source of X-ray diffraction', *Journal of Synchroton Radiation* 4: 366–79.

Huxley, J. S. 1940. 'Science in war' [review of *Science in War*], *Nature* 146: 112–13.

Ingram, V. 1956. 'A specific chemical difference between the globins in normal human and sickle-cell anaemia haemoglobin', *Nature* 178: 792–4.

1957. 'Gene mutations in human haemoglobin: the chemical difference between normal human and sickle-cell haemoglobin', *Nature* 180: 326–8.

IUPAC-IUB Commission on Biochemical Nomenclature 1970. 'Abbreviations and symbols for the description of the conformation of polypeptide chains', *Journal of Molecular Biology* 52: 1–17.

Jackson, D. A. and Stich, S. P. (eds.) 1979. *The Recombinant DNA Debate* (Englewood Cliffs, NJ: Prentice-Hall).

Jackson, L. 1991. *The New Look: Design in the Fifties* (London: Thames and Hudson).

Jacob, F. 1971. 'La belle époque', in J. Monod and E. Borek (eds.), *Of Microbes and Mice: Les microbes et la vie* (New York and London: Columbia University Press), pp. 98–104.

1972. 'Genetics of the bacterial cell. Nobel lecture, December 11, 1965', in *Nobel Lectures Including Presentation Speeches and Laureates' Biographies. Physiology or Medicine, 1963–1970* (Amsterdam: Elsevier Publishing Company), pp. 148–71.

1998. *Of Flies, Mice, and Men* (Cambridge, MA: Harvard University Press).

Jacq, F. 1995. 'The emergence of French research policy: methodological and historiographical problems (1945–1970)', *History and Technology* 12: 285–308.

Jones, R. A. 1997. 'The Boffin: a stereotype of scientists in post-war British films (1945–1970)', *Public Understanding of Science* 6: 31–48.

Jones, R. V. (ed.), 1975. 'A discussion on the effects of the two World Wars on the organisation and development of science in the United Kingdom', *Proceedings of the Royal Society of London A* 342 (1975), 439–586.

Jordanova, L. 1993. 'Gender and the historiography of science', *British Journal for the History of Science* 26: 469–83.

Judson, H. F. 1994. *The Eighth Day of Creation: Makers of the Revolution in Biology* (London: Penguin; first published 1979).

Kamminga, H. and de Chadarevian, S. 1995. *Representations of the Double Helix* (Cambridge: Wellcome Unit for the History of Medicine).

Kamminga, H. and Weatherall, M. W. 1996. 'The making of a biochemist: I. Frederick Gowland Hopkins' construction of dynamic biochemistry', *Medical History* 40: 269–92.

Katz, B. 1969. 'The EMBO question debated' [letter], *Nature* 224: 733.

1978. 'Archibald Vivian Hill, 26 September 1886 – 3 June 1977', *Biographical Memoirs of the Fellows of the Royal Society* 24: 71–149.

Kay, L. 1985. 'Conceptual models and analytical tools: the biology of physicist Max Delbrück', *Journal of the History of Biology* 18: 207–46.

1988. 'Laboratory technology and biological knowledge: the Tiselius electrophoresis apparatus, 1930–1945', *History and Philosophy of the Life Sciences* 10: 51–72.

1989. 'Selling pure science in wartime: the biochemical genetics of G. W. Beadle', *Journal of the History of Biology* 22: 73–101.

1993a. *The Molecular Vision of Life: Caltech, the Rockefeller Foundation, and the Rise of the New Biology* (Oxford: Oxford University Press).

1993b. 'Life as technology: representing, intervening, molecularizing', *Rivista di Storia della Scienza*: 85–103.

1995. 'Who wrote the book of life? Information and the transformation of molecular biology, 1945–55', *Science in Context* 8: 609–34.

2000. *Who Wrote the Book of Life? A History of the Genetic Code* (Stanford: Stanford University Press).

Kendrew, J. C. 1954. 'Structure of proteins', *Nature* 173: 57–8.

1961. 'The three-dimensional structure of a protein molecule', *Scientific American* 205 (December): 96–110.

1964. 'Myoglobin and the structure of proteins. Nobel lecture, December 11, 1962', in *Nobel Lectures Including Presentation Speeches and Laureates' Biographies. Chemistry, 1942–1962* (Amsterdam: Elsevier Publishing Company), pp. 676–98.

1966. *The Thread of Life: An Introduction to Molecular Biology Based on the Series of B.B.C. Television Lectures of the Same Title* (London: G. Bell and Sons).

1967. 'How molecular biology started', *Scientific American* 216: 141–4.

1968. 'EMBO and the idea of a European laboratory', *Nature* 218: 840–2.

1970. 'Some remarks on the history of molecular biology', in T. W. Goodwin (ed.), *British Biochemistry Past and Present* (London and New York: Academic Press), pp. 5–10.

Kendrew, J. C., Bodo, G., Dintzis, H. M., Parrish, R. G. and Wyckoff, H. 1958. 'A three-dimensional model of the myoglobin molecule obtained by X-ray analysis', *Nature* 181: 662–6.

Kendrew, J. C., Dickerson, R. E., Strandberg, B. E., Hart, R. G., Davies, D. R., Phillips, D. C. and Shore, V. C. 1960. 'Structure of myoglobin. A three-dimensional Fourier synthesis at 2 Å resolution', *Nature* 185: 422–7.

Kendrew, J. C., Watson, H. C., Strandberg, B. E., Dickerson, R. E., Phillips, D. C. and Shore, V. C. 1961. 'The amino-acid sequence of sperm whale myoglobin. A partial determination by X-ray methods, and its correlation with chemical data', *Nature* 190: 666–70.

[Kendrew Report] 1968. *Council for Scientific Policy. Report of the Working Group on Molecular Biology* (London: Her Majesty's Stationery Office), Cmnd. 3675.

Kevles, D. J. 1985. *In the Name of Eugenics: Genetics and the Use of Human Heredity* (New York: Alfred A. Knopf).

1987. *The Physicists: The History of a Scientific Community in Modern America* (Cambridge, MA: Harvard University Press).

Kevles, D. J. and Hood, L. (eds.) 1992. *The Code of Codes: Scientific and Social Issues in the Human Genome Project* (Cambridge, MA: Harvard University Press).

Kim, J. H. and van den Broeck, R. A. 1999. *The Renaissance of Antibodies*, Industry Report, Hambrecht & Quist Equity Research (March 15).

Kittler, F. 1993. *Draculas Vermächtnis: Technische Schriften* (Leipzig: Reklam Verlag).

Klug, A. 1968. 'Rosalind Franklin and the discovery of the structure of DNA', *Nature* 219: 808–10 and 843–4; reprinted in J. D. Watson, *The Double Helix: A Personal Account of the Structure of DNA. Text, Commentary, Reviews, Original Papers*, edited by G. Stent (New York and London: W. W. Norton & Company, 1980), pp. 153–8.

1983. 'From macromolecules to biological assemblies. Nobel lecture, 8 December, 1982', in *Les Prix Nobel 1982. Nobel Prizes, Presentations and Biographies* (Stockholm: Almqvist & Wiksell International), pp. 93–125.

Köhler, G. and Milstein, C. 1975. 'Continuous cultures of fused cells secreting antibody of predefined specificity', *Nature* 256: 495–7.

Kohler, R. 1977. 'Rudolf Schoenheimer, isotopic tracers, and biochemistry in the 1930s', *Historical Studies in the Physical Sciences* 8: 257–98.

1978. 'Walter Fletcher, F. G. Hopkins and the Dunn Institute of Biochemistry', *Isis* 69: 331–55.

1982. *From Medical Chemistry to Biochemistry: The Making of a Biomedical Discipline* (Cambridge: Cambridge University Press).

1991. *Partners in Science: Foundations and Natural Scientists 1900–1945* (Chicago: University of Chicago Press).

1994. *Lords of the Fly: Drosophila Genetics and the Experimental Life* (Chicago: University of Chicago Press).

1999. 'The constructivists' tool kit', *Isis* 90: 329–31.

Kornberg, A. 1964. 'The biological synthesis of deoxyribonucleic acid, Nobel lecture, December 11, 1959', in *Nobel Lectures Including Presentation Speeches and Laureates' Biographies. Physiology or Medicine, 1942–1962* (Amsterdam: Elsevier Publishing Company), pp. 665–80.

1989. *For the Love of Enzymes: The Odyssey of a Biochemist* (Cambridge, MA: Harvard University Press).

Kornberg, A., Horecker, B. L., Cornudella, L. and Oró, J. (eds.) 1976. *Reflections in Biochemistry in Honour of Severo Ochoa* (Oxford: Pergamon).

Krebs, H. 1981. *Reminiscences and Reflections* (Oxford: Clarendon Press).

[Krebs Report] 1969. *Biochemistry, 'Molecular Biology' and the Biological Sciences: Report of a Subcommittee of the Biochemical Society on the Report of the Working Group on Molecular Biology ('Kendrew Report') and the Present State of the Biological Sciences in this Country, with Particular Reference to Biochemistry* (London: Biochemical Society).

Krige, J. 1989. 'The installation of high-energy accelerators in Britain after the war: big equipment but not "big science"', in M. de Maria, M. Grilli and F. Sebastiani (eds.), *Proceedings of the International Conference on the Restructuring of Physical Sciences in Europe and the United States, 1945–1960. Università 'La Sapienza', Rome, Italy, 19–23 September 1988* (Singapore: World Scientific), pp. 488–501.

1997. 'The politics of European scientific collaboration', in J. Krige and D. Pestre (eds.), *Science and Technology in the Twentieth Century* (London: Harwood Academic Publishers), pp. 897–918.

1999. 'The Ford Foundation, European physics and the Cold War', *Historical Studies in the Physical and Biological Sciences* 29: 333–61.

forthcoming 'The birth of EMBO and EMBL', in S. de Chadarevian and B. Strasser (eds.), *Molecular Biology in Postwar Europe*, special issue *Studies in History and Philosophy of Biological and Biomedical Sciences*.

Krige, J. (ed.) 1996. *History of CERN*, vol. III (Amsterdam, NY: North Holland).

Krige, J. and Russo, A. 1995. *Europe in Space 1960–1973* (Noordwijk: ESA).

Krimsky, S. 1982. *Genetic Alchemy: The Social History of the Recombinant DNA Controversy* (Cambridge, MA: MIT Press).

1991. *Biotechnics and Society: The Rise of Industrial Genetics* (New York: Praeger).

Lambilliotte, M. (ed.) 1959. *Le Mémorial officiel de l'Exposition universelle et internationale de Bruxelles 1958. Vol. VI: Les Sciences* (Brussels: Etablissements généraux d'Imprimerie).

(ed.) 1961. *Le Mémorial officiel de l'Exposition universelle et internationale de Bruxelles 1958. Vol. VIII: Synthèse* (Brussels: Etablissements généraux d'Imprimerie).

Landsborough Thomson, A. 1973. *Half a Century of Medical Research. Vol. I: Origins and Policy of the Medical Research Council (United Kingdom)* (London: Her Majesty's Stationery Office).

1975. *Half a Century of Medical Research. Vol. II: The Programme of the Medical Research Council (UK)* (London: Her Majesty's Stationery Office).

Langridge, R. and Glusker, J. P. 1987. 'Computing in crystallography: the early days', in J. P. Glusker, B. K. Patterson and M. Rossi (eds.), *Patterson and Pattersons: Fifty Years of the Patterson Function* (New York and Oxford: Oxford University Press), pp. 382–407.

Latour, B. 1987. *Science in Action: How to Follow Scientists and Engineers through Society* (Cambridge, MA: Harvard University Press).

1988. *Pasteurization of France* (Cambridge, MA: Harvard University Press).

1990. 'Drawing things together', in M. Lynch and S. Woolgar (eds.), *Representation in Scientific Practice* (Cambridge, MA: MIT Press), pp. 19–68.

Latour, B. and Woolgar, S. 1986. *Laboratory Life: The Construction of Scientific Facts* (Princeton: Princeton University Press).

Lederberg, J. 1963. 'Biological future of man', in G. Wolstenholme (ed.), *Man and His Future* (London: J. & A. Churchill), pp. 263–3.

Lederman, M. and Burian, R. M. 1993. 'Introduction' [special section 'The right organism for the job'], *Journal of the History of Biology* 26: 235–7.

Leedham-Green, E. 1996. *A Concise History of the University of Cambridge* (Cambridge: Cambridge University Press).

Lenoir, T. 1997. *Instituting Science: The Cultural Production of Scientific Disciplines* (Stanford: Stanford University Press).

Lenoir, T. and Lécuyer, C. 1995. 'Instrument makers and discipline builders: the case of nuclear magnetic resonance', *Perspectives on Science* 3: 276–345.

Leslie, S. 1993. *The Cold War and American Science: The Military–Industrial–Academic Complex at M.I.T. and Stanford* (New York: Columbia University Press).

Lewenstein, B. V. 1992. 'The meaning of "public understanding of science" in the United States after World War II', *Public Understanding of Science* 1: 45–68.

Lewin, R. 1984. 'Why is development so illogical?', *Science* 224: 1327–9.

Liebenau, J. 1987. 'The British success with penicillin', *Social Studies of Science* 17: 69–86.

Lindee, S. 1994. *Suffering Made Real: American Science and the Survivors at Hiroshima* (Chicago: University of Chicago Press).

Lowe, R. 1988. *Education in the Post-War Years: A Social History* (London: Routledge).

  1993. *The Welfare State in Britain since 1945* (Basingstoke: Macmillan).

Löwy, I. 1996. *Between Bench and Bedside: Science, Healing, and Interleukin-2 in a Cancer Ward* (Cambridge, MA: Harvard University Press).

Lwoff, A. 1968. 'Truth, truth, what is the truth (about how the structure of DNA was discovered)?', *Scientific American* 219: 133–8; reprinted in J. D. Watson, *The Double Helix: A Personal Account of the Structure of DNA. Text, Commentary, Reviews, Original Papers*, edited by G. Stent (New York and London: W. W. Norton & Company), pp. 224–34.

Lwoff, A. and Ullmann, A. (eds.) 1979. *Origins of Molecular Biology: A Tribute to Jacques Monod* (New York: Academic Press).

Lynch, M. and Woolgar, S. (eds.) 1990. *Representation in Scientific Practice* (Cambridge, MA: MIT Press).

McGowan, H. D. 1942. 'Our purpose', *Endeavour* 1: 3

McGucken, W. 1978. 'On freedom and planning in science: the Society for Freedom in Science, 1940–46', *Minerva* 16: 42–72.

Mackenzie, M., Cambrosio, A. and Keating, P. 1988. 'The commercial application of a scientific discovery: the case of the hybridoma technique', *Research Policy* 17: 155–70.

Mackenzie, M., Keating, P. and Cambrosio, A. 1990. 'Patents and free scientific information in biotechnology: making monoclonal antibodies proprietary', *Science, Technology, and Human Values* 15: 65–83.

Mann, T. 1964. 'David Keilin 1887–1963', *Biographical Memoirs of the Fellows of the Royal Society* 10: 183–295.

Marwick, A. 1982. *British Society since 1945* (London: Penguin Books, 1982).

Mason, J. 1993. 'Women in Cambridge: some quandaries', *The Cambridge Review* (June): 67–74.

  1996. 'Honor Fell, 1900–1986', in E. Shils and C. Blacker (eds.), *Cambridge Women: Twelve Portraits* (Cambridge: Cambridge University Press), pp. 245–66.

Matthews, B. W., Sigler, P. B., Henderson, R. and Blow, D. M. 1967. 'Three-dimensional structure of tosyl-α-chymotrypsin', *Nature* 214: 652–6.

Medawar, P. 1986. *Memoir of a Thinking Radish: An Autobiography* (Oxford: Oxford University Press).

Megaw, H. 1951. 'The investigation of crystal structure', *The Architectural Review* 109: 236–9.

Meinel, C. forthcoming. 'Modelling a visual language for chemistry, 1860–1874', in S. de Chadarevian and N. Hopwood (eds.), *Displaying the Third Dimension: Models in the Sciences, Technology and Medicine* (Stanford: Stanford University Press).

Mellanby, E. 1943. 'Medical research in wartime', *British Medical Journal* 2: 351–6.

Meloun, B., Kluh, I., Kostka, V., Morávek, I., Prusík, Z., Vanecek, J., Keil, B. and Sorm, F. 1966. 'Covalent structure of bovine chymotrypsinogen A ', *Biochimica and Biophysica Acta* 130: 543–6.

Michaelson, M. G. 1972. 'Sickle cell anaemia: an "interesting pathology"', *Ramparts*: 52–8.

Milstein, C. 1985. 'From the structure of antibodies to the diversification of the immune response', in *Les Prix Nobel 1984. Nobel Prizes, Presentations, Biographies and Lectures* (Stockholm: Almqvist & Wiksell International), pp. 194–216.

Milward, A. S. 1984. *The Reconstruction of Western Europe, 1945–1951* (London: Methuen).

Monod, J. 1971. 'Du microbe à l'homme', in J. Monod and E. Borek (eds.), *Of Microbes and Mice: Les microbes et la vie* (New York and London: Columbia University Press), pp. 1–9.

Monod, J. and Borek, E. (eds.) 1971. *Of Microbes and Mice: Les microbes et la vie* (New York and London: Columbia University Press).

Morange, M. 1995. 'De la régulation bactérienne au contrôle du développement', *Bulletin de la Société d'Histoire et d'Epistémologie de les Sciences de la Vie* 2: 205–20.

  1996. 'Construction of the developmental gene concept. The crucial years: 1960–1980', *Biologisches Zentralblatt* 115: 132–8.

  1997. 'EMBO and EMBL', in J. Krige and L. Guzzetti (eds.), *History of European Scientific and Technological Cooperation. Florence, 9–11 November 1995* (Brussels: European Commission, European Science and Technology Forum), pp. 77–92.

  1998a. *A History of Molecular Biology* (Cambridge, MA: Harvard University Press).

  1998b. 'The transformation of molecular biology on contact with higher organisms, 1960–1980: from a molecular description to a molecular explanation', *History and Philosophy of the Life Sciences* 19: 425–49.

  2000. 'François Jacob's lab in the seventies: the T-complex and the mouse developmental genetic program', *History and Philosophy of the Life Sciences* 22 [issue delayed].

Morgan, K. O. 1990. *The People's Peace: British History 1945–1989* (Oxford: Oxford University Press).

Morrell, J. 1997. *Science at Oxford, 1914–1939: Transforming an Arts University* (Oxford: Clarendon Press).

Mott, N. 1986. *A Life in Science* (London and Philadelphia: Taylor & Francis).

Myers, G. 1989. 'The double helix as icon', *Science as Culture* 5: 49–72.

Needell, A. 1996. 'I. I. Rabi, Lloyd V. Berkner, and the American rehabilitation of European science, 1945–1954', in F. H. Heller and J. R. Gillingham (eds.), *The United States and the Integration of Europe: Legacies of the Postwar Era* (Basingstoke: Macmillan), pp. 289–305.

Neel, J. V. 1994. *Physician to the Gene Pool: Genetic Lessons and Other Stories* (New York: John Wiley & Sons).

Nelkin, D. and Lindee, M. S. 1995. *The DNA Mystique: The Gene as Cultural Icon* (London: W. H. Freeman).

Neuberger, A. 1939. 'Chemical criticism of the cycol and frequency hypothesis of protein structure', in T. Svedberg *et al.*, 'A discussion on the protein molecule', *Proceedings of the Royal Society of London B* 127: 25–6.

Nyhart, L. K. 1995. *Biology Takes Form: Animal Morphology and the German Universities, 1800–1900* (Chicago and London: University of Chicago Press).

Olby, R. 1970a. 'The origins of molecular biology at Cambridge and Caltech', in *Proceedings of the Conference on the History of Biochemistry and Molecular Biology sponsored by the American Academy of Arts and Science* (Brookline, MA: American Academy of Arts and Sciences), pp. 60–95.

1970b. 'Francis Crick, DNA, and the central dogma', *Daedalus* 99: 938–87.

1971. 'Schrödinger's problem: what is life?', *Journal of the History of Biology* 4: 119–48.

1979. 'Mendel no mendelian?', *History of Science* 8: 53–72.

1983. 'Practising without a license: discipline identity in biochemistry and molecular biology', paper presented at the Wellcome Institute in London.

1985a. 'The "mad pursuit": X-ray crystallographers' search for the structure of haemoglobin', *History and Philosophy of the Life Sciences* 7: 171–93.

1985b. 'Historical aspects of protein structure', paper presented at the Biochemistry Department in London.

1986. 'Structural and dynamical explanations in the world of neglected dimensions', in T. Horder, J. A. Witkowski and C. C. Wylie (eds.), *A History of Embryology* (Cambridge: Cambridge University Press), pp. 275–308.

1991. 'The molecular revolution in biology: what revolution?', paper presented at Stanford University.

1994. *The Path to the Double Helix: The Discovery of DNA* (New York: Dover Publications) [revised and enlarged edition with new subtitle; first published 1974].

Ophir, A. and Shapin, S. 1991. 'The place of knowledge. A methodological survey', *Science in Context* 4: 3–23.

Pauling, L. 1952. 'The hemoglobin molecule in health and disease', *Proceedings of the American Philosophical Society* 96: 556–65.

——— 1968. 'Reflections on the new biology. Foreword', *UCLA Law Review* 15: 267–72.

Pauling, L., Itano, H. A., Singer, S. J. and Wells, I. C. 1949. 'Sickle cell anemia, a molecular disease', *Science* 110: 543–8.

Pepinsky, R. 1947. 'An electronic computer for X-ray crystal structure analyses', *Journal of Applied Physics* 18: 601–4.

Perutz, M. F. 1940. 'The Crystal Structure of Horse Methaemoglobin'. Ph.D. dissertation, University of Cambridge.

——— 1948–9. 'An X-ray study of horse methaemoglobin. II', *Proceedings of the Royal Society A* 195: 475–99.

——— 1949. 'Recent developments in the X-ray study of haemoglobin', in F. J. W. Roughton and J. C. Kendrew (eds.), *Haemoglobin: A Symposium Based on a Conference Held at Cambridge in June 1948 in Memory of Sir Joseph Barcroft* (London: Butterworths Scientific Publications), pp. 135–47.

——— 1958. 'Some recent advances in molecular biology', *Endeavour* 17: 190–203.

——— 1962a. 'The MRC Unit for Molecular Biology', *New Scientist* 13: 208–9.

——— 1962b. *Proteins and Nucleic Acids: Structure and Function. Eighth Weizmann Memorial Lecture Series, April 1961* (Amsterdam and London: Elsevier).

——— 1964. 'X-ray analysis of haemoglobin. Nobel lecture, December 11, 1962', in *Nobel Lectures Including Presentation Speeches and Laureates' Biographies. Chemistry 1942–1962* (Amsterdam, London and New York: Elsevier).

——— 1965. 'Structure and function of haemoglobin. I. A tentative atomic model of horse oxyhaemoglobin', *Journal of Molecular Biology* 13: 646–68.

——— 1976. 'Fundamental research in molecular biology: relevance to medicine', *Nature* 262: 449–53.

——— 1978. 'Haemoglobin structure and respiratory transport', *Scientific American* 239: 92–125.

——— 1980. 'Origins of molecular biology', *New Scientist* 85: 326–9.

——— 1987. 'The birth of molecular biology', *New Scientist* 114: 38–41.

——— 1989. 'Enemy alien', in *Is Science Necessary? Essays on Science and Scientists* (London: Barrie & Jenkins), pp. 101–13.

——— 1990. 'How Lawrence Bragg invented X-ray analysis', *Royal Institution Proceedings* 62: 183–98.

——— 1996. 'The Medical Research Council, Laboratory of Molecular Biology', *Molecular Medicine* 2: 659–62.

——— 1998. 'How he saved the Laboratory of Molecular Biology', in E. A. Davis (ed.), *Nevill Mott: Reminiscences and Appreciations* (London: Taylor & Francis), pp. 133–4.

Perutz, M. F. and Lehmann, H. 1968. 'Molecular pathology of human haemoglobin', *Nature* 219: 902–9.

Perutz, M. F., Muirhead, H., Cox, J. M. and Goaman, L. C. G. 1968. 'Three-dimensional Fourier synthesis of horse oxyhaemoglobin at 2.8 Å resolution: the atomic model', *Nature* 219: 131–9.

Perutz, M. F., Rossmann, M. G., Cullis, A. F., Muirhead, H., Will, G. and North, A. C. T. 1960. 'Structure of haemoglobin. A three-dimensional Fourier synthesis at 5.5-Å. resolution, obtained by X-ray analysis', *Nature* 185: 416–22.

Pestre, D. 1992. 'The decision-making processes for the main particle accelerators built throughout the world from the 1930s to the 1970s', *History and Technology* 9: 163–74.

Phillips, D. 1979. 'William Lawrence Bragg, 31 March 1890 – 1 July 1971', *Biographical Memoirs of the Fellows of the Royal Society* 25: 75–143.

Pickering, A. 1992. *Science as Practice and Culture* (Chicago: University of Chicago Press).

   1993. 'Anti-discipline or narratives of illusion', in E. Messer-Davidow, D. R. Shumway and D. Sylvan (eds.), *Knowledges: Historical and Critical Studies in Disciplinarity* (Charlottesville: University Press of Virginia), pp. 103–22.

   1995. *The Mangle of Practice: Time, Agency, and Science* (Chicago: University of Chicago Press).

Pickstone, J. V. 1993a. 'Ways of knowing. Towards a historical sociology of science, technology and medicine', *British Journal for the History of Science* 26: 433–58.

   1993b. 'The biographical and the analytical. Towards a historical model of science and practice in modern medicine', in I. Löwy (ed.), *Medicine and Change: Historical and Sociological Studies of Medical Innovation* (Paris: Les Editions Inserm), pp. 23–47.

   1994. 'The place of the analytical/comparative in nineteenth-century science, technology and medicine', *History of Science* 32: 111–38.

Pickvance, S. 1976. ' "Life" in a biology lab', *Radical Science Journal* 4: 11–28.

Pinkerton, J. M. M., Hemy, D. and Lenaerts, E. H. 1992. 'The influence of the Cambridge Mathematical Laboratory on the LEO project', *IEEE Annals of the History of Computing* 14: 41–8.

Piper, A. 1998. 'Light on a dark lady', *Trends in the Biochemical Sciences* 23: 151–4.

Portugal, F. H. and Cohen, J. S. 1997. *A Century of DNA: A History of the Discovery of the Structure and Function of the Genetic Substance* (Cambridge, MA: MIT Press).

Pringle, J. W. S. 1963. *The Two Biologies: An Inaugural Lecture Delivered before the University of Oxford on 24 October 1963* (Oxford: Clarendon Press).

   1975. 'Effects of World War II on the development of knowledge in the biological sciences', *Proceedings of the Royal Society of London A* 342: 537–48.

Rabinow, P. 1996. *Making PCR: A Story of Biotechnology* (Chicago: University of Chicago Press).

Randle, P. 1990. 'Frank George Young, 25 March 1908–20 September 1988', *Biographical Memoirs of the Fellows of the Royal Society* 36: 581–99.

Rasmussen, N. 1996. 'Making a machine instrumental: RCA and the wartime origins of biological electron microscopy in America, 1940–1945', *Studies in History and Philosophy of Science* 27: 311–49.

——— 1997a. 'Midcentury biophysics: Hiroshima and the origins of molecular biology', *History of Science* 35: 244–93.

——— 1997b. *Picture Control: The Electron Microscope and the Transformation of Biology in America, 1940–1960* (Stanford: Stanford University Press).

——— 1999. 'Biotechnology before the "Biotech Revolution": life scientists, chemists and product development in 1930s–1940s America', paper presented at the conference Between Physics and Biology: Chemistry in the Twentieth Century, Munich, May 1999.

Raverat, G. 1952. *Period Piece: A Cambridge Childhood* (London: Faber and Faber).

Ravetz, J. R. 1971. *Scientific Knowledge and Its Social Problems* (Oxford: Clarendon Press).

Reader, W. J. 1975. *Imperial Chemical Industries: A History. Vol. II: The First Quarter Century 1926–1952* (London: Oxford University Press).

Reid, B. R., Koch, G. L. E., Boulanger, Y., Hartley, B. S. and Blow, D. M. 1973. 'Letter to the Editor. Crystallization and preliminary X-ray diffraction studies on tyrosyl-transfer RNA synthetase from *Bacillus stearothermophilus*', *Journal of Molecular Biology* 80: 199–201.

'Reports and documents. The emigration of British scientists' 1963. *Minerva* 1: 342–80.

Rettig, R. A. 1977. *Cancer Crusade: The Story of the National Cancer Act of 1971* (Princeton: Princeton University Press).

Rheinberger, H.-J. 1992. 'Experiment, difference, and writing. Part 1: Tracing protein synthesis' and 'Part 2: The laboratory production of transfer RNA', *Studies in the History and Philosophy of Science* 23: 305–31 and 389–422.

——— 1994. 'Konjunkturen: Transfer-RNA, Messenger-RNA, Genetischer Code', in M. Hagner, H.-J. Rheinberger and B. Wahrig-Schmidt (eds.), *Objekte, Differenzen und Konjunkturen: Experimentalsysteme im historischen Kontext* (Berlin: Akademie Verlag), pp. 201–31.

——— 1995. *Kurze Geschichte der Molekularbiologie. Preprint 24* (Berlin: Max-Planck-Institut für Wissenschaftsgeschichte).

——— 1996. 'Comparing experimental systems: protein synthesis in microbes and in animal tissue at Cambridge (Ernest F. Gale) and at the Massachusetts General Hospital (Paul C. Zamecnik), 1945–1960', *Journal of the History of Biology* 29: 387–416.

——— 1997. *Toward a History of Epistemic Things: Synthesizing Proteins in the Test Tube* (Stanford: Stanford University Press).

Rich, A. 1995. 'The nucleic acids: a backward glance', in D. A. Chambers (ed.), *DNA: The Double Helix. Perspective and Prospective at Forty Years.* Annals of the New York Academy of Sciences 758 (New York: New York Academy of Sciences), pp. 97–142.

Rich, A. and Crick, F. H. C. 1955. 'The structure of collagen', *Nature* 176: 915–16.

Rich, A. and N. Davidson (eds.) 1968. *Structural Chemistry and Molecular Biology: A Volume Dedicated to Linus Pauling by His Students, Colleagues, and Friends* (San Francisco and London: Freeman and Company).

Rich, A. and Watson, J. 1954a. 'Physical studies on ribonucleic acid', *Nature* 173: 995–6.

1954b. 'Some relations between DNA and RNA', *Proceedings of the National Academy of Sciences* 40: 759–64.

Richards, F. 1968. 'The matching of physical models to three-dimensional electron-density maps: a simple optical device', *Journal of Molecular Biology* 37: 225–30.

Richards, M. P. M. 1972. 'Consuming science: Rothschild, Dainton and the non-revolution in research policies', *Cambridge Review* 28: 75–7.

Richmond, M. 1997. 'A lab of one's own: the Balfour Biological Laboratory for Women at Cambridge University, 1884–1914', *Isis* 88: 422–55.

Roberts, L. 1990. 'The worm project', *Science* 248: 1310–13.

Robertson, J. M. 1953. *Organic Crystals and Molecules: Theory of X-Ray Structure Analysis with Applications to Organic Chemistry* (Ithaca, NY: Cornell University Press).

Robinson, P. and Spärck Jones, K. 1999. *EDSAC 99, 15–16 April 1999* (Cambridge: University of Cambridge Computer Laboratory).

Roche, J. 1994. 'The non-medical sciences, 1939–1970', in B. Harrison (ed.), *The History of the University of Oxford. Vol. VIII: The Twentieth Century* (Oxford: Clarendon Press), pp. 251–89.

Rook, A., Carlton, M. and Cannon, W. G. 1991. *The History of Addenbrooke's Hospital, Cambridge* (Cambridge: Cambridge University Press).

Rose, H. and Rose, S. 1971. *Science and Society* (Harmondsworth: Penguin).

Rosenhead, J. 1989. 'Operational research at the cross-roads: Cecil Gordon and the development of post-war operational research', *Journal of the Operational Research Society* 40: 3–28.

[Rothschild Report] 1971. 'The Organisation and Management of Government R. & D.', published as part of the Green Paper *A Framework for Government Research and Development* (London: Her Majesty's Stationery Office), Cmnd. 4814.

Roughton, F. J. and Kendrew, J. C. (eds.) 1949. *Haemoglobin: A Symposium Based on a Conference Held at Cambridge in June 1948 in Memory of Sir Joseph Barcroft* (London: Butterworths Scientific Publications).

Rowlinson, J. S. and Robinson, N. H. 1992. *Record of the Royal Society of London – Supplement to the Fourth Edition for the Years 1940–1989* (London: Royal Society).

Russell, B. 1968. *The Autobiography of Bertrand Russell, 1914–1944*, vol. II (London: Allen and Unwin).

Rutherford, E. 1915. 'Henry Gwyn Jeffreys Moseley' [Obituary], *Nature* 96: 33–4.

Sanger, F. 1981. 'Frederick Sanger' [Autobiography], in *Les Prix Nobel. Nobel Prizes. Presentations, Biographies and Lectures* (Stockholm: Almqvist & Wiksell International), pp. 141–2.

1988. 'Sequences, sequences, and sequences', *Annual Review of Biochemistry* 57: 1–28.

Santesmases, M.-J. and Muñoz, E. 1997a. *Establecimiento de la bioquímica y de la biología molecular en España* (Madrid: Fundación Ramón Areces. Consejo Superior de Investigaciones Científica).

1997b. 'Scientific organizations in Spain (1950–1970): social isolation and international legitimation of biochemists and molecular biologists on the periphery', *Social Studies of Science* 27: 187–219.

Sayre, A. 1975. *Rosalind Franklin and DNA* (New York: Norton & Company).

Schrödinger, E. 1992. *What is Life? The Physical Aspects of the Living Cell* with *Mind and Matter & Autobiographical Sketches* (Cambridge: Cambridge University Press); *What is Life?* originally published 1944.

Schroeder-Gudehus, B. and Cloutier, D. 1994. 'Popularizing science and technology during the Cold War: Brussels 1958', in R. W. Rydell and N. Gwin (eds.), *Fair Representations: World's Fairs and the Modern World* (Amsterdam: V.U. Press), pp. 157–80.

Schroeder-Gudehus, B. and Rasmussen, A. 1992. *Les fastes du progrès: le guide des expositions universelles 1851–1992* (Paris: Flammarion).

Schull, W. J. 1990. *Song among the Ruins* (Cambridge, MA: Harvard University Press).

1995. *Effects of Atomic Radiation: A Half-Century of Studies from Hiroshima and Nagasaki* (New York: Wiley-Liss).

Secord, J. (ed.) 1993. *The Big Picture*, special issue, *British Journal for the History of Science* 26: 387–483.

Segal Quince & Partners 1985. *The Cambridge Phenomenon: The Growth of High Technology Industry in a University Town* (Cambridge: Segal Quince & Partners).

Seymour-Ure, C. 1991. *The British Press and Broadcasting since 1945* (Oxford: Blackwell).

Shaffer Jr., P. A., Schomaker, V. and Pauling, L. 1946. 'The use of punched cards in molecular structure determinations: I. Crystal structure calculations', *The Journal of Chemical Physics* 14: 648–58.

Shapin, S. and Schaffer, S. 1985. *Leviathan and the Air Pump: Hobbes, Boyle, and the Experimental Life* (Princeton: Princeton University Press).

Shotton, D. M. 1969. 'The Structure of Porcine Pancreatic Elastase'. Ph.D. dissertation, University of Cambridge.

Shotton, D. M. and Watson, H. C. 1970. 'The three-dimensional structure of crystalline porcine pancreatic elastase', *Philosophical Transactions of the Royal Society of London B* 257: 111–18.

Silcock, B. 1982. 'The lab that makes geniuses', *Sunday Times*, 24 October, p. 17.

Silverman, J. B. and Wright Marino, L. 1998. 'Why everybody should love antibodies', BancBoston Robertson Stephens Healthcare Research (16 November).

Simon, B. 1991. *Education and the Social Order, 1940–1990* (London: Lawrence & Wishart).

Simpson, R. 1983. *How the PhD Came to Britain: A Century of Struggle for Postgraduate Education* (Guildford: Society for Research into Higher Education).

Sixty Penguin Years 1995 (Harmondsworth: Penguin).

Sloan, P. R. (ed.) 1999. *Controlling Our Destinies: Historical, Philosophical, Ethical, and Theological Perspectives on the Human Genome Project* (Notre Dame, IN: University of Notre Dame Press).

Snow, C. P. 1954. *The New Men* (London: Macmillan).

   1981. *The Physicists* (London: Macmillan).

   1993. *The Two Cultures with Introduction by S. Collini* (Cambridge: Cambridge University Press).

Söderqvist, T. (ed.) 1997. *The Historiography of Contemporary Science and Technology* (London: Harwood Academic Publishers).

Sparks, R. A. and Trueblood, K. N. 1983. 'Digital computers in crystallography', in D. McLachlan Jr. and J. P. Glusker (eds.), *Crystallography in North America* (New York: Crystallographic Association), pp. 228–32.

[Spinks Report] 1980. *Biotechnology: Report of a Joint Working Party of the Advisory Council for Applied Research and Development and Advisory Board for the Research Councils and the Royal Society* (London: Her Majesty's Stationery Office).

Star, S. and Griesemar, R. 1988. 'Institutional ecology. "Translations" and boundary objects: amateurs and professionals in Berkeley's Museum of Vertebrate Zoology 1907–1939', *Social Studies of Science* 19: 387–420.

Steitz, J. A. 1986. 'Shaping research in gene expression: role of the Cambridge Medical Research Council Laboratory of Molecular Biology', *Perspectives in Biology and Medicine* 29: S90–S95.

Stent, G. S. 1968. 'That was the molecular biology that was', *Science* 160: 390–5.

   1980. 'The DNA double helix and the rise of molecular biology', in J. D. Watson, *The Double Helix: A Personal Account of the Discovery of the Structure of DNA. Text, Commentary, Reviews, Original Papers*, ed. G. Stent (New York and London: W. W. Norton & Company), pp. xi–xxii.

Strasser, B. J. in press. 'Microscopes électroniques, totems des laboratoires et réseaux transnationaux: L'émergence de la biologie moléculaire à Genève (1945–1960)', *Revue d'Histoire des Sciences*.

Strickland, S. P. 1972. *Politics, Science and Dread Disease: A Short History of United States Medical Research Policy* (Cambridge, MA: Harvard University Press).

   1989. *The Story of the NIH Grants Programs* (Lanham and New York: University Press of America).

Sullivan, W. 1980. 'A book that couldn't go to Harvard', in J. D. Watson, *The Double Helix: A Personal Account of the Discovery of the Structure of DNA. Text, Commentary, Reviews, Original Papers*, ed. G. Stent (New York and London: W. W. Norton & Company), pp. xxiv–xxv (repr. from *New York Times*, 15 February 1968, pp. 1 and 4).

Sulston, J. E. and Brenner, S. 1974. 'The DNA of *Caenorhabditis elegans*', *Genetics* 77: 95–104.

Sulston, J. E., Schierenberg, E., White, J. G. and Thomson, J. N. 1983. 'The embryonic cell lineage of the nematode *Caenorhabditis elegans*', *Developmental Biology* 100: 64–119.

Sulston, J. *et al.* 1992. 'The *C. elegans* genome sequencing project: a beginning', *Nature* 356: 37–41.

Svedberg, T. *et al.* 1939. 'A discussion on the protein molecule, 17 November 1938', *Proceedings of the Royal Society of London B* 127: 1–40.

Svedberg, T. and Fåhraeus, R. 1926. 'A new method for the determination of the molecular weight of the proteins', *Journal of the American Chemical Society* 48: 430–8.

Svedberg, T. and Pederson, K. O. 1940. *The Ultracentrifuge* (Oxford: Oxford University Press).

Swann, B. and Aprahamian, F. (eds.), 1999. *J. D. Bernal: A Life in Science and Politics* (London: Verso).

Swann, J. P. 1983. 'The search for synthetic penicillin during World War II', *British Journal for the History of Science* 16: 154–90.

Szilard, L. 1968. 'Reminiscences', in D. Fleming and B. Bailyn (eds.), *The Intellectual Migration: Europe and America, 1930–1960* (Cambridge, MA: Belknap Press of Harvard University Press), pp. 94–141.

Tansey, E. M. Catterall, P. P., Christie, D. A., Willhoft, S. V. and Reynolds, L. A. (eds.) 1997. *Wellcome Witnesses to Twentieth Century Medicine*, vol. 1 (London: Wellcome Trust).

Teich, M. 1975. 'A single path to the double helix?', *History of Science* 13: 264–83.

Thackray, A. (ed.) 1998. *Private Science: Biotechnology and the Rise of the Molecular Sciences* (Philadelphia: University of Pennsylvania Press).

Thomas, M. H. 1951. *The Souvenir Book of Crystal Designs: The Fascinating Story in Colour of the Festival Pattern Group* (Colchester: Benham and Company Limited).

Timmins, N. 1995. *The Five Giants: A Biography of the Welfare State* (London: HarperCollins).

Timoféeff-Ressovsky, N. W., Zimmer, K. G. and Delbrück, M. 1935. 'Über die Natur der Genmutation und der Genstruktur', *Nachrichten der Gesellschaft der Wissenschaften zu Göttingen. Mathematisch-physikalische Klasse, Fachgruppe 6* 1: 189–245.

Todd, A. 1983. *A Time to Remember: The Autobiography of a Chemist* (Cambridge: Cambridge University Press).

Tooze, J. 1981. 'A brief history of the European Molecular Biology Organization', in *The Embo Journal: An Announcement* (Oxford: IRL Press), pp. 1–6.

—— 1986. 'The role of European Molecular Biology Organisation (EMBO) and European Molecular Biology Conference (EMBC) in European molecular biology (1970–1983)', *Perspectives in Biology and Medicine* 29: (Spring): S28–S46.

[Trend Report] 1963. *Report of the Committee of Enquiry into the Organisation of Civil Science, under the Chairmanship of Sir Burke Trend* (London: Her Majesty's Stationery Office), Cmnd. 2171.

Tucker, A. 1968. 'Science council dissatisfied with "trivial" research', *Guardian*, 10 July, p. 4.

Turney, J. 1998. *Frankenstein's Footsteps: Science, Genetics and Popular Culture* (New Haven and London: Yale University Press).

Uchida, H. 1993. 'Building a science in Japan: the formative decades of molecular biology', *Journal of the History of Biology* 26: 499–543.

Varcoe, I. 1974. *Organizing for Science in Britain: A Case-Study* (Oxford: Oxford University Press).

Vaughan, J. 1987. 'Honor Fell, 1900–1986', *Biographical Memoirs of Fellows of the Royal Society* 33; 235–59.

Veldman, M. 1994. *Fantasy, the Bomb and the Greening of Britain: Romantic Protest* (Cambridge: Cambridge University Press).

Vig, N. J. 1968. *Science and Technology in British Politics* (Oxford: Pergamon Press).

von Neumann, J. 1986. 'First draft of a report on EDVAC' [Moore School of Electrical Engineering, University of Pennsylvania, 30 June 1945], in W. F. Aspray and A. W. Burks (eds.), *Papers of John von Neumann on Computing and Computer Theory*. Charles Babbage Institute Reprint Series for the History of Computing 12 (Cambridge, MA: MIT Press and Los Angeles: Tomash Publishers), pp. 17–82.

Waddington, C. H. 1969. 'Some European contributions to the pre-history of molecular biology', *Nature* 221: 318–21.

—— 1973. *O.R. in World War 2: Operational Research against the U-Boat* (London: Elek Science).

Walker, P. M. B. 1972. 'Cancer: the relevance of certain fundamental research', *The Lancet*, 24 June: 1379–81.

Watson, E. L. 1991. *Houses for Science: A Pictorial History of Cold Spring Harbor Laboratory* (Cold Spring Harbor, NY: Cold Spring Harbor Laboratory Press).

Watson, H. 1969. 'The stereochemistry of the protein myoglobin', *Progress in Stereochemistry* 4: 299–333.

Watson, J. D. 1964. 'The involvement of RNA in the synthesis of proteins. Nobel Lecture, December 11, 1959', in *Nobel Lectures Including Presentation Speeches and Laureates' Biographies. Physiology or Medicine, 1942–1962* (Amsterdam: Elsevier Publishing Company), pp. 785–808.

1966. 'Growing up in the phage group', in J. Cairns, G. S. Stent and J. D. Watson (eds.), *Phage and the Origins of Molecular Biology* (Cold Spring Harbor, NY: Cold Spring Harbor Laboratory of Quantitative Biology), pp. 239–45.

1968. *The Double Helix: A Personal Account of the Discovery of the Structure of DNA* (New York: Atheneum and London: Weidenfeld and Nicolson).

1980. *The Double Helix: A Personal Account of the Discovery of the Structure of DNA. Text, Commentary, Reviews, Original Papers*, ed. G. Stent (New York and London: W. W. Norton & Company).

1987. 'Minds that live for science', *New Scientist* 114: 63–6.

Watson, J. D. and Crick, F. H. C. 1953a. 'Molecular structure of nucleic acids', *Nature* 171: 737–8.

1953b. 'Genetical implications of the structure of deoxyribonucleic acid', *Nature* 171: 964–7.

Watson, J. D. and Tooze, J. 1981. *The DNA Story: A Documentary History of Gene Cloning* (San Francisco: W. H. Freeman).

Weart, S. 1988. *Nuclear Fear: A History of Images* (Cambridge, MA: Harvard University Press).

Weatherall, M. and Kamminga, H. 1992. *Dynamic Science: Biochemistry in Cambridge 1898–1949* (Cambridge: Cambridge Wellcome Unit Publications).

1996. 'The making of a biochemist: II. The construction of Frederick Gowland Hopkins' reputation', *Medical History* 40: 415–36.

Weaver, W. 1970. 'Molecular biology: origin of the term', *Science* 170: 581–2.

Weiner, J. 1999. *Time, Love, Memory: A Great Biologist and His Quest for the Origins of Behavior* (London: Faber and Faber).

Werskey, G. 1988. *The Visible College: A Collective Biography of British Scientists and Socialists of the 1930s* (London: Free Association Books).

Wheeler, D. J. 1992. 'The EDSAC Programming Systems', *IEEE Annals of the History of Computing* 14: 34–40.

Wheeler, J. M. 1992. 'Applications of the EDSAC', *IEEE Annals of the History of Computing* 14: 27–33.

White, J. G., Southgate, E., Thomson, J. N. and Brenner, S. 1986. 'The structure of the nervous system of the nematode *Caenorhabditis elegans*', *Philosophical Transactions of the Royal Society London B* 314: 1–340.

Whittemore, G. F. 1975. 'World War I, poison gas research, and the ideals of American chemists', *Social Studies of Science* 5: 135–63.

Wigglesworth, V. 1983. 'John William Sutton Pringle, 22 July 1912 – 2 November 1982', *Biographical Memoirs of the Fellows of the Royal Society* 29: 525–51.

Wilkes, M. V. 1985. *Memoirs of a Computer Pioneer* (Cambridge, MA: MIT Press).

1992. 'EDSAC 2', *IEEE Annals of the History of Computing* 14: 49–56.

Wilkie, T. 1991. *British Science and Politics since 1945* (Oxford: Blackwell).

Wilkins, F. 1987. 'John Turton Randall, 23 March 1905 – 16 June 1984', *Biographical Memoirs of the Fellows of the Royal Society* 33: 491–535.

Williams, S. 1971. 'The responsibility of science', *The Times: Saturday Review*, 27 February, p. 15.

Wilson, A. J. C. 1951. 'Summarized proceedings of a conference on the development and application of Fourier methods in crystal-structure analysis – London, November 1950', *British Journal of Applied Physics* 2: 61–70.

Winstanley, M. (1976). 'Assimilation into the literature of a critical advance of molecular biology', *Social Studies of Science* 6: 545–9.

Wolstenholme, G. (ed.) 1963. *Man and His Future* (London: J. & A. Churchill).

Wood, W. B. and the Community of *C. elegans* Researchers (eds.) 1988. *The Nematode* Caenorhabditis elegans (Cold Spring Harbor, NY: Cold Spring Harbor Laboratory Press).

Woodham, J. M. 1990. *Twentieth-Century Ornament* (London: Studio Vista).

Wright, S. 1994. *Molecular Politics: Developing American and British Regulatory Policy for Genetic Engineering, 1972–1982* (Chicago: University of Chicago Press).

Wrinch, D. M. 1937. 'The cyclol hypothesis and the "globular" proteins', *Proceedings of the Royal Society A* 161: 505–24.

Yockey, H. P. 1958a. 'Some introductory ideas concerning the application of information theory in biology', in H. P. Yockey (ed.), *Symposium on Information Theory in Biology. Gatlinburg, Tennessee, October 29–31, 1956* (New York: Pergamon Press), pp. 50–9.

Yockey, H. P. (ed.) 1958b. *Symposium on Information Theory in Biology. Gatlinburg, Tennessee, October 29–31, 1956* (New York: Pergamon Press).

Yoxen, E. 1978. 'The Social Impact of Molecular Biology'. Ph.D. dissertation, University of Cambridge.

   1979. 'Where does Schroedinger's "What is Life?" belong in the history of molecular biology?', *History of Science* 17: 17–52.

   1981. 'Life as a productive force: capitalising the science and technology of molecular biology', in L. Levidow and R. Young (eds.), *Science, Technology and the Labour Process. Marxist Studies I* (London: CSE Books and Atlantic Highlands, NJ: Humanities Press), pp. 66–122.

   1982. 'Giving life a new meaning: the rise of the molecular biology establishment', in N. Elias, H. Martins and R. Whitley (eds.), *Scientific Establishments and Hierarchies. Sociology of the Sciences Yearbook 6* (Dordrecht: Reidel Publishing Company), pp. 123–43.

   1983. *The Gene Business: Who Should Control Biotechnology?* (London: Pan in conjunction with Channel Four Television Company).

   1985. 'Speaking out about competition. An essay on "The Double Helix" as popularisation', in T. Shinn and R. Whitley (eds.), *Expository Science: Forms and Functions of Popularisation. Sociology of the Sciences Yearbook 9* (Dordrecht: Reidel Publishing Company), pp. 163–81.

Zallen, D. T. 1992. 'The Rockefeller Foundation and spectroscopy research. The programs at Chicago and Utrecht', *Journal of the History of Biology* 25: 67–89.

1999. 'From butterflies to blood: human genetics in the U.K.', in M. Fortun and E. Mendelsohn (eds.), *The Practices of Human Genetics*. Sociology of the Sciences Yearbook 21 (Dordrecht: Kluwer), 197–216.

[Zuckerman, S.] 1940. 'Men of science and the war', *Nature* 146: 107–8.

Zuckerman, S. 1972. *Cancer Research* (London: Cabinet Office and Her Majesty's Stationery Office).

1978. *From Apes to Warlords: Autobiography (1904–1946)* (London: Hamish Hamilton).

1988. *Monkeys, Men and Missiles: An Autobiography 1946–1988* (London: Collins).

Index